A Game of Clans

collectores venatoresque,

agricolae pastoresque

Carlos Quiles

A Game of Clans

collectores venatoresque,

agricolae pastoresque

Carlos Quiles

A Song of Sheep and Horses

Book 1

ACADEMIA PRISCA

2019

A SONG OF SHEEP AND HORSES:

EURAFRASIA NOSTRATICA, EURASIA INDOURALICA

Book One: A Game of Clans: *collectores venatoresque, agricolae pastoresque.*
Book Two: A Clash of Chiefs: *rex militaris, rex sacrorum.*
Book Three: A Storm of Words: *vetera verba, priscae linguae.*
Book Four: A Feat of Crowds: *hic sunt leones, hic sunt dracones.*

Version: 1.2 (first printed edition), June 1st, 2019
ISBN-13: 978-1-072-00322-9 paperback

© 2019, 2017 by Carlos Quiles cquiles@academiaprisca.org
ACADEMIA PRISCA
Avda. Sta. María de la Cabeza, 3, E-LL, Badajoz 06001, Spain.
European Union

Work based on Quiles, Carlos (2017). Indo-European demic diffusion model (3rd ed.). Badajoz, Spain: Universidad de Extremadura. DOI: 10.13140/RG.2.2.35620.58241

Official site: <https://indo-european.eu/asosah/>
Publisher's site: <https://academiaprisca.org/>
Full text and latest revisions: <https://indo-european.info>
Blog for images, discussion, and new findings: <https://indo-european.eu/>

Cover image *Die Glockenbecherleute* by Gerhard Beuthner, appeared in Erdal-Bilderreihe Jugensteinzeit Nr. 117 Bild 5 (1930).

Table of Contents

Guide to the reader

Abbreviations

AASI: Ancient Ancestral South Indian

AEA: Ancient East Asians

AHG: Anatolian Hunter-Gatherer

AME: Ancient Middle Easterner

ANA: Ancient North African

ANE: Ancient North Eurasian

ANI: Ancestral North Indian

ANS: Ancient North Siberian

ASI: Ancestral South Indian

AP: Ancient Palaeosiberian

BA: Bronze Age

BBC: Bell Beaker culture

BMAC: Bactria and Margiana Archaeological Complex

BE: Basal Eurasians

CA: Copper Age

CHG: Caucasus Hunter-Gatherer

CWC: Corded Ware culture

CWE: Common West Eurasian

ENA: Eastern non-Africans

EWE: Early West Eurasians

EN: Early Neolithic

FBA: Final Bronze Age

IA: Iron Age

IE: Indo-European

IN: Iranian Neolithic

LBA: Late Bronze Age

LBK: Linearbandkeramik

LCA: Late Chalcolithic

LN: Late Neolithic

MBA: Middle Bronze Age

MCW: Multi-Cordoned Ware

MLBA: Middle–Late Bronze Age

MN: Middle Neolithic

NWAN: North-West Anatolian Neolithic

PCA: Principal Component Analysis

PIE: Proto-Indo-European

CEU: Central European

GAC: Globular Amphora(e) culture

EBA: Early Bronze Age

EEA: Early East Asian

EEBA: Early European Bronze Age

EEF: Early European Farmer

EH: Early Helladic

EHG: Eastern Hunter-Gatherer

EIA: Early Iron Age

PU: Proto-Uralic

SNP: Single Nucleotide Polymorphism

SRBW: Simple-Relief-Band Ware

TMRCA: Time to Most Recent Common Ancestor

TRB: Funnel Beaker culture

WHG: Western Hunter-Gatherer

WSHG: Western Siberian Hunter-Gatherer

Symbols

(x[SNP])	denotes "negative for [SNP]"
[SNP]+	marks an unofficial or probabilistic [SNP] call
[SNP]*	implies that the sample is of a "basal" [SNP] subclade
[N%]+	denotes an unofficial result of "N% ancestry"
[SNP]+	denotes "positive for [SNP]"
*[word]	denotes a reconstructed form

Conventions used in this book

This text is not a simple essay anymore. Even though I conceived it initially as a mere fourth revision at the end of 2017, it grew rapidly out of hand, as I intended to include as much relevant information as possible on published (and reported) population movements supported by genetic investigation. The association of genetic data with potential prehistoric ethnolinguistic communities required in turn the addition of all potentially relevant archaeological data.

The first two volumes of this series must be understood as a *detailed supplement* of the main work, which is the third volume concerning linguistic data. This order of relevance is not only related to this series' emphasis on languages over prehistoric cultures or genetics, but to the actual nature of the matter at hand: this a comprehensive work on reconstructed languages and the peoples who might have spoken them.

The work follows simple rules in its aim to achieve clarity and coherence.

It is an encyclopaedia-like text, free and organised in more or less isolated linguistic, archaeological, and genetic sections organised to facilitate future revisions by anyone, incorporating the latest research.

Unlike armchair work in linguistics or bioinformatics, where results and interpretations can be reviewed with knowledge and proper access to data, it is impossible to be an *armchair archaeologist* without ample experience in the specific field investigated. Therefore, secondary archaeological sources, giving proper interpretation and synthesis of primary research and fieldwork, are preferred over primary sources in the archaeological section, with little or no personal additions in this part, although primary sources and proper connections of the data have also been added whenever necessary. All archaeological summaries included are properly referenced, with the main author or authors behind the content of each paragraph properly cited—at least the author of the secondary source, often more relevant than primary sources— to allow for proper identification of the original text and for further reading.

A chronologically and regionally organised structure has been given to the full text, to allow for an easy searching of the content, and for the reading of the text in either a linear or non-linear manner.

Names of samples, their cultures or groupings, ancestries, or clusters do not necessarily follow the nomenclature systems used by the different authors, papers, research labs, or archaeological teams, but are made to fit into the coherent picture of this book (Eisenmann et al. 2018).

Haplogroup (hg.) will be frequently used to refer to Y-chromosome haplogroups, unless otherwise expressly stated. Y-DNA haplogroups and subclades will also be referred to as *line* or *lineage*, whereas common admixture components defined in recent papers will be referred to as *ancestry*. The preferred nomenclature system of haplogroups is *X-Y*, where *X* is the standard name by ISOGG (2018), and *Y* is one or more SNP mutations defining the haplogroup, using whenever possible the one preferred by YFull. An asterisk *X-Y** is used to represent a *basal lineage*, commonly understood as a subclade with different mutations from the most common, 'successful' ones.

Additional positive *Y*+ reported online in non-peer reviewed publications are represented in this text as *X-Y⁺*. The originally published haplogroup for the samples, other reported positive and negative SNPs, as well as the author or authors of the additional information, can be found in online supplementary materials of this book.

For the sake of consistency, only YFull estimates for year formed and time to most recent common ancestor (TMRCA) of Y-chromosome haplogroups have been used[1], unless other sources are expressly stated. Years before present (ybp) have been approximated to BC assuming 2,000 years of difference, to round out estimates. Estimates were obtained by Vladimir Tagankin by applying the method published in Adamov et al. (2015) to the data received

---

[1] Dates were retrieved from the website <https://www.yfull.com/> during October-December 2018.

from voluntary users[2]. Also for the sake of consistency, dates expressed as years before present (YBP) have been simplistically approximated to BC.

TMRCA dates are used as gross approximations to expansions of Y-DNA lineages (see Figure 1). They can offer an inaccurate idea of the lineage evolution because a) the actual rate of mutation is unknown, and b) TMRCA estimates are based on the lineages that survived, which may obviate other previous expansions in the same trunk.

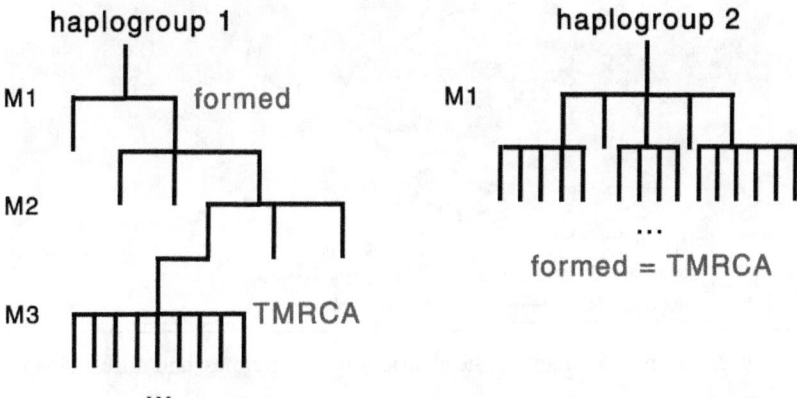

Figure 1. Simplistic example of SNP mutations in a haplogroup. Lines represent diverging male lineages. Haplogroup 1 is only successful after the third mutation, and is thus defined by mutations M1, M2, and M3, with M1 representing its formation date, and M3 its TMRCA. Haplogroup 2 is successful during its formation, and thus M1 defines it, as well as its formation and TMRCA date.

Modern physical maps are used to illustrate potential expansion routes of ancient cultures, peoples, and languages, even though they pose a significant danger to the development of a sound model, since they almost invariably involve "a concatenation of weakly supported links that corporately form an 'arrow' of dispersion" (Mallory 2014). Map routes are only depicted as a visual help to add movement to the otherwise stationary maps of ancient cultures,

---

[2] For details on the specific methodology used, see <https://www.yfull.com/faq/what-yfulls-age-estimation-methodology/>.

peoples, languages, and ancient DNA obtained from scattered burials. Eurasian biomes (Figure 2) and Suppl. Fig. 19) are commonly referenced to in this book to delimit cultural groups and migration routes.

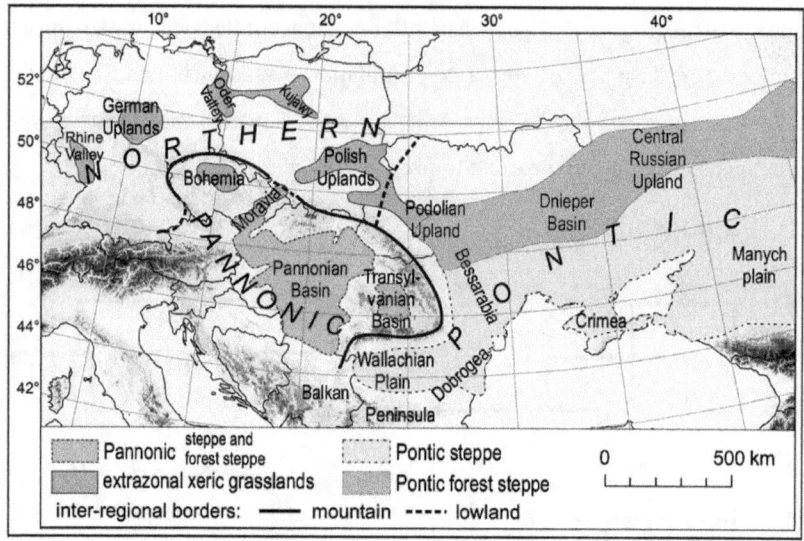

Figure 2. Simplified map of the distribution of steppes and forest-steppes (Pontic and Pannonian) and xeric grasslands in Eastern Central Europe (with adjoining East European ranges). Modified from Kajtoch et al. (2016).

List of Supplementary Figures

The following is an ordered list of supplementary graphics with their description and links to online sites for download. All supplementary data is also contained in the fourth volume of this series, *A Feat of Crowds*.

Maps

A list of available maps can be found at <https://indo-european.eu/maps/>.

Suppl. Fig. 1. Map of out-of-Africa migrations of anatomically modern humans (before ca. 35000 BC) with Y-DNA haplogroups, ADMIXTURE, and mtDNA haplogroups. <https://indo-european.eu/maps/out-of-africa/>.

Suppl. Fig. 2. Map of Upper Palaeolithic cultures (ca. 35000–20000 BC), with Y-DNA haplogroups, ADMIXTURE, and mtDNA haplogroups. <https://indo-european.eu/maps/palaeolithic/>.

Suppl. Fig. 3. Map of Epipalaeolithic cultures (ca. 20000-10000 BC), with Y-DNA haplogroups, ADMIXTURE, and mtDNA haplogroups. <https://indo-european.eu/maps/epipalaeolithic/>.

Suppl. Fig. 4. Map of Early Mesolithic cultures (ca. 10000–7500 BC), with Y-DNA haplogroups, ADMIXTURE, and mtDNA haplogroups. <https://indo-european.eu/maps/mesolithic-early/>.

Suppl. Fig. 5. Map of Late Mesolithic cultures (ca. 7500–6000 BC), with Y-DNA haplogroups, ADMIXTURE, and mtDNA haplogroups. <https://indo-european.eu/maps/mesolithic/>.

Suppl. Fig. 6. Map of Neolithic and hunter-gatherer pottery expansion (ca. 6000–5000 BC), with Y-DNA haplogroups, ADMIXTURE, and mtDNA haplogroups. <https://indo-european.eu/maps/neolithic/>.

Suppl. Fig. 7. Map of Early Eneolithic cultures (ca. 5000–4000 BC), with Y-DNA haplogroups, ADMIXTURE, and mtDNA haplogroups. <https://indo-european.eu/maps/eneolithic-early/>.

Suppl. Fig. 8. Map of Late Eneolithic cultures (ca. 4000–3300 BC), with Y-DNA haplogroups, ADMIXTURE, and mtDNA haplogroups. <https://indo-european.eu/maps/eneolithic/>.

Suppl. Fig. 9. Map of Final Eneolithic / Chalcolithic expansions (ca. 3300–2600 BC), with Y-DNA haplogroups, ADMIXTURE, and mtDNA haplogroups. <https://indo-european.eu/maps/copper-age/>.

Suppl. Fig. 10. Yamna – Bell Beaker evolution <https://indo-european.eu/maps/yamna-bell-beaker/>.

10.A) Map of attested Yamnaya pit-grave burials in the Hungarian plains; superimposed in shades of blue are common areas covered by floods before the extensive controls imposed in the 19th century; in orange, cumulative thickness of sand, unfavourable loamy sand layer. Marked are settlements/findings of Boleráz (ca. 3500 BC on), Baden (until ca. 2800 BC), Kostolac (precise dates unknown), and Yamna kurgans (from ca. 3100/3000 BC on).

10. B) Map of Yamna – Bell Beaker migrations in Central Europe (ca. 2800–2300 BC).

10.C) Tentative map of fine-scale population structure during steppe-related expansions (ca. 3500–2000 BC), including Repin–Yamna–Bell Beaker/Balkans and Sredni Stog–Corded Ware groups. Data based on published samples and pairwise comparisons tested to date. Notice that the potential admixture of expanding Repin/Early Yamna settlers in the North Pontic area with the late Sredni Stog population (and thus Sredni Stog-related ancestry in Yamna) has been omitted for simplicity purposes, assuming thus a homogeneous Yamna vs. Corded Ware ancestry.

Suppl. Fig. 11. Map of Chalcolithic expansions (ca. 2600–2200 BC), with Y-DNA haplogroups, ADMIXTURE, and mtDNA haplogroups. <https://indo-european.eu/maps/chalcolithic/>.

Suppl. Fig. 12. Map of Early Bronze Age cultures (ca. 2200–1750 BC), with Y-DNA haplogroups, ADMIXTURE, and mtDNA haplogroups, as well as a tentative ethnolinguistic identification of archaeological cultures <https://indo-european.eu/maps/early-bronze-age/>.

Suppl. Fig. 13. Map of Middle Bronze Age cultures (ca. 1750–1250 BC), with Y-DNA haplogroups, ADMIXTURE, and mtDNA haplogroups. <https://indo-european.eu/maps/middle-bronze-age/>.

Suppl. Fig. 14. Map of Late Bronze Age cultures (ca. 1250–750 BC), with Y-DNA haplogroups, ADMIXTURE, and mtDNA haplogroups. <https://indo-european.eu/maps/late-bronze-age/>.

Suppl. Fig. 15. Map of Early Iron Age cultures (ca. 750–250 BC), with Y-DNA haplogroups, ADMIXTURE, and mtDNA haplogroups. <https://indo-european.eu/maps/iron-age/>.

Suppl. Fig. 16. Map of Late Iron Age cultures (ca. 250 BC – AD 250), with Y-DNA haplogroups, ADMIXTURE, and mtDNA haplogroups. <https://indo-european.eu/maps/classical-antiquity/>.

Suppl. Fig. 17. Map of peoples and cultures in Antiquity (ca. AD 250–750), with Y-DNA haplogroups, ADMIXTURE, and mtDNA haplogroups. <https://indo-european.eu/maps/antiquity/>.

Graphics

A list of available graphics may be found at <https://indo-european.eu/pca/>.

Introduction

This project began as a short essay called "Indo-European Demic Diffusion Model", published in April 2017 in the Department of Anatomy, Cell Biology, and Zoology of the University of Extremadura, in which I contended that recent genetic investigation suggested that the expansion of Indo-European languages from the steppe was linked to the expansion of R1b1a1b-M269 lineages in Eurasia. In particular, genetic data recovered from ancient individuals seemed to support that the expansion of R1b1a1b1-L23 lineages in Europe was associated with Yamna migrants, and thus also subsequently with the expansion of East Bell Beakers as North-West Indo-European-speakers in Europe, whereas the spread of the Corded Ware culture likely represented the expansion of Uralic speakers.

Some researchers had already expressed doubts on the traditional association of Corded Ware with the Indo-European expansion, although none of them had given an alternative model consistent with the current data, explaining the role of R1a1a1-M417 lineages spreading with Uralic speakers (Horváth 2014), or the recently described "Yamnaya ancestry" peaking among Uralic individuals (Heggarty 2015; Klejn et al. 2017) as the result of mixed Indo-European–Uralic communities.

The theory laid in this text takes dialectal evolution as its stable framework, as the core which should underlie any Indo-European expansion model, and uses genetic investigation (of ancient and modern DNA samples) and its potential relationship with archaeological cultures to establish an expansion model step by step. It also takes into account that there are complex problems found in correlations of languages with archaeological cultures (Meier-Brügger 2003) and human genetics (Campbell 2015).

Even though phylogenetic methods became popular in the early 2000s, and have been used intermittently since then, especially by non-linguists (Ringe, Warnow, and Taylor 2002; Anthony and Ringe 2015), it seems more reasonable to avoid such methods in scientific publications, due to their controversial pseudoscientific nature and questionable results (Pereltsvaig and Lewis 2015)[3]. Historical linguistics can only provide a relative historical framework for individual proto-languages and their relationships, though.

Archaeology works with the concept of culture, and as such it is able to determine timelines. When these timelines complement relative chronologies and wide guesstimates of proto-languages beautifully, both are able to provide a contextualised historical explanation of linguistic frameworks (Vander Linden 2015; Hänsel and Zimmer 1994). The model of Indo-European migrations set forth by Marija Gimbutas (Gimbutas 1963, 1977) has been impressively corrected and expanded recently by Volker Heyd (Heyd 2004; Harrison and Heyd 2007; Heyd 2007; Heyd 2011), Valentin A. Dergachev (2007), David W. Anthony (Anthony 2007; Anthony and Brown 2011; Anthony 2013), James P. Mallory (2013), or Christopher Prescott (Prescott and Walderhaug 1995; Prescott 2012), among others. Similarly, the models of early

---

[3] Kallio (2015): "Rather than the standard comparative method the most popular tool has recently been the probabilistic approach (…). To put it crudely, all this means that instead of comparing two reconstructed protolanguages one only compares two 50-item or 100-item wordlists. Call me a grumpy old man, but I fail to see how such laziness would be progress, although it is of course faster and easier, requiring hardly any knowledge of the languages themselves. It is therefore no wonder that limited wordlists are particularly popular among nonlinguists applying phylogenetic methods to languages (see now Pereltsvaig & Lewis 2015 for an excellent review)."

Uralic–Indo-European contacts (Koivulehto 1991; Koivulehto 2003) and Uralic migrations from eastern Europe recently advocated by Kallio (Kallio 2002, 2014) or (Parpola 2013) have paved the way for a clearer understanding of the cultures and peoples involved in steppe-related migrations, and constitute the basic starting point of this book.

Language and culture expansion have been usually explained by two main alternative models: the demic diffusion model, which involves mass movement of people; and the cultural diffusion model, which refers to cultural impact between populations, and involves limited genetic exchange between them. Language transfer since ancient times seems to be associated with an expansion of people (Mikhailova 2015).

Ancestry of any selected population is likely to be a mixture of several ancient groups, which is reflected on the genetic structure (Haak et al. 2010; Skoglund et al. 2012; Malmström et al. 2009; Lazaridis et al. 2014). However, the genetic landscape for ancient populations is limited by the number of ancient DNA samples and ancient populations studied (Hellenthal et al. 2014). Population expansions are often accompanied by a significant replacement of patrilineal lineages (see Figure 3), due to the formation of expanding kin groups and to the intergroup competition among them (Zeng, Aw, and Feldman 2018). This reduction in variability and Y-chromosome haplogroups is exacerbated by migration and exogamy practices, and also by violent conflicts involving mainly males.

Male-biased population expansions cannot explain all language expansions and replacements, though, and exceptions are frequent throughout history. Archaeological research combined with population genomics is showing how bilingualism and multilingualism was common among different prehistoric groups which interacted closely and often became symbiotically integrated with each other. This is usually associated with chiefdom-like systems and long-lasting exogamy practices, such as those found between Abashevo and Sintashta-Potapovka-Filatovka peoples, between Bell Beakers and El Argar population, or among cultures of the Baltic Sea and expanding Akozino warrior-traders. Beyond that, population bottlenecks including founder effects often obscure the original expansion of a language with male-driven migrations, as happened in Finns with N1a1a1a1a-L392 subclades, and Basques with R1b1a1b1a1a2-P312 lineages.

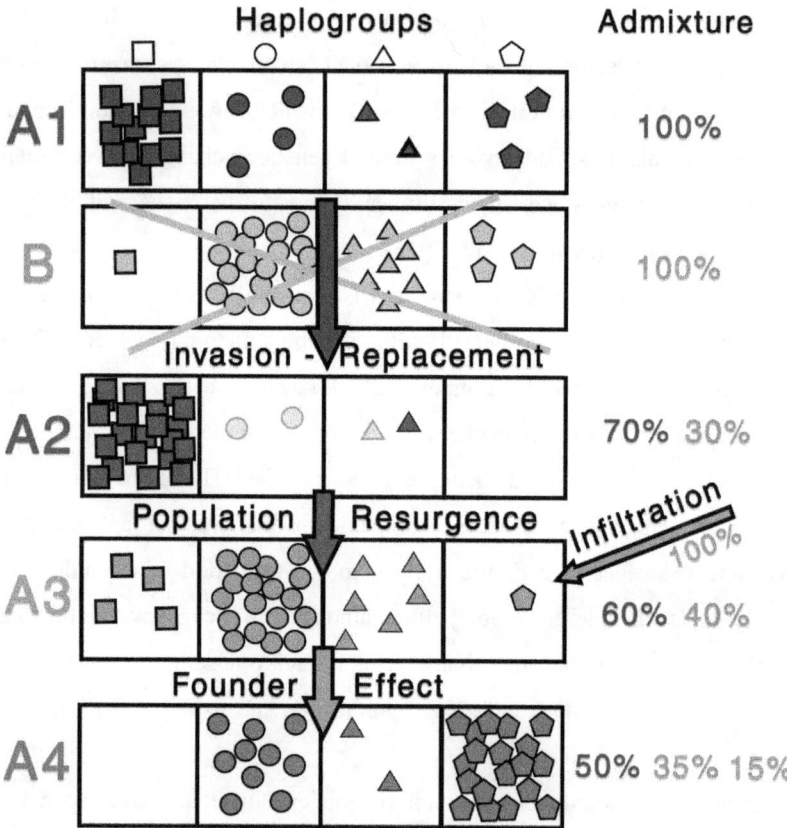

Figure 3. Simplified model of evolution of an ethnolinguistic community in terms of Y-chromosome haplogroups and admixture: Community A1 invades the territory of Community B, eventually replacing most 'indigenous' haplogroups at the same time as certain paternal lineages of A1 have more reproductive success, developing Community A2 through exogamy (evidenced by 30% of admixture from Community B). A language change is very likely to have taken place in the territory of Community B. No language change is necessary, however: (1) in the case of population resurgence (such as the change in haplogroups and admixture from Community A2 to Community A3); (2) after the incorporation of Y-DNA haplogroups and admixture from neighbouring populations; or (3) through founder effects (represented in the evolution from

Community A3 to Community A4). Graphic inspired by images in Zeng, Aw, and Feldman (2018).

Ancient DNA (aDNA) investigation allows us to disentangle complex human history (Slatkin and Racimo 2016). The most recent breakthrough in Indo-European migrations obtained thanks to population genomics, concerned with general population movements of Eurasians westwards from the steppe (Haak et al. 2015; Allentoft et al. 2015; Mathieson et al. 2015), suggested that a common so-called "Yamnaya ancestry" represented by individuals of the Yamna culture could also be found in Corded Ware, Bell Beaker, and Únětice in descending proportions, coherent with their radiocarbon dates, apparently connecting them to succeeding migrations. However, the strong reliance on ancestry to derive conclusions on potential population movements, suited for gross interpretations of Palaeolithic and Mesolithic movements over thousands of years based on few assessed samples, has proven much trickier when deriving models of ethnolinguistic change from movements of neighbouring and closely interacting populations lasting just hundreds of years. Proper assessment and interpretation of Y-DNA, which does not change with generations of admixture, has been demonstrated to be key when investigating the connection of certain groups, as is clear from the Iberian (Olalde et al. 2019) and South Asian cases (Narasimhan et al. 2018).

The massive migration of people of Yamna origin with the Bell Beaker culture some four hundred years after the expansion of Corded Ware peoples (a group probably originating close to the north Pontic area), witnesses thus the latest language shift before the start of the Early European Bronze Age. All data put in common, there is little space (if any at all) to relate the expansion of Corded Ware peoples to any Indo-European dialect surviving into historical times, because the partial genetic link of Corded Ware peoples with Yamna is probably earlier than the expansion of Proto-Anatolian, and there is a strong connection of surviving Corded Ware groups with Finno-Ugric populations. Even the 2015 papers on Indo-European migrations showed with their published data that haplogroups R1b1a1b-M269 and R1a1a1-M417 were absent from central and western Europe until after the expansion of Eurasian pastoralists. This data help thus trace most modern European languages to the Eneolithic Pontic–Caspian steppes, and therefore to a massive expansion starting at nearly the same time from eastern Europe after around 3000 BC. In these studies, R1a1a1-M417 was prevalent in Corded Ware samples and was absent from samples of the Yamna horizon, most of which belonged to haplogroup R1b-M269. Further publications on early European (Mathieson et al. 2018), Bell Beaker (Olalde et al. 2018), Turan/South Asian (Narasimhan et al. 2018), and ancient Eurasian samples (de Barros Damgaard, Marchi, et al. 2018; de Barros Damgaard, Martiniano, et al. 2018) further confirm a surprisingly long-lasting clear-cut division of patrilineal lineages among Eneolithic steppe communities, and recent studies on the prehistoric Caucasus (Lazaridis et al. 2018; Wang et al. 2019) are helping reconstruct the different fine-scale population structure of Corded Ware and Yamna peoples.

The recent genetic revolution is helping thus support the mainstream view of a natural evolution of reconstructed languages, including their dialectal stages, with concrete prehistorical communities defined in time and space (Lehmann 1992). Population genomics has therefore cleaned up the comparatist's desk, dismissing almost all models of cultural diffusion proposed to date, especially flexible frameworks such as the "constellation analogy" (Clackson 2007, 2013) of loosely interconnected prehistoric communities, sharing language and culture through unending waves of areal contact among dialect *continua*. These were probably the result of fashionable linguistic trends akin to the 'pots, not people' paradigm prevalent in archaeology since the mid–20th century, and most of them need be rejected, even though there is still an ongoing controversy over many details of the potential expansion of peoples with certain cultures (Kristiansen et al. 2017; Heyd 2017; Sørensen 2017; Furholt 2017), and there is growing concern about the need for fine-scale studies (Lazaridis 2018; Veeramah 2018).

Even more interesting than the general genetic revolution is the more specific one regarding Indo-European and Uralic migrations. Recently published data is helping reject previously popular theories concerning the dialectal and cultural evolution of Late Proto-Indo-European, such as the Anatolian homeland (Renfrew 1987), or the prevalent identification of Corded Ware with Indo-Europeans (Gimbutas 1977; Kristiansen 1989; Anthony 2007). In this sense, David Reich's words regarding the dismissal of the Anatolian homeland theory by genetic data have proven premonitory for the dismissal of their preferred model of Corded Ware as expanding Indo-European dialects:

"A great lesson of the ancient DNA revolution is that its findings almost always provide accounts of human migrations that are very different from preexisting models, showing how little we really knew about human migrations and population formation prior to the invention of this new technology" (Reich 2018).

Ancient DNA is helping locate different peoples in a very specific place, time, and route of expansion, supporting in turn the most appropriate models of dialectal splits, which closes the circle of interrelated connections between linguistics, archaeology, and genetics, and turns anthropological investigation into a shrinking helix that points more and more precisely to the true ancient ethnolinguistic picture.

While the picture is clearer today than it was just a year ago, the most recent genetic research is also correcting, not just old technology and ancient anthropological interpretations of the 1990s or 2000s, but also genetic methods and results of just months or years ago. Some data that we believed could be breakthroughs in the field have been demonstrated with time to be most likely wrong, either in the radiocarbon dating (due to mixed archaeological layers) or due to errors in technique or recent improvements in technology.

So, for example, the finding of haplogroup N1a-F1206 in the Comb Ware culture (Chekunova et al. 2014) becomes more and more unlikely with each new paper, like the finding of R1a1a-M198 around Lake Baikal during the

Early Neolithic (Mooder et al. 2005; Moussa et al. 2016). Similarly, reports based on modern populations, such as the estimated origin R1b1a1b-M269 in Neolithic Europe (Myres et al. 2011), or R1a-L146 in South Asia (Underhill et al. 2015), and many others have been proven repeatedly wrong with ancient DNA. Even today, errors in cultural attribution, radiocarbon dates, and estimated haplogroups or subclades are bound to happen—in addition to technical errors involving the processing and assessment of samples (very difficult to test without the resampling of specimens) —as we have seen most recently in samples from Hajji Firuz in Narasimhan et al. (2018), a huge investigation including scattered Asian samples and necessitating an international collaboration of many different archaeological teams.

More than *Kosinna's smile* (Heyd 2017) of equating prehistoric culture to population in a general sense, the most recent genetic investigation should probably represent the joy of Starostin's Nostratic Eurasian Epipalaeolithic, Kortlandt's Eurasiatic northern Eurasian Mesolithic, Vennemann's and Villar's Vasconic Mediterranean and western (and possibly central) European Neolithic, Wiik's Uralic northern and eastern European Chalcolithic, and Krahe's Old European Early Bronze Age. Beyond petty sociopolitical and ethnolinguistic grievances of neighbouring Eurafrasian populations, and beyond the infinite pet theories on the potential ancestral population or language *homelands*, genetics is cutting up to the chase and dismissing all theories but a few (usually related ones), or even just one, despite the obstinate defence of traditional theories by many academics.

Sadly, the field is plagued with unending setbacks: on one hand, the eternal search for academic authority, and the need to publish and to collect as much citations and publications in journals of high impact factor as possible, are provoking all kinds of reactionary views, to fit previous models with the clear-cut picture emerging in some cases from genetic investigation. On the other hand, modern political and ethnolinguistic views burden this field, ranging from modern Indian politics in favour of an "indigenous Sanskrit" opposed to the so-called "Aryan invasion theory"; through the interest of modern Russian politics in supporting "indigenous Slavs", opposed to the known history of colonisation and Slavicisation of essentially all of their modern territory; to the interest of certain western European groups in supporting an "indigenous" Palaeolithic Vasconic-speaking population. Reactionary views and 'nativist' ethnolinguistic trends are slowly eroding this new anthropological subfield of population genomics, and I would not be surprised if some education systems would reject it as a useful anthropological discipline, for one or other reason.

No one is free of personal or professional bias, and mine is clear: at Academia Prisca, Fernando López-Menchero and I have invested years supporting the reconstruction of North-West Indo-European as a Late Proto-Indo-European dialect, which puts a clear red line in this series of books to any interpretation of the data that challenges this dialectal scheme. Also, I am of haplogroup R1b1a1b1a1a2a-DF27, like many in south-western Europe. On the other hand, we have been publishing texts about Proto-Indo-European since 2005, and I knew my haplogroup since 2008, but until 2015 I supported the spread of North-West Indo-European with Corded Ware and a later Old European dialect *continuum* centred on the pan-European Únětice culture (Quiles 2012). These cultures were thought to be dominated by R1a-M420 lineages, so that R1b-M343 lineages (probably Vasconic speakers) would have acquired the language by way of cultural diffusion in western Europe, maybe by Bell Beakers along the Rhine.

Only after 2015—when, paradoxically, genetic papers seemed to support my preferred model—did I realise that Corded Ware may not have been linked to the expansion of Proto-Indo-European, and Volker Heyd's theories seemed to take the lead, with R1a-M420 lineages potentially expanding Indo-Uralic through North Eurasia, but not Indo-European from the steppe, which would have been hitchhiked by R1b-M343 lineages which expanded Afroasiatic from Anatolia into south-eastern Europe (Quiles 2017).

After the most recent papers of 2017 and 2018, it seems more and more unlikely that the early arrival into eastern Europe and lack of expansions of R1a-M420 lineages could be associated with the spread of Indo-Uralic or Eurasiatic through North Eurasia, and therefore R1b-M343 lineages, with a likely origin in (and multiple expansions from) eastern Europe, seem like the most appropriate lines to follow most of the time for the spread of Pre-Indo-European languages.

These are simplistic assessments, and it should be obvious to anyone involved in the field that 1) the current picture shown by available ancient DNA research is clearly shifted towards Europe and R1b-M343 samples, for different reasons, which may be distorting our view of ancient population movements; 2) uniparental markers cannot be linked in a simplistic way to assess ethnolinguistic communities and their movements, because other relevant linguistic, archaeological, and genetic data must be assessed in order to obtain proper migration models; and 3) stages before Indo-Uralic are at best speculative, and are used only to give a coherent account of migrations coupled with reconstructed languages.

Even if all potential biases seem to be under control, a word of caution is due: This book tries to reflect the state of the art of linguistics and archaeology coupled with the available information of population genomics as of the day of its publishing. There is little in science that can be called definitive, and ethnolinguistic identification of prehistoric cultures is not even close to those discoveries and conventions that we could consider firmly established. I have no intention to invest myself into the defence of lost causes, so I would not mind changing any of my interpretations as new data is published: e.g. to argue that Ancient North Eurasian ancestry and Q1a2-M25 represent the Eurasiatic expansion; or to argue that R1a-M420 and Ancient North Eurasian or East Asian ancestry in eastern Europe connects all the necessary dots for the Indo-Uralic expansion, if the new data supports this.

David W. Anthony is a great example of an academic who has invested a lot of time and effort supporting an idea, and has nevertheless changed it as necessary: from a non-Indo-European Corded Ware culture unrelated to Indo-European-speaking Yamna, with certain neighbouring groups adopting the language through "patron–client relationships" (Anthony 2007; Anthony and Ringe 2015); to a Corded Ware culture that expanded with Yamna peoples from the steppe, based on the (then) recently described "Yamnaya ancestry" of genetic papers (Anthony and Brown 2017); to a Corded Ware culture that expanded from Yamna peoples in Hungary, at roughly the same time as it evolved into Bell Beaker, based on the R1a/R1b Y-chromosome bottleneck (Anthony 2017); this last one probably in need of a thorough revision today, as new data has appeared clearly contradicting it. Against this example of a dynamic researcher, there are dozens of known academics unwilling to change one iota of their previous theories, trying to adapt genetic data to their own models. I don't have much doubts about my intentions or interpretations today, but I do hope that I will be able to change what needs to be changed in the future, like Anthony; but also, to distinguish what is wrong from what is not and needs to be defended in spite of what is fashionable, comfortable, or politically correct.

I. Palaeolithic

I.1. Modern humans

Initial Upper Palaeolithic industries associated with the spread of anatomically modern humans could have begun as early as 48000 BC, with Emirian lithics found in the Negev Desert ca. 45000–43000 BC. This first expansion was followed by another successful one from the Levant, represented by the Early Ahmarian (Near East), Kozarnikian (eastern Balkans), and Proto-Aurignacian (south-west and south-central European) lithic cultures.

Admixture with Neanderthals probably took place in the levant during these early population movements out of Africa, as seen in an Upper Palaeolithic Siberian—Ust'Ishim ca. 43000 BC (Fu et al. 2014), of hg. K-M9(xLT)—and an early Upper Palaeolithic East Asian population—Tianyuan ca. 40500 BC (Yang et al. 2017). The so-called Basal Eurasians, not yet sampled, do not show this admixture, which indicates that they formed part of another expanding group, probably located somewhere in the Near East.

By ca. 39000 BC, modern humans had spread into southern Europe, with transitional industries from Middle Palaeolithic Mousterian style (typical of Neanderthals) to Upper Palaeolithic cultures found widespread in Europe: Uluzzian in northern and southern Italy and Greece; Châtelperronian in northern Spain and western and central France; Szeletian in the Czech Republic and Hungary; and the Lincombian–Ranisian–Jerzmanowician from east to west across the Northern European plain.

These pioneer populations in Europe, represented by Ust'Ishim in Siberia and by the Oase1 individual from Romania ca. 38000 BC, who shows a recent Neanderthal contribution less than six generations back in his family tree, did not contribute detectably to any present-day European population, which suggests that later population expansions have wiped out most of their genetic contribution. It has been speculatively proposed that these populations were affected by the eruption of the Archiflegreo volcano ca. 37000 BC, like Neanderthals, and were thus more easily replaced by newcomers.

Populations that began to diverge 40,000 years ago or earlier in Eurasia may thus be simplistically divided into geographic regions—without care for sub-structured populations and gene flow—as (Suppl. Fig. 1):

- Ancestral North Africans (ANA): a deeply splitting ghost population without Neanderthal admixture, assumed to be present during the Upper Palaeolithic in northern Africa.

- Basal Eurasians (BE): another "deep" ghost population, not participating in the Neanderthal admixture, assumed to have diverged from other non-African Eurasian populations ca. 67,400–101,000 years ago. It experienced most of the common bottleneck of non-Basal Eurasians, which suggests their common involvement in their further migration to the Levant (Lazaridis et al. 2018).

- Upper Palaeolithic Siberians: represented by Ust'Ishim and Oase1, they stem from a common Main Eurasian population which admixed with Neanderthals.

- Early East Asians (EEA): represented by Tianyuan [4], mainly contributed to by a common source to Upper Palaeolithic Siberians

---

[4] Tentative SNP call O-M175 or P-295, obtained with Yleaf, from Wang et al. (2019) supplementary materials.

(ca. 98%). It also represents the ancestral population of eastern non-Africans (ENA) —i.e. modern Papuans and Onge, and a contribution of some Native Americans—which in turn contributes to present-day South and East Asians.

- Early West Eurasians (EWE): with the earliest representative samples being far eastern Europeans Kostenki14 and Sunghir3, their parent population—probably widespread through Eastern Europe, the Caucasus and the Near East—contributed for thousands of years to different Upper Palaeolithic and Mesolithic migrations, and their ancestry is found in present-day Europeans and Near Easterners.

I.2. Upper Palaeolithic

The Aurignacian lithic cultural complex appeared around 38000 BC in central Europe, and ca. 33000 BC in western Europe, replacing the earlier Middle and Upper Palaeolithic stone toolmaking styles under a unifying trend. The Goyet cluster (defined by GoyetQ116-1, ca. 33330 BC) appeared associated with the Aurignacian cultural complex, showing that a more homogeneous population accompanied the expanding culture. This genetic homogeneity is probably the result of inbreeding due to small population sizes, as can be inferred from the current estimates for the mean 1,500 (ca. 800–3,300) persons for western and central Europe (Schmidt and Zimmermann 2019).

The GoyetQ116-1 individual is more closely related to the EEA and ANS than any other subsequent European population. Its mitochondrial haplogroup M (found mainly among East Eurasian, Oceanian, and Native American populations) is probably related to this ancestral link (Yang et al. 2017). The ancestry related to Aurignacian samples would continue to contribute to the European stock for thousands of years after this culture's demise.

The Gravettian complex (ca. 31000–15000 BC) succeeded the Aurignacian. This culture is known for its Venus figurines, typically made from ivory or limestone carvings. The Věstonice genomic cluster (represented by Vestonice16, ca. 28060 BC) was also genetically quite homogeneous in samples studied from Italy, Austria, the Czech Republic, and Belgium. This population was related to an ancient hunter-gatherer population sampled in far eastern Europe (Kostenki and Sunghir) ca. 35500–30000 BC (Sikora et al. 2017). Samples from Goyet shared no ancestry with Aurignacian samples from the same site, supporting that this cultural change was also associated with a population replacement.

Although it is likely that this population came originally from the east, the primary centre of expansion of the Gravettian culture (and thus the Věstonice cluster), based on archaeological data, was located in the Middle Danube Basin, spreading to the Upper Danube Basin and into the mid–Atlantic France (Middle Rhine Group, Maisierian group) and south-western Europe (western Gravettian), as well as to eastern Europe (Dniester and Prut Basins) and the Russian Plains (Kostenki–Avdeyevo group).

Chronologically coincident with the Gravettian were the Ancient North Siberians, represented by two Yana RHS samples (ca. 29600 BC), of P1-M45 lineage, ancestral to haplogroups Q-M242 and R-M207 (Sikora et al. 2018). They share the most genetic drift with Ancient North Eurasians (ANE), represented by three individuals from Upper Palaeolithic south-central Siberia (from Mal'ta–Buret' ca. 22350 BC to Afontova Gora ca. 16000 BC), spanning the duration of the Late Glacial Maximum (LGM). Even though Mal'ta–Buret' had Venus figurines with potential similarities to Gravettian ones, the ANE ancestry proper of this cluster is not found among Europeans until the Eneolithic steppe expansions.

The Mal'ta boy's paternal lineage diverged from haplogroup R-M207* shortly before its split into R1-M173 and R2-M479 subclades (Raghavan et al. 2014), which is estimated to have happened ca. 26200 BC. His so-called Ancient North Eurasian (ANE) ancestry contributed substantially to the genetic ancestry of Siberians, Native Americans, and Bronze Age Yamna individuals (Lazaridis et al. 2014), being close to modern-day Native Americans, Kets, Mansi, Nganasans, and Yukaghirs (Flegontov et al. 2016).

The Magdalenian culture spread after the LGM (ca. 23000–17000 BC) from a refuge in southern Iberia, chasing the retreating ice sheet expanding in a northeast direction with El Mirón genomic cluster. From the Late Magdalenian assemblages of Germany and Poland, the expansion of the Hamburgian and Creswellian technocomplexes ca. 13000 BC spread across the Northern European Lowlands south of the Scandinavian Ice sheet.

Relevant known populations before the Final Upper Palaeolithic period may be simplistically divided into geographic regions—without care for sub-structured populations and gene flow—as (Suppl. Fig. 2):

- Goyet cluster: stemming from an EWE source, they expand first with the Aurignacian, and then again from an Iberian refuge during the Magdalenian. During this initial expansion, and also later during the Gravettian, hg. C1a2-V20 (formed ca. 43000 BC, TMRCA ca. 41500 BC) dominates the ancient DNA record.

- Common West Eurasians (CWE): ghost population stemming from an EWE source, probably spread somewhere around the Black Sea. It contributes to different West Eurasian populations to the north (eastern Europe) and south of the Caucasus (Fertile Crescent) during the Palaeolithic.

- Ancient Middle Easterners (AME): Represented by two samples from the Dzudzuana cave in the southern Caucasus (ca. 25000–22000 BC), they show contribution from CWE (ca. 72%) and a BE or earlier BE-like population (ca. 28%). Their ancestral population contributed to Epipalaeolithic Levantine and Anatolian populations, so they are a likely proxy for a contemporary population spanning the whole Fertile Crescent (Lazaridis et al. 2018).

- Věstonice cluster: formed probably in far eastern Europe by an admixture of an EWE source close to CWE (ca. 62%), and from another EWE source close to the Goyet cluster (ca. 38%), it expanded west with the Gravettian expansion. The spread of haplogroup I-

M170 (formed ca. 40900 BC, TMRCA ca. 25500 BC) in Europe is probably associated with the expansion of certain groups of the Věstonice cluster.

- Ancient North Siberians (ANS): It diverged ca. 36000 BC from EWE (soon after this population split from East Asians), with further contributions from EEA (ca. 25%) also soon after the split EEA–EWE.

- Ancient North Eurasians (ANE): contributed to by ANS, an EWE source close to the Goyet cluster (ca. 75%) and an EEA population (ca. 25%). This component contributed substantially to the genetic ancestry of Siberians, Native Americans, and Bronze Age Yamna individuals (Lazaridis et al. 2014), being close to modern-day Native Americans, Kets, Mansi, Nganasans, and Yukaghirs (Flegontov et al. 2016). Their wide distribution of a consistent Initial Upper Palaeolithic technology in Siberia and neighbouring territories up to north Mongolia seems to be coincident with the eastward expansion of peoples through the steppe belt (Zwyns and Lbova 2018).

- El Mirón cluster: it derives most of its ancestry from Goyet (63%), and it has not been found in the previous Gravettian period. This is thus the first described case of a 'resurgence' of a population from an (as of yet) unsampled pocket that survived a previous population turnover, and persisted for a long time after their assumed disappearance. The main haplogroup associated with its expansion is I-M170 also found later in the Villabruna cluster, which suggests that Gravettian lineages admixed with a population of the Goyet cluster, and remained in isolation (most likely in Iberia) until the Magdalenian expansion.

I.3. Epipalaeolithic

Starting ca. 12000 BC with the first strong warming period after the last ice age, known as the Bølling-Allerød interstadial, a new migration replaced part of the European population, helped by the melting of the Alpine glacial wall that divided west and east Europe. Individuals associated with diverse Epipalaeolithic cultures (ca. 12000–5000 BC)—transition to the Epigravettian in southern Europe, and Magdalenian–to–Azilian transition in western Europe—fall into a newly emerged Villabruna genomic cluster, which displaced previous hunter-gatherer populations.

The population ancestral to the Villabruna cluster separated from the ancestors of contemporary populations found in the Near East. It is during this time that western European hunter-gatherers become much more closely related to modern Near Easterners, proving that the new migration likely happened from the Near East into Europe. The defining sample comes from an Epigravettian individual from Villabruna, Italy (ca. 13000 BC), of hg. R1b1-L754, and this lineage is also found in Loschbour (ca. 9775 BC), along with an individual of hg. I2-M438.

Other individuals include from Bichon, Switzerland (ca. 11700 BC), hg. I2a1a1b1-L286; Loschbour, Luxembourg (ca. 6100 BC), hg. I2a1a2-M423; as well as samples from La Braña, Iberia (ca. 5865 BC), hg. C1a2-V20; and Körös (Hungary ca. 5710 BC), hg. G-M201. Ancient individuals from France, Sicily, Croatia, France, and Germany share this ancestry, which suggests that the Villabruna cluster was widely distributed in Europe for at least six thousand years, and probably expanded from a south-eastern European refugium following the last Ice Age ca. 13000 BC (Mathieson et al. 2017).

Of the fifteen samples studied, four individuals from central and central-west Europe show a distinct component found in modern East Asians, particularly Loschbour and La Braña, which indicates gene flow from a population related to modern-day East Asians into some groups of the Villabruna cluster, consistent with gene flow between populations related to

East Asians (Fu et al. 2016). This supports the potential arrival of R1b1-L754 lineages from Asia associated with a male-biased migration of an eastern population.

Based on the most recent data of modern populations, an origin of the split into R1b-M343 and R1a-M420 is estimated ca. 20800 BC, with a TMRCA ca. 18400 BC for R1b-M343, and ca. 16200 for R1a-M420. The formation of R1b1-L754 is estimated ca. 16900 BC, with a time to MRCA ca. 15100 BC, suggesting successive migration events, starting probably near Siberia in Asia, based on the Mal'ta sample of hg. R-M207.

Hunter-gatherers from the Iron Gates prove the regional continuity of haplogroup R1b1-L754 (xR1b1a1-P297, xR1b1a1b-M269). These samples were probably from branches that have not survived in modern populations, and they cover an extensive period spanning from the first half of the 10th millennium to the first half of the 6th millennium BC, with the latest samples showing already Middle East farmer ancestry (Mathieson et al. 2017; González-Fortes et al. 2017).

More individuals possibly related to these ancient branches are found later in Ukraine, Iberia, and central European Neolithic in Quedlinburg as R1b1-L754 (xR1b1a1b-M269) ca. 3590 BC (Haak et al. 2015). These samples, coupled with individuals of hg. R-M207 found in Ganj Dareh (Iranian Neolithic) in the first half of the 9th millennium BC might suggest a southern Eurasian migration route for R1b1-L278 lineages, through the Iranian plateau.

The samples of basal R1b-M343* lineages in modern populations of southern Kazakhstan (Myres et al. 2011) and Iran (Grugni et al. 2012) give further support to the southern migration route into Europe. Basal R1b1-L278* lineage was found in five individuals—3 Italians, 1 West Asian, 1 East Asian—out of 5,326 samples studied (Cruciani et al. 2010) , which also point to a potential ancestral migration into Europe.

During the Bølling-Allerød interstadial, various divergent populations coexisted in Eurasia and Africa (Suppl. Fig. 3):

- Epipaleolithic Iberomaurusians: represented by samples from Taforalt (ca. 18000–8000 BC), of hg. E1b1b1a1-M78 (formed. ca. 17600, TMRCA ca. 11300 BC) they derive their ancestry from ANA (ca. 45%) and a mix of CWE (ca. 40%) and other "deep" ancestry (ca. 15%). They contributed mainly to Early Neolithic populations from Morocco, and also to the Natufian population.

- Epipalaeolithic Natufians: represented by samples from the Raqefet Cave (ca. 11300–10800 BC), probably all of hg. E1b1b1a1-M78, are a Levantine population of hunter-gatherers who lived in permanent dwellings and managed local wild plants. They show contribution from AME (73%), but also from ANA (ca. 27%), consistent with the spread of morphological features and artefacts into the Near East, as well as Y-chromosome haplogroup E.

- Anatolian hunter-gatherers (AHG): represented by an individual from Pınarbaşı in Northern Anatolia (ca. 13350 BC), of hg. C1a2-V20, mtDNA k2b, whose ancestry descends mostly from AME (>95%), with small contributions from an ENA/ANE source, from Villabruna, and possibly from the Levant (Feldman et al. 2019). It contributed to Early Anatolian Neolithic populations.

- El Mirón: While a Mesolithic individual from Chan in north-west Iberia (ca. 7200 BC) shows continuity with El Mirón ancestry, the La Braña brothers from ca. 1,300 years later, were closer to central European hunter-gatherers like the Hungarian Körös, with an even more extreme shift ca. 700 years later in Canes, also from the Cantabrian region, reflecting a gene flow from Villabruna affecting north-west Iberia but not the south-east or south-west, where individuals remained close to El Mirón. The incursion of Villabruna ancestry in Iberia is dated to at least 12000 BC, when a sample from Balma Guilanya in north-east Iberia shows it. Nevertheless, all these Mesolithic samples show still higher Goyet ancestry than non-Iberian hunter-gatherers. One example of a late El Mirón-like individual from the south-east comes from Cueva de la Carigüela (ca. 9700–5500 BC) potentially of hg. I1-M253 (formed ca. 25500 BC, TMRCA ca. 2600 BC), and a certain I1-M253 is found in an older sample from Balma Guilanya, which could mean that this lineage expanded from Iberia to the north with the Magdalenian expansion or later population movements (Olalde et al. 2019; Villalba-Mouco et al. 2019).

- West European hunter-gatherers (WHG): derived from CWE, they are represented by the Villabruna cluster in Europe. Due to its common root with AME—relative to which it lacks BE-like contribution—they are supposed to represent a population in or near Anatolia that expanded to central Europe probably from a region near the Black Sea. It is possibly part of a big AME transitional cline that connected WHG and AHG during the Palaeolithic, since south-eastern European hunter-gatherers show extra Anatolian admixture, just like AHG shows small WHG admixture. This ancestry dominates over most European hunter-gatherer populations until the arrival of the Neolithic ca. 6000 BC. Samples like Villabruna in northern Italy (ca. 12000 BC), OrienteC in Sicily (ca. 12000 BC), Bichon in Switzerland (ca. 11700 BC), Croatia Mesolithic (ca. 7200 BC), Loschbour in north-west Europe (ca. 6100 BC), La Braña 1 in north Iberia (ca. 5900 BC), or Körös in Hungary (ca. 5700 BC), all form a close WHG cluster spanning 6,000 years from the Atlantic façade to Sicily in the south and to the Balkan peninsula in the south-east (Mathieson et al. 2018).

- ANE: represented in this late Upper Palaeolithic period by the Afontova Gora 3 sample from Lake Baikal, tentatively classified as of haplogroup Q1a-F1096 (formed ca. 24000 BC, TMRCA ca. 23900

BC) or possibly R1-M173[5] (formed ca. 26200 BC, TMRCA ca. 20800 BC). The creation of EHG ancestry (basically a WHG:ANE cline, see below) in Eastern Europe and the Caucasus was most likely associated with the westward migration of groups of ANE ancestry, probably mainly of hg. Q1a2-M25 (formed ca. 22400 BC, TMRCA ca. 14300 BC) and R1-M173 through North Eurasia. Samples of haplogroup Q-M242 found in a Baltic hunter-gatherer (ca. 6500 BC), and later in Eneolithic populations from the Caucasus, are likely remains of this early expansion. Modern-day Kets, Mansi, Native Americans, Nganasans and Yukaghirs show maximum ANE ancestry (Flegontov et al. 2016).

- Ancient East Asians (AEA): represented by hunter-gatherers from the Early Neolithic in Lokomotiv and Shamanka (ca. 5200–4200 BC) near Lake Baikal, they show predominantly East Asian ancestry closely related to ancient individuals from the Devil's Gate Cave (ca. 6000–5500 BC), and some ANE-related contribution (ca. 16%), representing thus another proxy for an ancestral ENA-like ancestry to compare with ANE (de Barros Damgaard, Martiniano, et al. 2018; Lazaridis et al. 2018; Sikora et al. 2018). They show one sample of haplogroup C2a1a1a-F3918 and other five probably N1a2-L666 (formed ca. 13900 BC, TMRCA ca. 6800 BC). A Jomon sample from Japan dated ca. 2,500 years ago, with close affinity with a 8,000-years old Hòabìnhian hunter-gatherer, and unaffected by ANE gene flow, supports a coastal route of the earliest wave of East Asian ancestry (Gakuhari et al. 2019), suggesting that the East Asian coast was a sink rather than a source during prehistoric population movements. The first appearance of AEA-related ancestry in Eastern Europe must have happened quite early, possibly later than the ANE expansion into

---

[5] Additional information: Q1a originally reported by YFull; tentative SNP call R1 obtained with Yleaf, from Wang et al. (2019) supplementary materials.

Europe, and possibly associated with the spread of R1a1-M459 (formed ca. 16200 BC, TMRCA ca. 12000 BC) from Siberia into the Pontic–Caspian area.

- Eastern hunter-gatherers (EHG): it can be modelled as an admixture of ANE (ca. 63%) with WHG (34%), with additional ancestry related to AEA (ca. 7%), but without additional AME contribution. It is represented by hunter-gatherers from eastern Europe, each with its specific contributions of different components, which suggests that it formed a quite stable east European cline between more WHG-like populations from the west and more ANE-like populations from the east. The presence of a Q1a2-M25 sample ca. 6500 BC in Zvejnieki points probably to the resurgence of this lineage that had spread ANE ancestry to eastern Europe, although R1a1-M459 lineages may have been involved in the creation of the cline, too. The first EHG sample is an individual from Sidelkino, from the Samara region (ca. 9300 BC), of mtDNA U5a2—with mtDNA U5 being a constant in prehistoric eastern European populations. The appearance of hg. J1-M267 in two early EHG samples from Karelia (ca. 6300 BC) may be related to an early expansion from the south, possibly of J1b1-Y6034 (formed ca. 12900 BC, TMRCA 9700 BC) from the Caucasus with ANE ancestry, creating the described cline of ANE:WHG:CHG ancestry (ca. 60:25:15) in these samples (Sikora et al. 2018).

- Caucasus hunter-gatherers (CHG): represented by samples from Satsurblia (ca. 11300 BC, haplogroup J1-M304), and Kotias Klde (ca. 7800 BC, haplogroup J2a-M410), both sites near Dzudzuana. They show AME ancestry (ca. 56-64%) and a contribution of ENA/ANE-like populations, apart from a small "deep" ancestry.

- Iran Neolithic (IN): represented by a Mesolithic child from the Belt Cave (ca. 12000–8000 BC), of hg. E1b1-P2, and individuals from Ganj Dareh in the Zagros Mountains (ca. 9000–8000 BC), of hg. R2-

M479. They form an EHG:ANE cline similar to CHG, and thus form likely an ancestral CHG/IN population formed mainly by AME ancestry (ca. 50–58%), where IN shows a slightly higher contribution of ANE, with a statistically significant greater contribution of "deep" ancestry likely from the south.

- Ancient Palaeosiberians (AP): represented by the Kolyma1 individual (ca. 7800 BC), they derive their ancestry from a mixture of EEA and ANS ancestry similar to that found in Native Americans (but with greater EEA contribution, 75% vs. 63%), with a closer relation to Mal'ta than to Yana RHS. The divergence of AP/Native Americans and present-day East Asians (Han Chinese) is estimated to have happened ca. 22000 BC, with AP/Native Americans showing further contribution (ca. 18000 BC) related to ANS (Sikora et al. 2018).

- Ancient Ancestral South Indian (AASI): hypothesised South Asian hunter-gatherer ancestry deeply to present-day indigenous Andaman Islanders (Mallick et al. 2016), in particular the sampled Onge population, mainly of hg. D-M174 (Thangaraj et al. 2003), and mtDNA M2 and M4 (Reich et al. 2009; Moorjani et al. 2013).

i.3. Nostratians

Based on the most recent data of modern populations, an origin of the split into R1b-M343 and R1a-M420 is estimated ca. 20800 BC, with an expansion of R1b-M343 (TMRCA 18400 BC), then R1b1-L278 (formed ca. 18400, TMRCA ca.16900 BC). The finding of intrusive haplogroup R1b1-L754 (formed ca. 16900 BC, TMRCA ca. 15100 BC) with a homogenous WHG ancestry in Europe, and the consistent presence of mtDNA hg. U5b in different samples of the Villabruna cluster, support a male-biased migration coincident with the Bølling-Allerød interstadial.

The Mal'ta individual and the finding of the other main R1b-M343 subclade, R1b2-PH155 (TMRCA ca. 5200 BC), among early Xiongnu individuals in East Asia—and later accompanying Turkic peoples—supports a split of R1b-L278 in eastern Europe or central Eurasia, and was most likely associated with the expansion of ANE and/or EHG ancestry from Asia, possibly in successive waves of expansion that also accompanied haplogroups Q1a2-M25 and R1a-M459 to the area.

This eastern origin may justify the presence of East Asian ancestry among some samples from the Villabruna cluster associated with expanding R1b1-L754 lineages. Whether R1b-M343 lineages traversed the Middle East or expanded from the Pontic–Caspian region into Europe is unclear, although the high variability of ancient subclades found to date in eastern Europe and the Caucasus supports the regions on both sides of the Urals as the most likely cradle of R1b-M343 expansions.

Sampled hunter-gatherers from south-eastern Europe show the long-term regional continuity of haplogroup R1b1-L754 (xR1b1a1-P297, xR1b1a1b-M269), found in the Iron Gates—and also later in Mesolithic and Neolithic populations from Ukraine, the Balkans, Central Europe and Iberia. They have WHG (87%) and EHG (13%) ancestry, show mtDNA K1 and H (not present in WHG and EHG individuals), and many of these samples have been confirmed as of subclade R1b1b-V88 (formed ca. 15100 BC, TMRCA ca.

9700 BC), which must have split at the same time as R1b1-L754 was expanding into Europe. European R1b1b-V88 lineages cover thus an extensive period spanning from the first half of the 10th millennium to the first half of the 3th millennium BC, and are found widespread from Iberia in the west to the north Pontic area in the east (González-Fortes et al. 2017; Mathieson et al. 2018).

The samples of basal R1b-M343* lineages in modern populations of southern Kazakhstan (Myres et al. 2011) and Iran (Grugni et al. 2012) give further support to an eastern origin in Central Asia. Basal R1b1-L278* lineages were found in five cases out of 5,326 cases studied – three Italians, one West Asian, one East Asian (Cruciani et al. 2010) –, which also point to a potential ancestral migration into Europe. Nevertheless, population movements after their initial expansion may have obscured the original migration route, and an expansion through Anatolia cannot be excluded.

Tracing backwards potential Eurasiatic and Afroasiatic movements, and based on male-driven population expansions, the clearest link to an expanding Nostratic-speaking community is represented by the expansion of R1b1-L754 lineages, starting probably after ca. 16000 BC through the North Pontic area into south-eastern Europe, acquiring along the way the characteristic CWE-like ancestry of the Villabruna cluster.

The presence of R1b1b-V88 lineages widespread among European hunter-gatherers point to a likely early "southern Nostratisation" of Europe from east to west. The expansion of R1b1b-V88 subclades within Africa is most likely linked to the spread of Proto-Afroasiatic (see below *§ii.4. Early Afrasians*). The expansion of R1b1a1-P297 into north-east Europe, later emerging with post-Swiderian cultures, marks the clearest trace of the potential Eurasiatic expansion (see *§ii.1. Eurasians*). Even though the precise origin of expansions of R1b1-L754 subclades remains unclear, the regions surrounding the Pontic–Caspian area are the best candidates at this moment.

While the formation of hg. R2-M479 was quite early (ca. 26200 BC), its lineages survived probably somewhere in Asia until its successful expansion (based on its TMRCA ca. 14300 BC), and should probably be identified with the additional ENA/ANE contribution to Iranian Neolithic (and possibly CHG) ancestry, since they are found in samples from Ganj Dareh during the 9th millennium BC. Because one sample is R2a-M124 (formed ca. 14300 BC, TMRCA ca. 9600 BC), Iran Neolithic individuals are probably close descendants from this haplogroup's successful expansion. Haplogroup R2a-M124 seems to be prevalent among ancient and modern Dravidians (see *§viii.20. Dravidians and Indo-Aryans*), and is also found in the Caucasus (Huang et al. 2017).

A connection of Dravidian with R1b-M343 is not straightforward, then, lacking fitting ancient DNA samples. Nevertheless, the likely initial expansion of R1b1-L754 lineages with ANE ancestry, as well as early expansions through the Caucasus or Turan, may have contributed to the development of other Nostratic communities in the Near East. Similarly, there is no clear connection between this haplogroup and Kartvelian, although the complex evolution of multiple small communities in the Caucasus probably allowed for many ethnolinguistic changes in the region, associated with different haplogroup expansions.

The expansion of R1b1a1-P297 lineages apparently associated with Eurasians (see below *§ii.1. Eurasians*) and the later emergence of R1b1a2-V1636 lineages (TMRCA ca. 4700 BC) in the Pontic–Caspian steppe region (see below *§iv.2. Indo-Anatolians*) supports the expansion of their upper clade R1b1a-L388 (TMRCA ca. 13600 BC) from far eastern Europe, having separated (ca. 15100 BC) with sister clade R1b1b-V88 from the ancestral R1b1-L754 trunk.

This early split of R1b1a-L388 may account for the separation of Eurasians from Pre-Kartvelians, who would have expanded close to the Caucasus with R1b1a2-V1636 lineages, while Eurasians expanded through the north. The early separation of R1b1b-V88 from the eastern European *cradle* of hg. R1b1-L754 and of ANE/EHG ancestry expansions would support a closer connection of ancestral Eurasiatic, Kartvelian, and possibly also Dravidian communities with each other than with Afroasiatic. The presence of basal R1b1a-L388 subclades in modern individuals from Turkey, Bulgaria, and Italy would also suggest eastern European routes of expansion for this lineage, rather than southern routes through the Caucasus or West Asia.

The timing of expansion and separation of these lineages from the common R1b-M343 trunk (Suppl. Graph. 19) coupled with known admixture events fit some of the previously published 'shape-shifting' Nostratic macro-languages (Campbell 1998), as well as roughly the dates published with help of language guesstimates coupled with archaeology (Beridze 2019) and statistical models (Pagel et al. 2013), which essentially predict an earlier separation of Dravidian, followed by that of Kartvelian from the common Eurasiatic superfamily (Figure 4).

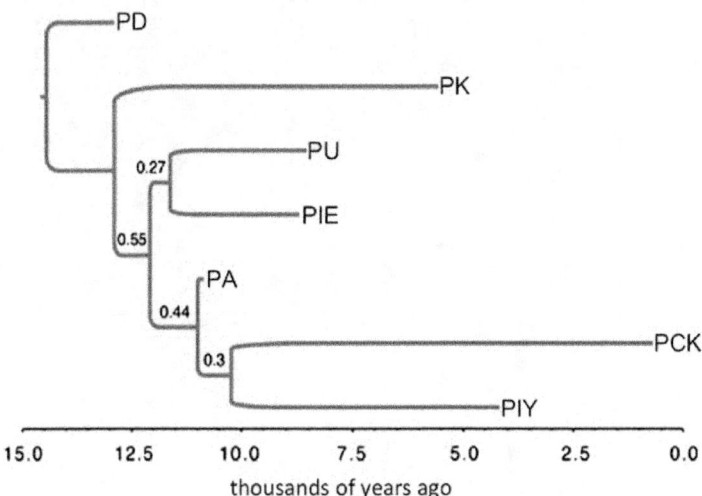

Figure 4. Consensus phylogenetic tree of Eurasiatic superfamily rooted tree with estimated dates of origin of families and of superfamily by Pagel et al. (2013). P proto followed by initials of language family: PD proto-Dravidian, PK proto-Kartvelian, PU proto-Uralic, PIE proto–Indo-European, PA proto-Altaic, PCK proto–Chukchi-Kamchatkan, PIY proto–Inuit-Yupik. Consensus tree rooted using proto-Dravidian as the outgroup. The age at the root is 14.45 ± 1.75 kya (95% CI = 11.72–18.38 kya) or a

slightly older 15.61 ± 2.29 kya (95% CI = 11.72–20.40 kya) if the tree is rooted with Proto-Kartvelian.

II. Mesolithic

II.1. North-Eastern Technocomplex

Blade production by pressure technique is a marker of a particular craft tradition that emerged in the Mongolian area ca. 20,000 years ago, and spread from east to west during the last glacial maximum, reaching the Baltic Sea and Scandinavia in the Mesolithic.

In the north, the population of the final Palaeolithic Swiderian culture of deer hunters, which had developed in Poland on the sand dunes left behind by retreating glaciers, migrated during the Palaeolithic–Mesolithic transition at the turn of the 11th–10th millennium BC to the north-east following the retreating tundra (Terberger et al. 2018), which is evidenced by a centuries-long settlement break before a new populations arrived (Kobusiewicz 2002). Morphological similarities in the tanged points of the East European Swiderian points with Kunda and Butovo cultures supports groups migrating north during the Late Pleistocene / Early Holocene (Zaliznyak 1999).

Post-Swiderian cultures developed in particular the surface pressure flaking technology further, and technical differences with Swiderian cultures include single platform cores, pressure blade debitage, inset technology, etc. (Darmark 2012). The culture expanded in the Baltic and in the east European forest zone, north of the (then unstable) Pontic–Caspian area, especially during the Early Mesolithic (ca. 9000–8300 BC) and Middle Mesolithic (ca. 8300–6000 BC), although the earliest date for similar material is currently set at the end of the 11th millennium BC. The Kunda culture developed around the eastern Baltic, from the Polish Plains to the Gulf of Finland; the Butovo culture in the Volga and Oka regions; and the Veretye culture in the eastern part of Lake Onega (Suppl. Fig. 4).

During the Early Mesolithic period, human settlement shifted from the major river valleys to the inland lake regions, and changes are seen in the extraction and processing of lithic raw materials, technology, and tool morphology. There is a rich bone and antler inventory—harpoons with large, widely spaced barbs, slotted and needle-shaped points, daggers, etc.—and a less diverse lithic inventory—flint end-scrapers and blade inserts, rarely tanged points (Damlien et al. 2018).

A typical feature is the use of imported high-quality Cretaceous flint, originating from areas to the south, in the forest zone of western Russia. In the Middle Mesolithic, settlements concentrate in inland lake basins, the most extensively excavated site being Zvejnieki II, in the northern region of the Kunda culture. Compared to the previous period, there is a richer lithic inventory, dominated by side- and end-scrapers, inserts, and some burins, and mainly local raw material is used (Damlien et al. 2018).

Mesolithic arrowheads from Butovo (Figure 5) show that they were made using a standard operation chain, with sophisticated technology, and some of them were treated with special care. They were mainly used for hunting, for a short time, but use-wear and traces of repair in some specimens suggest they were used for a long time. These arrowheads show more differences than common traits with Dubensee and Maglemose cultures of western and central Europe. Slotted bone points with flint inserts appearing in Denmark and Scania in the second part of the Boreal period differ in the position and morphology from East European artefacts (Zhilin 2017).

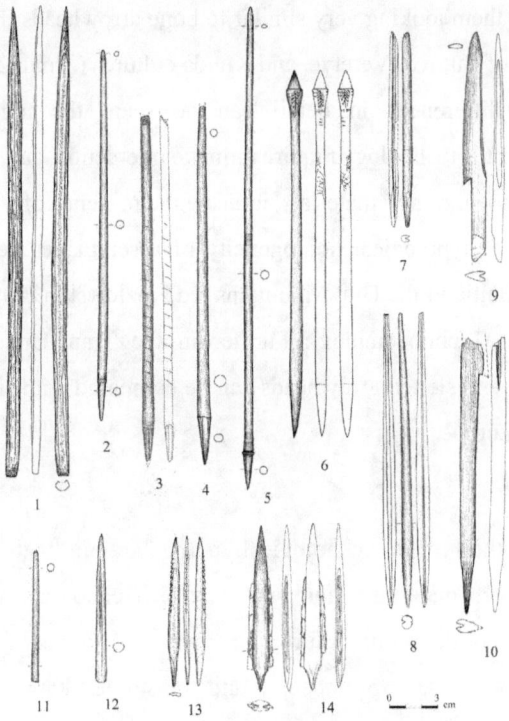

Figure 5. Bone arrowheads from the early Mesolithic layer IV. From Zhilin (2017).

On the other hand, the technology and typology of early Mesolithic arrowheads from the Eastern Baltic to the Upper Volga area show similarities that indicate regular communication and the existence of social networks among populations of these regions. This is further confirmed by the spread of Pulli type flint arrowheads, some types of retouched inserts, and specific types of flint raw materials. Further developments in technology during the Boreal period show a resemblance between Veretye and Butovo cultures, differing from the Kunda culture, but still with contacts between the regions (Zhilin 2017).

The Kunda–Butovo complex is linked together thus by lithic provenience from hundreds of km apart, and by a similar technology. Peat bog sites of the Trans-Urals area also produced bone arrowheads: needle-shaped, narrow flat slotted, one-winged with barbs, paddle-shaped and arrowheads with thickened head, some of them looking very similar to bone arrowheads from Mesolithic peat bog sites of Butovo, Veretye, and Kunda cultures (Zhilin 2017).

Although differences in detail can be seen, the transmission and maintenance of the technology require intimate interaction, and, in conjunction with the movement of raw materials, indicate the presence of a social network in the area. The typological homogeneity of assemblages between regions spanning the Baltic to the Ural Mountains led Kozłowski (2009) to call it the 'North-Eastern Technocomplex'. Further in the Trans-Urals and Siberia, needle-shaped and slotted arrowheads can be compared with similar artefacts from Eastern Europe.

ii.1. Eurasians

Following the French technological approach, along with the concept of *chaîne opératoire*, material culture such as lithic technology can be argued to represent a manifestation of culturally transmitted knowledge that is learned and shared among a group of people and transmitted between generations, thereby reflecting social traditions. The specific combinations of tehnology (including operational sequence, combination of raw materials, tools, gesture, etc.) are part of a craft tradition, a knowledge and know-how shared by a specific social group. While single technological elements can be transmitted easily within a generation or peer group, complete production concepts are more likely to be learned and passed unchanged through many generations (Damlien et al. 2018).

The early successful expansion of haplogroup R1b1a1-P297 (TMRCA ca. 11300 BC) may be associated with the spread of post-Swiderian cultures in north-eastern Europe, linked thus to the North-Eastern Technocomplex and potentially to the expansion of Eurasiatic, whereas the earlier successful expansion of its parent haplogroup R1b1a-L388 (TMRCA ca. 13600 BC) in eastern Europe may have brought R1b1a1-P297 in contact with the disintegrating Swiderian culture. Samples from the Kunda and the succeeding Narva cultures show an ancestry intermediate between WHG (ca. 70%) and

EHG (ca. 30%). The earliest samples of Kunda in Zvejnieki during the 8th–6th millennium BC are of hg. R1b1a1-P297, and there is continuity of this lineage until the end of the 4th millennium in the Baltic (Jones et al. 2017; Mathieson et al. 2018).

Other samples of hg. R1b1a1-P297 are also found (at least since the 6th millennium BC) in the Volga–Ural region. Subclades R1b1a1a-M73 found in Siberia (see *§v.8. Palaeosiberians*) and later in Asian populations, as well as the expansion of R1b1a1b-M269 much later from the Volga–Ural area (see below *§iii.5. Early Indo-Europeans and Uralians* and *§viii.21.1. Yukaghirs*) support this early expansion of R1b1a1-P297 lineages through eastern Europe into the Trans-Urals area up to the Altai Mountains. The finding of R1b1a1-P297 subclades in modern East Asian territories from Russia, China, or Japan further support the later association of some eastern European or central Asian groups with these lineages.

II.2. Colonisation of Scandinavia

The ice sheet retracted from northern Europe allowed for the colonisation of the Scandinavian Peninsula from about 9700 BC, according to the archaeological record, both in southern and northern Scandinavia, while ice still dominated the interior.

The peninsula seems to have been colonised first from the south by peoples from central Europe, related to late-glacial lithic technology (direct blade percussion technique), which brought WHG ancestry with them. Komornica traditions from western Latvia show technological similarities with the western Baltic region, shared with Maglemose cultures across the Polish Plain to the islands of eastern Denmark and southern Sweden, where the core platforms were generally kept unprepared, and formal microliths were an integrated element of the lithic tool tradition, suggesting that the pressure blade technology was not adopted completely (Damlien et al. 2018).

An invasion from the north-east (through Finland) by post-Swiderian Mesolithic groups from east Europe is evidenced by the arrival of EHG ancestry with them. Their technology is represented by their pressure blade technology in central and western Scandinavia, where the core platform was formed and repeatedly rejuvenated by detachments of core tablets, and by systematically faceting the platform surface. Formal microliths are generally absent, and blade inserts dominate. This is clearly documented in Zvejnieki II, in northern Latvia (Damlien et al. 2018).

This regional variation in pressure blade technology is an expression of two different culturally derived traditions that existed synchronously in the Northern European Plains and around the Baltic Sea during the Middle Mesolithic (Figure 6). The adoption of the earliest form of pressure blade technology in Maglemose/Komornica (dating to the early Middle Mesolithic in post-Swiderian cultures) points to the adoption of technology by western groups, and thus to the existence of two main distinct cultural, economic, social, and communication networks in northern Europe (Damlien et al. 2018).

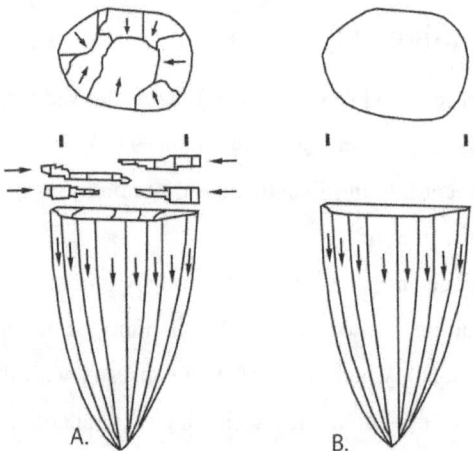

Figure 6. Different types of core platform rejuvenation and preparation. A: Platform preparation by systematic faceting and repeated rejuvenation. B: Unprepared platforms (modified after Sørensen et al. 2013, fig. 1). Image from Damlien et al. (2018).

These population movements left a paradoxical pattern of increased EHG ancestry in northern and western Scandinavia, and WHG ancestry in eastern and central Scandinavia, which correlates with Baltic samples (Suppl. Graph. 3). It has also been shown that selection drove the unique combination of light skin and hair and varied blue to light-brown eye colour, as part of the adaptation to a different environment, contrasting with WHG who had the specific combination of blue eyes and dark skin (Günther et al. 2017).

ii.2. Northern Europeans

Individuals from Mesolithic Scandinavia, which define the so-called Scandinavian hunter-gatherer (SHG) ancestry, show exclusively I-M170 subclades, probably resurged from the previous Magdalenian-Hamburgian expansions, and mtDNA haplogroups U5 and U2 (Mathieson et al. 2017). The first sample of haplogroup I1-M253 is probably found in Gotland ca. 6950 BC, which points to the survival of these lineages in pockets of Scandinavia until the formation of the Nordic Late Neolithic.

The expansion of the pressure blade technology proper of post-Swiderian cultures was also driven by I2-M438 lineages, which probably 'resurged' in certain Baltic communities at the end of the 7th millennium, side by side with hg. R1b1a1-P297. For example, Zvejnieki hunter-gatherers show six individuals of hg I2-M438 (ca. 6000–4775 BC), with two I2a1a-P37.2, two I2a1b1a2-CTS10057 (formed ca. 8500 BC, TMRCA ca. 8500 BC), and two I2a1a2a1-S2639 (formed ca. 5000 BC, TMRCA ca. 4700 BC).

The assemblages of Motala hunter-gatherers from south-western Sweden, for example, have been described as having general traits in common with Maglemosian cultures (e.g. microblade technology), but differ in other respects (e.g. an absence of geometrical microliths), with further similarities and differences with the "quartz and pecked axe complex" found in north-east and central Sweden. Because of that, it is considered a transition zone, a mixture of cultural contacts (Eriksson et al. 2018).

The east–west and north–south mixture of ancestry and lineages in Scandinavia is thus further complicated by cultural contacts. Motala hunter-gatherers, sampled since the 8th millennium up to the 6th millennium BC, show mainly I2a1a2-M423 lineages (formed ca. 16400 BC, TMRCA ca. 9200 BC). An individual of eastern pressure blade technology from Huseby Kiev (ca.

7800 BC) shows mainly WHG ancestry, belonging to the *south-eastern* SHG group, in spite of its culture belonging to the expansion from the north-east, which may support cultural diffusion, but also (probably more likely) the gradual admixture during the population expansion, and later population replacement from the south (Kashuba et al. 2019).

The same WHG-related ancestry is found in another sample of Maglemosian culture from Stora Förvar (ca. 6600 BC), and in an individual associated with handle core technology from Motala (ca. 5640 BC), which supports continuity of this ancestry in the region. Furthermore, it seems that EHG ancestry did not reach southern Denmark before the Neolithic (Jensen et al. 2018).

II.3. Pontic–Caspian zone

The end of the last Ice Age ca. 14000–12000 BC brought instability to the Pontic–Caspian area: meltwater flew torrentially from the northern glaciers and the permafrost into the Khvalynian Sea (the Caspian Sea is a small remaining part of it). A shoreline between the Middle Volga and the Ural rivers restricted east–west movements south of the Ural Mountains (Anthony 2007).

By 11000–9000 BC water may have poured into the Black Sea (Major et al. 2006; Ryan 2007), enlarging it and creating the Sea of Azov. Although the magnitude and rapidity of this flow remains controversial (Yanko-Hombach, Gilbert, and Dolukhanov 2007), it is agreed that meltwater created unstable shores in eastern Europe (Patton et al. 2017).

Deglaciation and palaeoclimatic changes were probably more important in their potential for environmental, cultural, social and historical changes of this region, though. A significant deterioration is found first during the Younger Dryas, a severe cold spell that lasted between 10800 and 9500 BC. Then a warming trend began, with climate aridisation and reduction of overall biomass density seen in the region during the transition to the pre-Boreal period, with large group segmentation, local population dispersion, increase in population mobility, and decrease in population density (Smyntina 2016).

In the Boreal period, the Pontic–Caspian steppe became stable with an increase in climatic humidity, and a growth of biomass density. Hunters, probably incoming from eastern and western regions, settled there and population density increased (Anthony 2007).

II.3.1. North Pontic steppes

In the Pontic steppes, Crimea shows new technology of flint blade production from a Near Eastern inspiration in the Murzak–Koba culture (ca. 10th–8th millennium BC). Reindeer hunters of the western region, in the Dniester valley, became deer hunters and riverine fishers during the Early Mesolithic. In the Dnieper–Donets steppes, bison and horse hunters of the Late Palaeolithic became deer and horse hunters of the Early Mesolithic. Early Mesolithic horse hunters also appeared in the Dniester–Prut steppes (Biagi and Kiosak 2010).

The Near East component is still visible in the Epigravettian Kukrek and Grebeniki cultures. Kukrek appeared as heir of this Near Eastern tradition in the late 8th millennium BC, and included the eastern part of the north Pontic area, expanding to the north (Biagi and Kiosak 2010).

The north-western Pontic region was thus exploited by intertwined communities distinguished by different material cultures: the non-geometric Anetivka variant of Kukrek, and the geometric Grebeniki material culture, offspring of Early Mesolithic Tsarinka flint knapping tradition, continuing thus earlier western Pontic traditions. Two basic inventions are attributed to this period: the first attempts at aurochs domestication, in the Lower Danube region, and a significant intensification of the use of wild plants, fish, and other river resources (Smyntina 2016).

The next development was the transition to the stage of food-producing economy, with two cultures appearing in the north Pontic area: the Azov culture in the east, and the Bug–Dniester culture in the west. Both have pre-pottery stages starting around 7500–6000 BC, with the Azov culture being a continuation of the Kukrek tradition. The Bug–Dniester culture (emerging ca.

6500 BC) was in close contact with the Criş–Starčevo culture from the Balkans, and existed for about a thousand years, until they became integrated into the Early Trypillian ethnocultural complex (Telegin et al. 2015).

In the Dnieper rapids—represented by the Vasylivka site—the Kukrek tradition was replaced by the Surskii culture—with stone vessels characteristic of the Neolithic of Asia Minor—and then by the Middle Neolithic Dnieper–Donets I (Anthony 2006), which replaced it also ca. 5500 BC. The earliest Mariupol-type cemeteries of the Dnieper–Donets culture (Figure 7), such as Vasylivka II and Marivka, as well as the early horizons of the Rakushechny Yar sites in the Lower Don area and the monuments of the Kaia-Arsy type in the Crimea, all date approximately to 6500-5500 BC (Telegin et al. 2015).

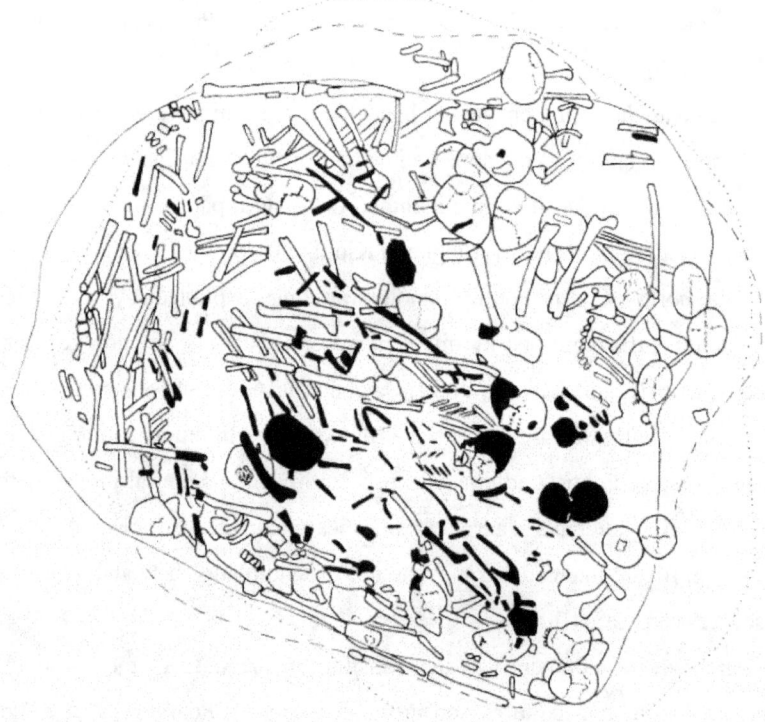

Figure 7. Dnieper–Donets culture collective burials in pit 3 at Nikolskoye. Burnt bones are shown in black. After Telegin. Image modified from Whittle (1996).

In the forest-steppe area, the Mesolithic is represented by the Zimovniki culture (in its different stages, since the Early Mesolithic), from north-east Azov to the Don–Volga interfluve, with a material culture that has been linked to the lithic industry of the Caucasus. This culture is probably related to the emergence of the Late Mesolithic–Neolithic Donets culture, in the Don Right Bank valley. The dispersal of these Mesolithic materials, possibly up to the Volga–Oka interfluve, attest to possible trans-regional ancient population migrations (Fedyunin 2015).

II.3.2. North Pontic forest

In the north Pontic forest area, the Swiderian culture had been replaced by local groups which grew from its tradition. Early west Mesolithic forest hunters from flooded areas between Britain and Scandinavia migrated to the east into Middle European lowlands, forming the Duvensee cultural unit (Maglemose culture) in the 8th–7th millennium BC. The only group to reach into the north Pontic forest-steppe region comes from the Komornica culture, from the Vistula river basin (Zalizniak 2016).

The eastward migration of west European Late Mesolithic groups, as well as the appearance of post-Maglemosian cultures, have been suggested as the result of further floodings in western Europe, specifically the flooding of Doggerland, ca. 6200 BC. In the southern part of the east European forest zone, between the Vistula and the Dnieper, Janislawice became the dominant post-Maglemosian culture. There is no influence of the Caucasus here (Zalizniak 2016).

Two main groups developed in the area: the Tacenki group around the Kyiv Polesia, and its offshoot the Kudlaievka in the valley of the upper Dnieper and Desna, under influence from the Kukrek tradition. Similar findings in the River Donets and to the north are thought to derive from a migration of population from the Kudlaievka culture in both directions. The adaptation of Janislawice to Caucasian–Pontic traditions, including animal husbandry, happened late, and only after the expansion of this culture into the forest-steppe (Zalizniak 2016).

II.3.3. North Caspian steppes

To the east, the Early Mesolithic forager camps (ca. 8000–7000 BC) inhabited the North Caspian depression, then filled with lakes, in a cooler and moister climate. The economy of these groups, like that of north Pontic groups, depended on equid hunting, and their camps probably represented single families or small hunting parties. Their garbage dumps contain almost exclusively bones of onagers, and their flint tool inventories mainly geometric microliths (Anthony 2006).

During the Late Mesolithic (ca. 7000–6000 BC), the North Caspian area became increasingly dry, with pollen evidence suggesting that desert conditions spread precisely during this time, with a peak dry episode ca. 6000 BC (related to the cold oscillation ca. 6200 BC in the northern hemisphere), after which humid conditions returned. Camps reached larger sizes, hunters preferred a grey flint different from the previously prevalent yellow-grey flint, and geometric microliths diminished (Anthony 2006).

Toolmaking traditions recognised in these sites include many different lowland steppe foraging groups, from the North Caspian area (Istai IV type), West Caspian steppes (Kharba type), and Azov–Crimean steppes (Kukrek and Zimovniki). When pottery-making emerged in this area, Early Neolithic cultures of this Azov–Caspian region (Surskii, Kairshak III) made pots with similar shapes and similar patterns of decoration. Because of that, Vasiliev advanced the idea that the dry Azov–Caspian steppe constituted a 'cultural region' during the Late Mesolithic and Early Neolithic, a network of interacting forager bands. Different from this common material culture was that one seen east of the Caspian Sea and east of the Urals, which belonged to distinct social networks (Anthony 2006).

II.3.4. Hunter-gatherer pottery

The oldest known pottery appeared in East Asia, with findings in the Yangze Basin (20th–17th mllennium BC), and further examples in northern China (11th millennium), the Korean Peninsula (11th–10th millennium BC). In Japan, pottery was apparently introduced during the incipient Jōmon phase (16th–13th millennium BC), and was probably related to the introduction in North Asia from the east, with findings spanning ca. 15th–12th millennium BC (Piezonka 2016).

In the Trans-Baikal area, the first findings are dated ca. 13th–11th millennium BC, but it seems to have been a still-stand zone, because pottery appeared to the west only millennia later, in west Siberia and the Trans-Uralian regions. While some initial findings can be dated to the 10th–8th millennium BC, the start of a westward cultural wave can be dated only to the first half of the 8th millennium BC, with widespread findings in cultures of the North Asian regions dating to the 7th and 6th millennium BC (Piezonka 2016). On the other hand, the period between 11th-8th millennium seems to be one of expansion from east to west.

Before the arrival of farmers to the western frontier of the Pontic–Caspian steppe, pottery was produced in the Volga region by the Elshanian culture ca. 7000–6500, with the oldest pottery in Europe coming from the Samara region. This material culture was probably derived from the Eastern Asian tradition of the Late Pleistocene, which arrived through a path following Siberia and the Trans-Urals (Piezonka 2015). The earliest Trans-Uralian sites include Yurtobor and Lake Andreevskoe, in the Tobol-Ishim region, of the late Mesolithic Boborykino culture—featuring round- and flat-based bots with incised and impressed decoration—and Sumpanya in the Konda River Basin (Gibbs and Jordan 2013). The scarce sites investigated and variable dates obtained from them (ca. 12000–8000 BC) make it impossible to ascertain a precise origin of eastern European pottery, though.

From this source, the first Neolithisation wave reached hunter-gatherer groups of the Pontic–Caspian steppe, starting with the Lower Volga region (Suppl. Fig. 5), whose oldest pottery is dated ca. 6200 BC, with similarities to those of the Kairshak culture in the northern Caspian steppes, whose first sites with the oldest pottery appeared ca. 6500 BC (Vybornov 2016). From the north-western Caspian region pottery spread south- and westward into north Pontic societies, appearing in different steppe regions simultaneously ca. 6200–6000 BC, including the Bug–Dniester culture (Zaitseva et al. 2009).

Pots were made of a clay-rich mud collected from the bottoms of stagnant ponds, and they were formed by the coiling method and baked in open fires at 450–600° C. These pots are bottom-tappered and non-decorated in the Middle Volga (Elshanian culture); non-decorated and with scoring marks with round and flat bottoms in the North Caspian steppe; non-decorated in the Lower Don; and non-decorated or sparsely decorated in the Dvina–Lovat' region (Zaitseva et al. 2009).

Sparsely decorated pottery (decorated with little punctures, and having a tapered bottom) dispersed north into the forest zone ca. 6000 BC or slightly earlier, from the Upper Volga and Dvina–Lovat' regions to the east (into the Dvina–Pechora region) and west (into the eastern Baltic), reaching the Upper Volga, Serteya, and Valday cultures, and later the Narva culture. It reached the Bug–Dniester culture in the Southern Buh valley ca. 6200–6000 BC, just before the western pottery type was adopted in the Middle Dniester valley ca. 5900–5700 BC (Vybornov, Kosintsev, and Kulkova 2015).

Pottery production increases at the end of the 7th millennium and beginning of the 6th millennium, with the second stage of Elshanian pottery appearing ca. 5750–5500 BC, with pottery assemblages also found in the North Caspian region. More pottery decorated with triangular impressions can be traced to the second half of the 6th millennium in the region (Mazurkevich and Dolbunova 2015).

In the forest-steppe of the Don River region, sites appear in the upper parts of the first terrace above the floodplain, and sometimes on bedrock shores. Pottery of the Karamyshevo type appear through direct contacts with the population of the Elshanian culture, which had the skills and arrived in the Don region probably at the end of the 6th millennium BC, with Karamyshevo culture prevalent throughout the first half of the 5th millennium BC, eventually produced from sanded silty clay containing natural inclusions, like the pottery of the Middle Don culture (Smolyaninov, Skorobogatov, and Surkov 2017).

The Middle Don culture, in turn, appears to be contemporaneous with early Karamyshevo, with the southern periphery of the Upper Volga region also coincident with early stage of both, hence connecting it to the Valday and Volga–Oka groups that formed the Upper Volga culture. It is difficult to pinpoint exactly the intertribal communication networks between Upper Don River basin and those of the Upper Volga, as is the relationship to the communities of the Dnieper–Donets culture around the mid–5th millennium BC (Smolyaninov, Skorobogatov, and Surkov 2017).

A second expansion of eastern pottery reached the eastern Baltic region ca. 5500 BC, expanding from the Dnieper region to the north-west, generating the sparsely decorated Dubičiai pottery (later evolving into the Neman culture), and influencing the north European regions from the Narva to the Ertebølle cultures (Piezonka 2015).

A third expansion of eastern pottery spread from the Volga–Kama region to the east ca. 5000 BC, connected to influences from beyond the Urals, showing a more elaborately decorated ware (with bands of pits and impressions made from comb stamps), spreading north and west in the Sperrings and Säräisniemi 1 cultures (Piezonka 2015).

ii.3. Indo-Uralians

The successful expansion of haplogroup R1a-M459 (formed ca. 20800 BC, TMRCA ca. 16200 BC) may have been associated with waves of migration of ANE/EHG ancestry into eastern Europe from northern Eurasia. An individual from Vasylivka ca. 8700 BC, of hg. R1a1-M459 (Jones et al. 2017), supports this haplogroup as one of the prevalent Epipalaeolithic lineages of north Pontic hunters. The spread of R1b1b-V88 lineages (TMRCA ca. 9700 BC) in the North Pontic area, with the first sample found in Vasylivka ca. 7050 BC (Mathieson et al. 2018), may be linked to remnants of ancestral populations expanding westward, but they could be more directly associated with the eastward expansion of the Grebeniki tradition, of west Pontic origin, close to Balkan hunter-gatherers of the same haplogroup (see *§i.3. Nostratians*). A sample of hg. I2a1a-P37.2 in Vasylivka ca. 8100 BC may also represent either an earlier Epipalaeolithic population, or a recent migration of Iron Gates hunter-gatherers or post-Swiderian migrants.

Mesolithic samples from the north Pontic area are intermediate between EHG and SHG, with further contribution of WHG ancestry (Mathieson et al. 2018), which—together with their position in PCA—support close contacts of post-Swiderian and Maglemosian cultures in the region. The expansion of eastern ancestry detected in north Pontic samples (from the Middle Volga) and in a Middle Volga sample (from further east) also support westward migration waves coinciding with the expansion of hunter-gatherer pottery (see below *§iii.5. Early Indo-Europeans and Uralians*).

A sample of a Mesolithic hunter-gatherer at Lebyanzhinka in the Samara region, dated ca. 5600 BC, shows hg. R1b1a1-P297[+] (Mathieson et al. 2015), while R1b1a1a-M73 is found later in Donkalnis, in the Baltic region (ca. 5200 BC), belonging to the Narva culture (Mittnik, Wang, et al. 2018). Similarly, expanding Indo-Anatolians show hg R1b1a1b-M269 (see below *§iv.2. Indo-Anatolians*), apart from other likely local haplogroups. These lineages are compatible with the previous expansion of R1b1a1-P297 lineages with the North-Eastern Technocomplex (see *§ii.1. Eurasians*).

Nevertheless, the presence of R1b1a1-P297 late in the Baltic, after the resurgence of I2-M438 lineages, could also suggest a back-migration of certain groups associated with the westward expansion of hunter-gatherer pottery from the Trans-Urals into the Pontic–Caspian and north-eastern European regions. This is supported by the presence of AEA ancestry in two individuals, from Samara and Karelia (Lazaridis et al. 2018), and the appearance of haplogroup R1b1a1-P297 in Karelia ca. 5600 BC (Mathieson et al. 2015). The presence of R1b1a1-P297 lineages in Latvia since the Mesolithic until the arrival of Corded Ware does not help distinguish between continuity, resurgence, and back-migration events.

Samara and Karelia hunter-gatherers also show in 'speculative' estimates further contribution of El Mirón ancestry compared to the older Sidelkino sample (ca. 9300 BC), and higher than the one found in other EHG samples from Karelia, in south-eastern Europe, or in the north Pontic region during the Mesolithic or Neolithic (Lazaridis et al. 2018). This contribution is thus compatible with the expansion of populations related to northern Mesolithic Europe, and thus post-Swiderian cultures.

The ancestry of later Eneolithic individuals of the Khvalynsk culture (5th millennium BC) is also found in two Sintashta outliers from Kamenyi Ambar (ca. 2000–1650 BC), one of hg. R1b1a1-P297, the other R1b1a1a-M73 (Narasimhan et al. 2018), both probably related to local groups of the southern Urals and Trans-Urals region, remnant populations of these groups expanding hunter-gatherer pottery to the west. The same haplogroup R1b1a1a-M73 is found further east among the Botai, supporting the widespread distribution of these groups beyond the Urals (see *§v.9. Pre-Tocharians* and *§v.8. Palaeosiberians*). Like these late outliers, Khvalynsk-related individuals on the Cis-Urals probably hosted some West Siberian Hunter-Gatherer (WSHG) ancestry, and were thus most likely part of an ancestral EHG:WSHG cline from the Cis-Urals to the Trans-Urals and West Siberian region.

R1a1-M459 lineages continued probably in groups of the Pontic–Caspian region, evidenced by one individual from Deriïvka (ca. 6900 BC), a hunter-gatherer from Karelia (ca. 6300 BC), and an individual of the Khvalynsk culture from Samara (ca. 4600 BC). The expansion of R1a1a-M198 lineages (formed ca. 12000 BC, TMRCA ca. 6600 BC) and the finding of sister clade R1a1b-YP1272 (TMRCA ca. 5300 BC) in a Maikop outlier from the northern Caucasus, and in an individual from Kudrukŭla, Estonia (Saag et al. 2017), of the Comb Ware culture (ca. 3000 BC), suggest—together with their early split date—an early expansion of R1a1-M459 subclades from within eastern Europe, probably associated with the spread of some groups of hunter-gatherer pottery within the forest zone[6]. the with later expansions likely associated with specific groups of hunter-gatherer pottery.

If post-Swiderian cultures are associated with the expansion of Eurasiatic languages into eastern Europe and the Trans-Urals region, the westward expansion of hunter-gatherer pottery from the Elshanian culture in the Volga–Ural region—probably originally associated with R1b1a1-P297 lineages in the Volga–Ural area, and continued after the resurgence of local R1a1a-M198 lineages mainly in the forest zone—can be linked to the expansion of Indo-Uralic from the east. The emergence of the elk as an animal of great symbolic value in the east European forest zone, appearing in rock art, as well as in elk-head staffs (see below Figure 12) and other elk-head sculptures since the 7th

---

[6] The finding of two isolated R1a1a-M198 samples in Lokomotiv (Mooder et al. 2005) associated with Kitoi material culture (dated ca. 6000-5200 BC), among sixteen reported haplogroups, with the rest being K-M9 (likely N-M231) and one C2-M21, in the western shores of Lake Baikal (Moussa et al. 2016), may suggest an expansion of these lineages from the Trans-Urals region. However:

  a) the earliest reported R1a1-M459 lineages are all from eastern Europe (Mathieson et al. 2018);
  b) there are no calibrated dates for those samples, paternal lineages for Kitoi seem too heterogeneous, and there are known later eastward expansions of hg. R1a1a1b-Z645 to the region (see *§viii.20.3. Turkic peoples and Mongols*);
  c) recently reported samples from the region during the Early Neolithic have not obtained any R1a-M420 subclade, but are consistent with the presence of N-M231 and C2-M217 (de Barros Damgaard, Martiniano, et al. 2018);
  d) the obtained strontium isotope ratios that serve to confirm their local origin (Moussa et al. 2016) cannot differentiate between earlier and later periods; and
  e) based on the study of modern populations, basal R1a-M420* and R1a1-M459* subclades have only found around the Caucasus, not among Asian or Siberian populations (Underhill et al. 2015; Karafet et al. 2018).

The most likely interpretation for both samples, until confirmed in different ones with proper radiocarbon dates, is that they belong to later archaeological layers, or to contamination.

millennium BC, may have also been associated with this cultural expansion, which may have included also the spread of symbolic red ochre in paintings and graves (Norberg 2019), eventually appearing among Comb Ware groups.

II.4. North Africa

During the African Humid Period (AHP), starting ca. 12800–10000 BC, hunter-gatherer groups repopulated the Sahara, marked landscapes with rock art, and occasionally created cemeteries.

The Typical Capsian culture (ca. 7500–6000 BC) is characterised by large tools, mainly burins, scrapers made on blades, with blade production involving knapping schemes derived from simple or complex core preparation. These indicate the use of soft and hard hammer percussion for blades and flakes, which are then retouched to produce a variety of tools. Sites show accumulation of land snail shells, ash, burnt rocks, knapped flint, worked (human and not human) bones, and mammalian faunal remains. Capsian sites are concentrated on the high plateaus of eastern Algeria and southern Tunisia, although the culture shows variants to the west and east into Lybia (Rahmani and Lubell 2012).

Cord-wrapped roulettes are among the earliest cord-based decorative tools in Africa, appearing at Aïr Massif in Niger, later in the Sahara region between Algeria and Libya ca. 8th–6th millennium, as well as on the Atlantic coastline and Mauritania's interior dating from the 6th–5th millennium BC. Vessels with flat bases are also found along the western coast up to pre-Saharan Morocco. These features appear in Morocco's Middle Neolithic pottery, which suggests contacts between the populations of the interior potentially related to the end of the AHP.

Before the expansion of Neolithic farmers, clear links can be seen between both shores of the Mediterranean in Mesolithic hunter-gatherer blade and trapeze industries from the North African Upper Capsian, the Castelnovian in Italy and southern France, and the Geometric Mesolithic in the Iberian

Peninsula. This north–south network of connections is also seen later during the Mediterranean expansion of the Neolithic (Guilaine 2017).

ii.4. Early Afrasians

Proto-Afroasiatic is proposed to have emerged in the southern fringe of the Sahara in an "upside-down" view (Bender 2007), which would put the Afroasiatic homeland near Megalake Chad during the African Humid Period (ca. 9000–3000 BC). This period was probably marked by a Sahelo-Sudanian palaeoenvironment of isolated wetlands and small lakes, with a pale-green and discontinuously wet Sahara forming north–south and east–south pathways (and a continuous east–west corridor in the southern half of the current desert) that allowed for human migrations (Quade et al. 2018).

This corridor would have allowed for a sizeable population expansion in south-central Saharan territory, for an eastward expansion of Cushitic and Omotic, and for a migration of Hamito-Semitic speakers (including possibly Berbero-Semitic) to the north-east. This model agrees with Chadic languages being the most divergent of the Afroasiatic group, excluding Omotic, whose population has been shown to be mainly of sub-Saharan ancestry, in contrast to other Afroasiatic peoples (Baker, Rotimi, and Shriner 2017).

R1b1b-V88 lineages are found widespread in north and central Africa, mainly among Chadic-speaking peoples, but also in modern populations of northern and eastern Africa (Figure 8), with a "star-like" topology estimated to have begun ca. 5850 BC (Figure 9)—roughly at the same time as Saharan populations adopted pastoralism—and tracing a trans-Saharan axis (D'Atanasio et al. 2018):

- R1b1b2a2-Y8447, the main African subclade (formed ca. 5500, TMRCA ca. 5000 BC).
- R1b1b2a2a-V1589/Y7771 (formed ca. 5000 BC, TMRCA ca. 3100 BC) and most other rare African R1b1b-V88 lineages (not considered below) are distributed widely, in descending proportion, in central Sahel (around the prehistoric Megalake Chad), in eastern Libya and

north-west Egypt, in south Egypt and north Sudan, and in central Sudan and east Sahara.

- o R1b1b2a2aX-V1589 (TMRCA ca. 3700 BC), with a distribution similar to the parent haplogroup R1b1b2a2a-V1589/Y7771.

- o R1b1b2a2aX-V4759 (TMRCA ca. 2700 BC) peaking in eastern Libia and north-west Egypt, and spread also through the Maghreb and south-western Africa.

- o R1b1b2a2a1-V69 (formed ca. 3000 BC, TMRCA ca. 2600 BC), peaking around lake Chad, and also found in south Egypt, Sudan, and eastern Africa and Arabia.

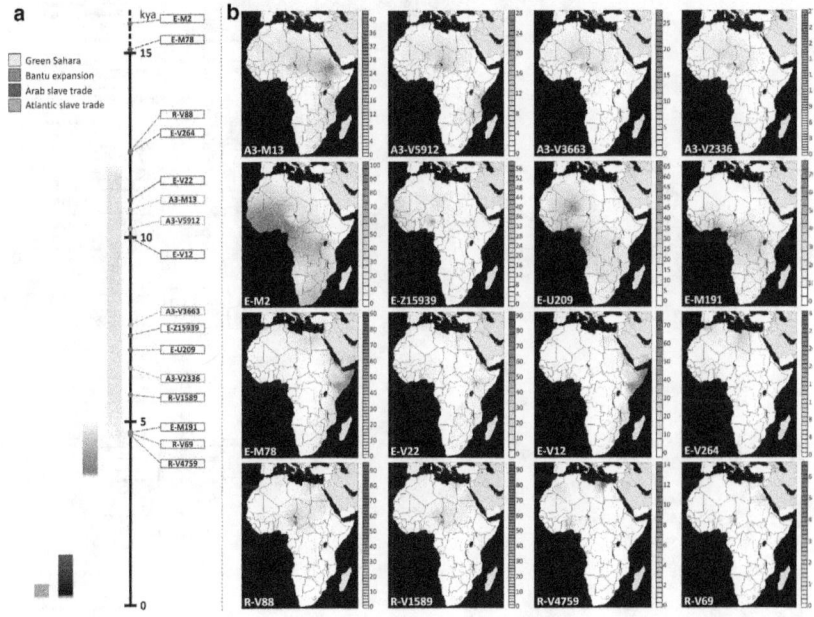

Figure 8. Image modified from D'Atanasio et al. (2018): "Time estimates and frequency maps of the four trans-Saharan haplogroups and major sub-clades. a) Time estimates of the four trans-Saharan clades and their main internal lineages. To the left of the timeline, the time windows of the main climatic/historical African events are reported in different colours (legend in the upper left). b) Frequency maps of the main trans-Saharan clades and sub-clades. For each map, the relative frequencies (percentages) are reported to the right."

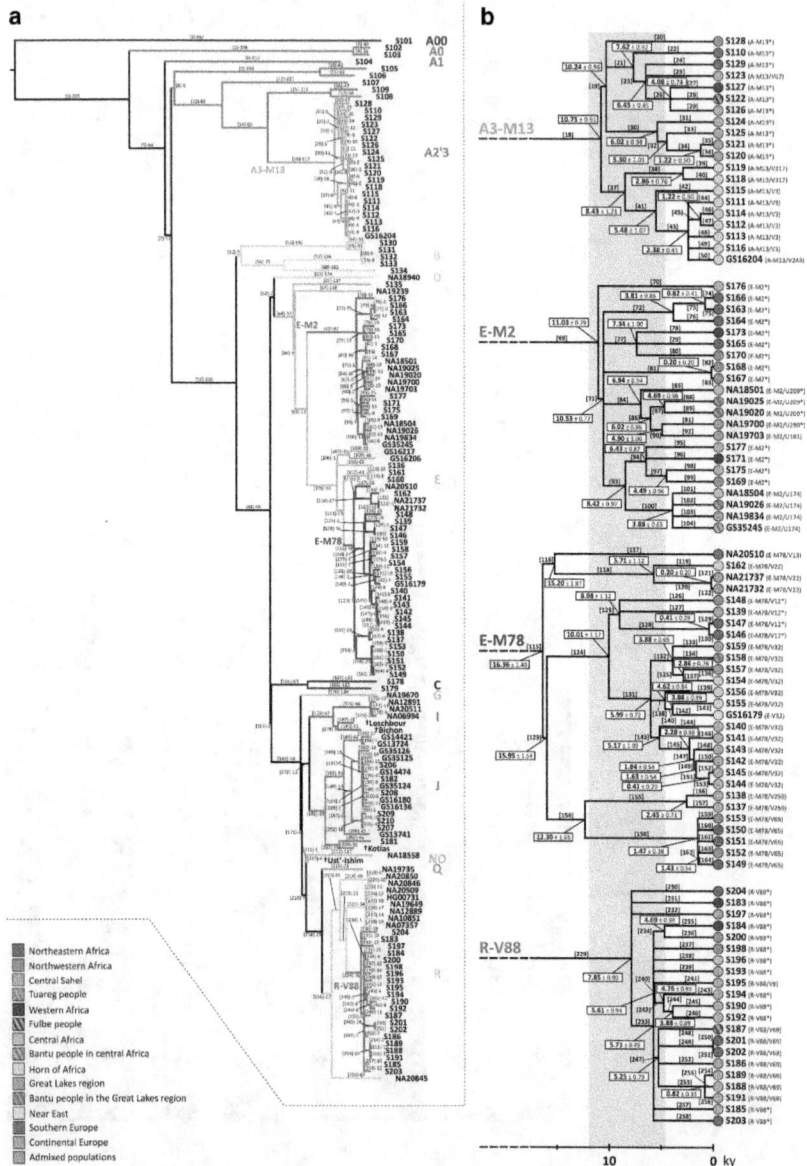

Figure 9. Image modified from D'Atanasio et al. (2018): "Maximum parsimony Y chromosome tree and dating of the four trans-Saharan haplogroups. a) Phylogenetic relations among the 150 samples analysed here. Each haplogroup is labelled in a different colour. The four Y sequences from ancient samples are marked by the dagger symbol. b) Phylogenetic tree of the four trans-Saharan haplogroups, aligned to the timeline (at the bottom). At the tip of each lineage, the ethno-geographic affiliation of the corresponding sample is represented by a circle, coloured according to the legend

(bottom left). The last Green Sahara period is highlighted by a green belt in the background."

R1b1b-V88 lineages related to the Villabruna cluster probably crossed the Mediterranean into northern Africa, most likely through the southern Italian Peninsula, eventually expanding with the Capsian tradition. This is supported by rare subclades found in southern Europe, like R1a1a1a2-M18 (split ca. 10300 BC), present mainly in Corsica and Sardinia; R1b1b2a-Y8451/V2197 (formed ca. 7500 BC, TMRCA ca. 5500 BC) and R1b1b2a1-V35 (split ca. 6700 BC), present mainly in Sardinia (Cruciani et al. 2010). The presence of the rare mtDNA haplogroup J/T in Epipalaeolithic samples from Afalou, as well as in modern populations of northern Morocco (1.8%), Sicily (1.8%), and other Italian populations (1.6%) also point to ancient contacts of west Africa with Italy (Kefi et al. 2018).

A migration of these lineages through Iberia seems unlikely, because there is large genetic continuity of an endemic element related to ANA (found in Epipalaeolithic Iberomaurusians and in Early Neolithic samples) in north-west Africa until the Late Neolithic (Fregel et al. 2017), where only later does ancestry from Iberia appear associated with the expansion of European Neolithic, probably through the Strait of Gibraltar (Fregel et al. 2018). The presence of sub-Saharan admixture in Middle Neolithic and Chalcolithic samples, as well as the finding of mtDNA hg. L2a1 in a Bronze Age sample from south Iberia point to a north–south genetic structure in pre-Neolithic Iberia, with hunter-gatherers probably showing resemblance with sub-Saharan Africans (González-Fortes et al. 2019).

Alternatively, the finding of subclade R1b1b2a-V2197—prevalent among Sardinians and most African R1b1b-V88 carriers—in the Els Trocs individual and two ancient Sardinian individuals (ca. 3400–1000 BC) may be interpreted as part of the Early European Farmer expansion through the Mediterranean, with an infiltration of Mesolithic R1b1b-V88 lineages among Anatolian farmers in the Balkans. The coalescence times between the Sardinian and African R1b1b-V88 haplotypes, coupled with autosomal traces of Holocene admixture with Eurasians in ancient samples, may thus be interpreted as a more recent maritime wave of Cardial Neolithic migrants along the West Mediterranean coasts in the 5th millennium, and their subsequent movements across the Green Sahara (Marcus et al. 2019). This is also supported by the finding of R1b-M343 in a coastal site of Cueva de Chaves, likely from an Early Neolithic context (Villalba-Mouco et al. 2019). Nevertheless, the association of R1b1b-V88 lineages with the expansion of the Villabruna from eastern Europe, and its likely presence in the Villabruna individual himself (reported as R1b1-L754), suggests that the haplogroup might have expanded with Villabruna-like ancestry during the Epipalaeolithic, reaching north-east Iberia ca. 12000 BC (Villalba-Mouco et al. 2019).

There is a cline of ANA:Natufian ancestry in North Africa from east to west (Rodríguez-Varela et al. 2017), but a hint of Villabruna ancestry can also be inferred in all modern north African populations west of Libya, which is consistent with gene flow from Iberia during the Late Neolithic (Lazaridis et al. 2018), but could also be at least partially related to the expansion of R1b1b-V88 lineages through southern Italy in the 7th–6th millennium BC.

R1b1b-V88 is found at high frequencies among the Toubou population (34%) inhabiting Chad (Haber, Mezzavilla, Bergström, et al. 2016), who also show 20%–30% ancestry coming from outside of Africa, related to Eurasian herders (Schlebusch et al. 2017), potentially linked to this ancient migration, since a good proxy for this ancestry (apart from ancient Levantines) are present-day Sardinians (Pickrell et al. 2014).

From north-central Africa and through the Green Sahara, with gradual desiccation of the desert until ca. 4000–3000 BC (Drake et al. 2011), Afroasiatic speakers of R1b1b-V88 lineages could have expanded in all directions following for example the researched Fezzan–Chad–Chotts, and Chad–Chotts–Ahnet–Moyer pale green areas. The role of certain E1b1b1a1-M78 lineages (split ca. 13200 BC)—widespread in eastern Africa—in secondary expansions of Afroasiatic languages seems very likely:

E1b1b1a1b2-CTS9547/L677 (formed ca. 8000 BC, TMRCA ca. 2400 BC) in east and north-east Africa; E1b1b1a1a1-V12/Z1216 (formed ca. 9900 BC, TMRCA ca. 8000 BC) in the east, peaking in the Horn of Africa; and E1b1b1a1a2a1-CTS194 (formed ca. 2600 BC, TMRCA ca. 200 BC) in Libya and the Maghreb (D'Atanasio et al. 2018).

II.5. Caucasus Mesolithic

In the Caucasus, at least four regional Mesolithic traditions can be distinguished, linking the late Upper Palaeolithic traditions of foragers with the arrival of a farming economy: the north-east Pontic area, extending to the steppes of the northern foreland; the south-western Imeretian variant; the Trialeti highlands, where communities had access to nearby obsidian sources; and the Dagestan Mesolithic. The Trialetian Mesolithic is probably the best known, representing a wide-spread industry that reached into the Trans-Caspian region, eastern Anatolia and the Iranian Plateau, although there is a lack of absolute dating (Sagona 2017).

The Imeretian culture developed on the slopes of the south-western Caucasus, with an origin in the Gravettian cultures of the region, which maintained contacts with the areas of Syria–Palestine and the Zagros Mountains. The introduction of micro-burin technique and a Natufian retouch, together with the geometric microliths, which appear most frequently in North Africa and the Near East, show an intensification of contacts with the population of Natufian culture of the Levant during the Holocene. At the same time, these new elements reach communities of the north-western Caucasus (Sagona 2017).

Kotias Klde, a karstik cave above the Kvirila River in western Georgia, and the Darkvety rock shelter, forms part of the Trialetian tradition. The lithic industry in Kotias Klde is relatively homogeneous in both types and technology, and comparable industries are found in distant territories, in particular at Ali Tepe in the Elbruz region of Iran, and Hallan Çemi. While a chronology is difficult to establish, it seems that the Trialetian elements appear first at Ali Tepe (ca. 10500–8870 BC), slightly later at Hallan Çemi (ca. 8600–7600 BC) (Sagona 2017).

There is a 1,000–year gap between the Epigravettian tradition of the Dzuzuana Cave and the Mesolithic findings of Kotias Klde, a seasonal camp by then, which suggests the arrival of new peoples, hunters of wild boar and brown bear during the late spring and early summer. In the southern Caucasus, the Chokh variant apparently continues the earliest ceramic tradition, but in Georgia there is a clear distinct period with Neolithic geometrics and the trapeze microlithic. Their diet was mostly composed of mammals, mainly deer (ca. 50%), but also brown bear (ca. 34%) probably mainly for fur and symbolic reasons. Unlike their Upper Palaeolithic predecessors, they did not hunt other ungulates, such as aurochs, steppe bisons, Caucasian tur, and wild horses (Sagona 2017).

The Chygai rock shelter, in the northern foothills of the western Caucasus, represents the lithic industry of the north-western Caucasus region. Their technological change is gradual, until the abrupt appearance of geometric microliths, especially trapezes, which mirrors the technological change of Crimea (see *§II.3.1. North Pontic steppes*). Remains of small mammals like ovicaprids and deer species are predominant in the early period, while larger animals such as bison and wild pig are hunted later on (Sagona 2017).

ii.5. Caucasus hunter-gatherers

Different groups of Caucasus hunter-gatherers with strong differences in admixture emerging in the ancient DNA record support the existence of multiple small populations living in partial isolation, probably helped by the region's orography. This is evidenced e.g. by the signs of recent consanguinity of the Mesolithic individual from Satsurblia; by the existence of a prehistoric genetic and cultural barrier around the Caucasus Mountains (Wang et al. 2019); and by the partial continuity of ancestral Y-chromosome and mtDNA haplogroups into the modern population of the southern Caucasus (Jones et al. 2015).

Close cultural contacts of the Imeretian culture linking the southern Caucasus and the Zagros Mountains probably reflect an ancient, Upper Palaeolithic network that allowed for the spread of ANE ancestry in the mainly AME-like population of the region, possibly through expanding Q1a2-M25 lineages from West Siberia. This is supported by its modern distribution in eastern Europe and central Asia, with a higher frequency in the Iranian Plateau under Q1a2a-L712 subclade (formed ca. 14300 BC, TMRCA ca. 11300 BC), apart from ancient samples in Aleutian Islanders, ancient northern Athabaskans, and in a sample from Cukotkan Ust'Belaya culture (Flegontov et al. 2016), as well as its estimated expansion ca. 10000 BC (Grugni et al. 2019). The resurgence of a fully ANE-like individual of Q1a2-M25 lineage in the Lola culture during the Bronze Age proves the persistence of small, isolated ANE-like populations in the Caucasus since the arrival of these migrants.

II.6. Fertile Crescent

Epipalaeolithic Natufians were probably the first to exploit cereals occasionally, and there is proof of preparation and consumption of bread-like products ca. 12500–9500 BC, before agriculture was firmly established (Arranz-Otaegui et al. 2018).

The earliest pastoral systems developed in multiple regions across the Fertile Crescent within sedentary communities of hunter-gatherer-agriculturalists who began controlling and managing small populations of wild caprines and bovines by the early 9th millennium BC. This early stage of animal domestication during the 9th and early 8th millennium BC emerged first along the Upper Euphrates river, and was characterised by highly variable management techniques and relatively low herd productivity, with small controlled animal populations and wild phenotypes (Arbuckle and Hammer 2018).

At Ganj Dareh, the economy was focused on the pre-domestic management of goats, including the intensive slaughter of yearling males, probably designed to reduce the number of aggressive adult males in herds. This system was not copied elsewhere, and local domestication systems were the norm. Only after the mid–8th millennium were dramatic changes seen across south-west Asia, constituting a true "pastoral revolution": shift toward intensive caprine pastoralism, widespread appearance of domestic phenotypes, and spread of caprine husbandry outside the Fertile Crescent, with the standardisation of targeted culling of young male animals (Arbuckle and Hammer 2018).

The emergence of sedentism in central and southern Anatolia began in the 9th millennium BC, after a period of mobile groups in the Epipalaeolithic. Local building techniques, burial customs, and agriculture were gradually changing through contacts, especially with south-east Anatolia and the Levant. In south-east Anatolia, a gradual disintegration of the Aceramic Neolithic lifeway and a replacement by a society formally based on kinship appears starting with the Mature Aceramic III period (ca. 8000 BC), and continuing through the early part of the Pottery Neolithic (Rosenberg and Erim-Özdoğan 2011).

These social changes are intertwined with important economic changes involving the development of the full southwestern Asia domesticate complex, and technological advances like the widespread adoption of ceramic technology, although specifics of how they are related are not known (Rosenberg and Erim-Özdoğan 2011).

ii.6. Early agriculturalists

The presence of further contribution of "deep" ancestry in Iran Neolithic farmers compared to CHG, and the lack of such ancestry in Anatolian Hunter-Gatherers—who, on the other hand, seem to show Natufian-related ancestry—support the expansion of this "deep" ancestry into the Zagros Mountains and the Caucasus from a southern (probably south-western) population that did not undergo the Palaeolithic bottleneck with Neanderthal admixture (Lazaridis et al. 2018). Available samples with a similar cluster to Iran Neolithic include the following:

A sample from the Hotu Cave in the South Caspian region (dated ca. 6100 BC, although archaeologically attributed to the Mesolithic–Neolithic, ca. 9100–8600 BC) shows an ancestry similar to Iran Neolithic samples, but with contribution from EHG (ca. 10%), which supports the existence of an ancestral cline EHG:CHG (or WHG–ANE:CHG–IN) from Eastern Europe to the Middle East (Lazaridis et al. 2016). While this EHG/ANE contribution found in CHG/IN probably came from the north, with the expansion of haplogroup Q1a2-M25, the expansion of R2-M479 (and possibly that of R1b-M343 lineages before it) is likely the reason for the presence of slightly elevated ENA-like ancestry in Iran compared to CHG.

The five early Neolithic Aceramic Anatolian farmers sampled to date (ca. 8300–7800 BC) and later Ceramic farmers (ca. 7000–6000 BC) form a close cluster with AHG. Aceramic farmers derive most of their ancestry from AHG (ca. 90%), and the rest probably from a source near Iranian/Caucasus ancestry, which may have diffused via contacts through eastern Anatolia, hence allowing for the spread of cultural innovations. They show the emergence of haplogroup G2-P287 with a sample of G2a2b2b-PF3359 (formed ca. 12400 BC, TMRCA ca. 10200 BC)—found today mostly in Sardinia—and one sample of haplogroup C-M130, common in Palaeolithic populations (Feldman et al. 2019).

Later Ceramic farmers also derive their ancestry primarily from AHG (more than 75%), but there is gene flow from the Levant to Anatolia during the early Neolithic. In turn, Early Pre-Pottery Neolithic farmers (ca. 8300–6700 BC) from the southern Levant can be modelled as a two-way admixture of Natufians with Aceramic farmer contribution (ca. 18-21%), which supports reciprocal genetic exchange between the Levant and Anatolia during the early stages of the transition to farming (Feldman et al. 2019). Continuity in the Levant can be inferred also from the presence of haplogroup E1b1b1a1-M78 in three individuals, together with, two hg. H2-P96, one T1-L206 (xT1a1-L162, T1a2-L131) and one F-M89^{+} (Lazaridis et al. 2016).

III. Neolithic

III.1. Neolithic package

In the Anatolian plateau, full farming villages developed ca. 7500–7000 BC, at the same time as pottery. Around 7000–6400 BC, the full Neolithic package conquers the western part of Anatolia and the Greek region, reaching Crete ca. 7000 BC (in aceramic version), Thessaly ca. 6500 BC, and western Greece ca. 6400 BC, indicating a rapid diffusion process. Around 6500 BC, the majority of settlements in the Anatolian plateau were abandoned: new settlements appeared on trade routes, which suggests a trend to control long-distance exchange networks, which is also evidenced by the intensified contacts between regions seen in material culture and in the exchange of ideas (Özbaşaran 2011). This coincides with another pause in the expansion of Neolithic pottery lasting two to three centuries.

The rapid expansion of Neolithic farmers has been put in relation with changes in climate (ca. 6600–6000 BC), although the increase of intergroup violence associated with environmental and economic stress show only some variation in intensity (to profit from a weakened adversary), but do not correlate with all evidence of conflict. Instead, the true reason for emerging warfare is probably associated with the new essential aspect of territoriality for the economically superior (and demographically expanding) farming societies, and thus conflicts arise from either politically motivated expansion, or from political power imbalances, which may in turn be associated with drought, epidemics, or as a consequence of climatic stress (Clare and Weninger 2016).

The succeeding episodes of population growth and sudden collapse identified in the expansion of Neolithic farmers suggest that they correspond to the demographic signature of travelling waves (and not travelling wave-fronts, proper of classic demic diffusion, where the population of previously occupied territories is kept at carrying capacity after the wave-front passes through them). That is, the 'boom' is associated with the arrival of new people, whilst the 'bust' should be understood as outgoing migrants resuming their spread into a new region. Only the regions being passed by the travelling wave experience a noticeable demographic pressure, while the meta-population follows a neutral growth curve (Silva and Vander Linden 2018).

The Neolithic package covers a whole series of technical innovations that accompany the domestication of plants and animals, like ceramics and polished stone, changes in dynamics of territorial occupation and exploitation, organisation of domestic areas and forms of social production, social dynamics and reproduction. Settlements are variable depending on the geographical area, but they generally consist of traits signalling a more permanent occupation compared to foragers, such as domestic areas that include houses, grain storage, graves, hearths, and ovens. Tools and materials suggest that settlements are also areas of production and consumption (Guilaine 2017).

iii.1. Aegean farmers

Signs of the Neolithic package appear only after ca. 7000 BC in west Anatolia and the Aegean. Studies of human ancestry support the arrival Anatolian farmers into central and western Europe with the Neolithic expansion (Brandt et al. 2013; Olalde et al. 2015; Szecsenyi-Nagy et al. 2017), also signalled by an expansion of hg. G2a2-CTS4367 (formed ca. 15800, TMRCA ca. 14900 BC), found first in Anatolian Aceramic (ca. 8000 BC), and in most Anatolian Neolithic individuals spanning the whole 7th millennium BC.

Consistent with a migration to the west, north Aegean Neolithic individuals ca. 6350 BC from Barcın (north-west Anatolia) and Revenia (northern Greece) share the closest genetic relationship with central Anatolian Neolithic individuals from Boncuklu (Pre-Pottery Neolithic) and Tepecik (Pottery Neolithic), closer than these central Anatolian groups among each other, and a closer relationship than Anatolian Neolithic samples with Iran Neolithic farmers, Natufians, and WHG (Kılınç et al. 2017).

The contribution of Iran Neolithic ancestry and their close affinity with Levantines may be explained by a contribution from a similar Anatolian Neolithic group, close to the central samples studied from Boncuklu and Tepecik, receiving more of the known previous Middle Eastern gene flows, especially Iran Neolithic contributions from the east; the closer relationship of Aegean individuals with WHG points to the survival in Aegean populations of the ancestral AME cline formed between WHG–AHG (see *§I.3. Epipalaeolithic*).

This ancestral AME cline does not allow us to fully discard an initial hypothetical out-of-the-Aegean model, or a Neolithisation driven by coastal and interior interaction networks connecting populations through the Aegean and the Levant. Nevertheless, later samples from north-western Anatolia (Kumtepe ca. 5000 BC) and south Greece (Franchti Cave and Diros, ca. 4000 BC) lack noticeable WHG-like ancestry, which suggests that Neolithic migration waves fully replaced the local population (Kılınç et al. 2017), a fact that points to similar demographic dynamics in the previous centuries.

III.2. European Neolithic

III.2.1. Mediterranean

The coastal spread in the western Mediterranean was much faster than in central Europe, which may be explained by long-distance maritime travel and exploration combined with demic expansion in coastal areas and assimilation of local foraging communities (Rigaud, Manen, and García-Martínez de

Lagrán 2018). Through the Mediterranean, different chronological and geographical traditions appear.

The *Impressa* ware develop around 6000 BC in the Ionian Sea, and constitutes the first step in the expansion to the west. It replaces the monochrome pottery around the southern Adriatic, appearing first in west Greece and south-eastern Italy, and spreading fast to the north into Dalmatia, and to the west into southern Italy and Sicily (Suppl. Fig. 6). The primary distribution of the "leap frog" type of impressed ware led by small pioneering units with a well-established agropastoral economy is seen rapidly expanding in western Mediteranean sites ca. 6000–5600 BC, such as Sicily, Tuscany, Liguria, Provence, Languedoc, Valencia (Guilaine 2017).

The common material culture element is thus pottery impressed with diverse instruments, such as the *Cardium edule* shell in the western Mediterranean, from the Adriatic coast to Portugal. A secondary phase of generalisation and regional settlement starting ca. 5600/5500 BC sees the development of specific Early Neolithic cultures, such as Stentinello in Sicily and Calabria; Tyrrhenian Cardial in Latium-Tuscany-Sardinia-Corsica; regional groups of the Franco-Cantabrian Cardial (Provence, Catalonia, Valencia, Andalusia). Cardial groups spread also into Morocco, either from Sicily or Iberia (Guilaine 2017).

In Iberia and southern France, agriculture spreads rapidly as flora diversity increases with the arrival of farming/herding communities, transforming the – until then – intact temperate forest areas. For hundreds of years a clear increase in demography, technological developments and cultural practices unfold, and during the Middle Neolithic the proportion of dominant species changes to those resistant to constant cuts (deciduous *Quercus*, evergreen *Quercus*, *Olea*, *Pinus halepensis*), and open areas multiply. Cycles of land exploitation, associating wood cutting, farming and herding, followed by woodland regeneration are common (Guilaine 2017).

III.2.2. Central Europe

While economic practices changed, some cultural traits like funerary practices did not accompany the 'Neolithic package' acquired in the Balkans by local fisher-hunter-gatherers. It seems that farmers and their domestic animals spread fast, in ca. 10 human generations, from sub-Mediterranean Macedonia to the northern limits of the temperate Balkan Peninsula and the adjacent Carpathian Plain, which may have put serious difficulties for the spread of cattle until selective pressure could provide genetically-driven adaptations to harsh environments (Ethier et al. 2017).

The early 6th millennium pioneer settlers in the interior of the Balkans were probably the first to face these challenges. Among their response to unfamiliar ecological conditions, the easiest adaptation was to adjust the species mix in favour of crop and livestock taxa that reproduced best in the new environment. However, farmers from the southern Balkans (modern-day Bulgaria and northern Macedonia) chose a strategy of diversification, exploiting a very broad spectrum of crops, probably to reduce climate-related losses. With their expansion to the northern Balkans and the Carpathian Basin, farmers abandoned many leguminous crops, and reduced the spectrum of cultivated plants (Ivanova et al. 2018).

On the other hand, the faunal assemblages were clearly structured by climate: sheep, goats, and pigs were reduced, while cattle and wild spaces increased in frequency during the northward expansion of farming. This, in combination with dietary evidence – such as organic residues in ceramics and stable isotope values in human bones – suggests that animal husbandry and especially dairying became of key importance for the initial establishment of farming beyond the Mediterranean areas (Ivanova et al. 2018).

The Linearbandkeramik (LBK) culture brought (starting ca. 5700–5600 BC) the first farming settlements to central European uplands as well as the North European Plains along the Oder and Vistula rivers. It expanded by colonising habitats favourable to agriculture, through a progressive migration of farming peoples from the Danube Valley to the north and west. Its economy was based almost entirely on domesticated plants and animals, and settlements were concentrated on fertile loess soils along streams. Analysis of isotope in skeletons in the Rhine Valley suggest that peoples of local origin may have been involved in the establishment of these early farming communities (Midgley 2004), which suggests short-range migrations.

Early Neolithic warfare in central Europe shows violent organised conflict between independently acting (probably territorial) groups connected by kinship ties. Massacres seem to cluster near the end of the LBK period, in the decades before 5000 BC, which suggests profound changes that affected interlinked social and natural landscapes: climate-induced drops in agricultural production with mounting claims to inherited agricultural land and increasing hierarchical differentiation are likely factors for the rise of social tensions and lethal conflicts between local groups. Evidence of perimortem collective lethal violence, as well as possible torture, mutilation, execution, dismemberment and cannibalism, including children, shows violence was a major societal issue for later LBK populations (Meyer et al. 2018).

iii.2. Early European farmers

Neolithic farmers from Europe derive from a single Balkan population closely related to north-western Anatolians which split in two routes: one related to Danubian populations, represented by the Linearbandkeramik complex from central Europe; and another associated to the Impressa complex of Croatia and Epicardial Early Neolithic from Iberia. However, this north-west Anatolian Neolithic (NWAN) ancestry is distinct from the central Anatolian source found in Aegean Neolithic samples, being shifted away from CHG and towards WHG compared to them (Mathieson et al. 2018).

Neolithic populations from the Balkans during the 6th millennium BC cluster closely with north-western Anatolian Neolithic individuals, deriving most of their ancestry from them (ca. 98%) and the rest from WHG, consistent with archaeological evidence. An exception is found in individuals of the mid–

6th millennium BC from Malak Preslavets, in the west Pontic area south of the Danube, where higher contributions of WHG (ca. 15%) and EHG (4%) are found, possibly representing populations close to the highest density of hunter-gatherers (Mathieson et al. 2018).

The Iron Gates zone represents a region of interaction between groups in both ancestry and subsistence strategy, based on strontium and nitrogen isotope data: two individuals from Lepenski Vir (ca. 6200–5600 BC), of entirely NWAN ancestry, were migrants from outside of the region and ate primarily terrestrial diet; another (ca. 6070 BC) had a mixture of NWAN and hunter-gatherer-related ancestry and consumed aquatic foods; and a fourth, earlier individual from the same site (ca. 7850 BC), had entirely hunter-gatherer-related ancestry. Another individual from Padina (ca. 5950 BC) also shows a mixture of NWAN and hunter-gatherer-ancestry, confirming the approximate date and region of interaction of both groups (Mathieson et al. 2018).

Hunter-gatherer ancestry of expanding farmers is more similar to eastern WHG individuals—like Villabruna (ca. 12000 BC) and Körös (ca. 5700 BC)—in the east, and more similar to western WHG individuals—like La Braña 1 (ca. 5900 BC) and Loschbour (ca. 6100 BC)—farther west, which shows that their admixture derives from populations with which they lived in close proximity. In particular, LBK individuals show a greater affinity to Loschbour hunter-gatherers, whereas Iberian Early Neolithic populations have La Braña-related ancestry (Lipson et al. 2017).

The increase in hunter-gatherer ancestry after the Early Neolithic period is lower in Hungary than in LBK and Mediterranean farmers from Iberia, and closest to the more eastern WHG individuals (from Villabruna and Körös), with limited intra-population heterogeneity, which points to the relative isolation of this group, close to the original source of expanding LBK farmers. In Iberia, the average admixture date is estimated ca. 5650 BC, but probably closer to ca. 5900 BC when considering only the oldest individuals assessed, which suggests—given the start of farming in Iberia ca. 5500 BC—the presence of a small proportion of hunter-gatherer ancestry in earlier Cardial Neolithic populations acquired along their migration route (Lipson et al. 2017). El Mirón-like ancestry is found in Early Neolithic individuals in higher proportions than those outside Iberia, and especially to the south, suggesting a north–south cline of admixture with hunter-gatherers who carried mixed Upper Palaeolithic ancestry (Villalba-Mouco et al. 2019).

Y-chromosome haplogroup G2a2-CTS4367 becomes also increasingly mixed with I-M170 lineages with time, especially in the Iberian Peninsula. G2a2a-PF3147 lineages (formed ca. 14900, TMRCA ca. 9700 BC), are found in Anatolian Neolithic samples from Tepecik and Barcın (and also later in Maikop from Novosvobodnaya), also early in Croatia Impressa (ca. 5560 BC), and later in Iberia, but especially in LBK samples. Haplogroup E1b1b1a1b1-L618 (formed ca. 10000 BC, TMRCA ca. 6100 BC) is also found in Cardial from Croatia (ca. 5900 BC), in Iberian Epicardial (ca. 5000 BC), and in Late Neolithic Lengyel (ca. 4740 BC), which supports its presence—probably originally from a Levantine source—in expanding Anatolian farmers.

Haplogroup G2a2b-L30 (formed ca. 14900 BC, TMRCA ca. 12500 BC), present in Anatolian Neolithic groups from Boncuklu and Barcın, as well as Greece Neolithic, seems to be more restricted to south-eastern European groups (although it is also present in the Iberian Chalcolithic), which suggests different bottlenecks during the expansion of Early European farmers from an original population with more varied male lineages.

The two major Neolithic migration waves of Neolithic settlers, from Danubian and Mediterranean routes, did probably encounter in the Paris Basin LBK and Cardium groups during the Early/Middle Neolithic transition. Cultural exchanges are observed in the region at the end of the Early Neolithic between north-eastern LBK-derived cultures Rubané Récenet du Bassin Parisien (RRBP) and Villeneuve-Saint-Germain (VSG), and Cardial farmer groups from southern France. This interaction is supported by the admixture found in ancient mtDNA from the Gurgy necropolis (Rivollat et al. 2015).

Resurgence of hunter-gatherer ancestry is also seen in Middle Neolithic samples from France (Brunel 2018).

Population continuity is observed in ancient samples through the British and Irish Mesolithic, characterised by WHG ancestry sharing high genetic drift with contemporaries from France and Luxembourg. This continuity is broken with the arrival of agriculture, introduced after a millennium–long lag between the establishing of farming in the mainland by incoming continental farmers, evidenced by a massive shift in population structure, with little evidence for local admixture. There seems to be two geographically distinct entries of Neolithic farmers, the main one through the Mediterranean route of dispersal entering from north-western mainland Europe, given the more immediate affinities with Iberian Neolithic individuals (Brace et al. 2018; Cassidy 2018; Kador et al. 2018). This affinity of Middle Neolithic samples from Britain to Iberia may be explained as from a source close to La Braña and El Mirón, but also to other intermediate unsampled regions in western Europe with higher Goyet-like contribution (Villalba-Mouco et al. 2019).

While farmers and hunter-gatherers lived in settlements in close proximity during the Neolithic in the Balkans, in western, central and northern Europe, there are signs of long periods with minimal admixture (Mathieson et al. 2017). During the Middle Neolithic, a resurge of male-biased hunter-gatherer ancestry is seen in central Europe and Iberia. Persistent frontiers between hunter-gatherers and farmers are found in central and northern Europe, coincident with the loess belt of the northern European plain, to the north of which early farming techniques were probably not suitable. It is likely that new climates and environments led to the eventual breakdown of demic diffusion, and the spread of Neolithic traits by cultural diffusion (González-Fortes et al. 2017).

That resurgence of hunter-gatherer ancestry, with a ca. 4:1 WHG:EHG contribution, is found in the Balkan Neolithic in the territory of present-day Bulgaria, close to the Danube river. This suggests a heterogeneous landscape of farmer populations with different proportions of hunter-gatherer ancestry during the early Neolithic, probably due to pockets of hunter-gatherers surviving close to the coast and major rivers (Mathieson et al. 2018). There is no sign of increasing WHG ancestry in Britain as the Neolithic progresses, which discards a resurgence of hunter-gatherers (Brace et al. 2018), although there is evidence for local Mesolithic survival and introgression in southwestern Ireland, long after the commencement of the Neolithic, also implied in haplotypic-analysis (Cassidy 2018).

In Iberia, the resurgence of hunter-gatherer-related ancestry after 4000 BC occurs in higher proportions in groups from the north and centre, and is closely related to later north-western (Canes1-like) hunter-gatherers than to the El Mirón-like hunter-gatherers from the south-east (Olalde et al. 2019).

III.3. Caucasus

The Neolithic arrived in the Caucasus during the 7th millennium BC. Farmers began to settle in the Kura–Araxes interfluve, bringing with them traditions different to those found in Anatolia and the Zagros Mountains, developing regional cultures which shared affinities with traditions in northern Mesopotamia and north-western Iran. Three Neolithic traditions can be distinguished: the Shulaveri-Shomu culture, in the central and southern regions; the western Caucasus and east Pontic region, concentrated on the area of Colchis in the foothills and along the coast; and the central and northern Caucasus, from the Surami massif to the piedmont of the northern Caucasus (Sagona 2017).

Certain Georgian sites ca. 6th–5th millennium BC show a transitional material culture ("Pre-Pottery Neolithic"), different from both the Trialetian Mesolithic and Pottery Neolithic traditions. The Pottery Neolithic of central and southern Caucasus is represented by the Shulaveri-Shomu culture, which refers to three groups: the main one, centred in the middle reaches of the Kura River; another one between the Nakhichevan region, Mil Plain, and the Mugan steppes; and the third in the Ararat Plain. This culture is thought to have been started by immigrants from northern Mesopotamia or Iran, interacting with Late Mesolithic communities (Sagona 2017).

In the Kura corridor, the typical Caucasian Neolithic village (small hamlets averaging about 1–1.5 ha) consists of high-density, cell-like compounds of round or oval houses, measuring ca. 2.5–5 m in diameter, and linked by low walls. The building on top of old structures represents a conservative building code, a tradition strongly related to the own ancestors, and thus kinship organisation, with a mentality reminiscent of Çatalhöyük. Some settlements may have been left seasonally, and courtyards were important communal areas, enclosed by small storage cells and houses (Sagona 2017).

In the Ararat plain, similar settlement plans and round structures can be seen. In the southern Caucasus a shift to fully rectangular constructions can be seen in the second half of the 6th millennium BC, reflecting a change in the social organisation, from a more communal-based village to one orientated towards the nuclear family. The plain Neolithic ceramic wares represent a local production displaying affinities with the northern Iranian Plateau (Halaf) or Anatolia (Sagona 2017).

All Shulaveri-Shomu groups are characterised by the abundance and diversity of flake tools, large quantity of scrapers, the adoption of advanced blade techniques, and ground stone artefacts, with the lithic assemblage predominantly formed by obsidian. Their economy is based on animal husbandry, raising sheep, goats, pigs, and cattle, and complementing them with hunting and fishing. Farming included a diversity of crops greater than neighbouring regions (like Northern Iran or the eastern Caspian regions). The beginnings of the wine culture is associated with this culture (Sagona 2017).

The fully-fledged Neolithic farming communities from the Mil Plain, ca 5600–5400 BC, had strong ties to northern Iran and the south Caspian region, but interaction with Shulaveri-Shomu appears to be minimal. Communal co-operation is evident, with a sub-circular planning of mud-brick constructions, with ditched enclosures pointing to potential cattle corrals, enclosed marketplaces, fortifications, or even astronomical observatories with ritual values (Sagona 2017).

Burials do not show unity among the different cultures: most findings involve inhumations, many in a crouched position, positioned either on their back or on the left or right sides (no gender differentiation). Common is the use of figurines depicting humans. Metalworking begins in the Caucasus in the late 6th – early 5th millennium, with emerging jewellery industry, including bead production and far-flung connections with central Asia and its borderlands (Sagona 2017).

In central and northern Caucasus, the evolution of the 7th to the 6th millennium is marked by a change in lithic typology and stone resources, the appearance of Pottery Neolithic, and the addition of agriculture to the subsistence economy (Sagona 2017).

III.4. Africa

Levantine herders probably introduced domestic livestock in northern Arabia, with Natufian and Pre-Pottery Neolithic spreading across Arabia during the African Humid Period ca. 8000–4000 BC (with an unknown timing). Depictions of domestic goats in rock art points to caprine pastoralism typical of the Jordanian Badia, although cattle is found in early remains, too (Scerri et al. 2018).

The Upper Capsian culture followed the Typical Capsian phase after a change to bladelet production (ca. 6200 BC), normally produced by pressure technique that required preparation of sophisticated mitred cores. A consistent range of bladelet blanks are produced, which in turn allowed for the production of standardised tools. This technological change is synchronous with an environmental shift, probably under the influence from expanding Neolithic cultures (Rahmani and Lubell 2012).

Neolithisation along the Mediterranean fringe of the African continent is not well documented, but it seems to have happened late and simultaneously with the northern Mediterranean. In the Egyptian Delta, the introduction of an agropastoral economy from the neighbouring Middle East occurs ca. 6000 BC or earlier, and reaches later Libya over the course of the 6th millennium (Guilaine 2017). Along the coastline there is clearly a rapid extension, probably related to an expansion from the Levant in the east, too.

In the southern Mediterranean, bifacial pressure flaking follows farming westward during the first wave of distribution, unlike in the northern Mediterranean, where Impressed Ware is not accompanied by this technique (it appears in Iberia only in the mid–4th millennium). In north-west Africa, wheat and sheep—species of oriental origin—are evidenced in the Maghreb at the turn of the 6th to 5th millennium BC, with the introduction of bifacial arrowheads, associated with the expansion of the Neolithic of Capsian tradition. Bifacial thinning using pressure flaking associated with agriculture is known thus only in the Levant, Anatolia, and Northern Africa until ca. 3500 BC (Darmark 2012).

As herding economies entered the Sahara, monumental sites commenced with pastoralist cemeteries along the Nile and in the south-central Sahara (ca. 6000–5500 BC), including personal adornments in some burials. Slightly later (ca. 5400–4500 BC), rock art, platforms, and/or standing stones, as well as ritually interred cattle appear.

While population levels peaked between 6000–5500 BC, northern Africa underwent over the following millennium a population decline driven by a millennium-long dry event starting ca. 6000 BC. After ca. 5000 BC, domestic cattle, sheep and goat spread throughout northern Africa, which was followed by a second population boom lasting until ca. 3500 BC.

iii.4. Northern Africans

Large genetic continuity in north-west Africa can be inferred from the Early Neolithic samples of Ifri n'Amr or Moussa (ca. 5200 BC), which shows hg. E1b1b1b1a-M81 (formed ca. 11900 BC, TMRCA ca. 800 BC). Although their position in PCA is intermediate between Iberomaurusians and modern

North African populations, this has been interpreted as driven by isolation and genetic drift (Fregel et al. 2018).

While a population related to north-western Anatolian Neolithic farmers spread westward into Europe, farmers related to those of the Levant spread southward into north-east Africa (Lazaridis et al. 2016; Schuenemann et al. 2017).

III.5. Pontic–Caspian steppe

Contacts of north Pontic cultures with Criş settlers from the Starčevo–Kőrös–Criş culture about 5800 BC introduced domesticated cattle to the Bug–Dniester culture. Up to the mid–6th millennium BC, the only domesticated animal present in the steppe was the dog. From the mid–6th millennium BC, bone remains of domestic animals are found in the north Pontic steppes, and slightly later in the Volga–Ural steppes (Vybornov et al. 2016).

No signs of cultural assimilation have been found with the introduction of Neolithisation to the Bug–Dniester culture, though. The later invasion of Linear Pottery sites ca. 5500–5200 BC respected a similar cultural frontier, geographically coincident with the Dniester. Therefore, the language of western Neolithic settlers—assumed to come from the Middle East, if language accompanied the spread of Neolithic technology—was probably not transferred to north Pontic herders (Anthony 2007).

The economy of the Pontic–Caspian steppes did not diverge significantly from a settled hunter-fisher system, though, with elements of animal husbandry found only in the steppe Neolithic cultures, as evidenced by the numerous long-term Mariupol-type cemeteries in the Dnieper region. In the Volga region, a number of forest-steppe settlements provided material with approximately equal number of bones of domesticated and wild fauna, with a considerable proportion represented by horse (Rassamakin 1999).

Neolithic period settlements and temporary encampments had an essential hunting economy, as proved by the overwhelming predominance of wild fauna. The shift to imported food-production economy in native foragers seems to have happened first during the Early Neolithic, possibly in the Bug–Dniester settlements through contacts with Criş settlers, ca. 5800–5500 BC. From the Bug–Dniester culture domesticated cattle, sheep, and goats spread quickly from about 5200 BC east- and northward into Pontic–Caspian sites, reaching Khvalynsk and the Samara region about 5100 BC (Anthony 2007).

The domesticated horse must have appeared at roughly the same time, having almost disappeared form the archaeological record during the previous Mesolithic period. Examples of horse remains include also sites on the steppe margins, such as the Pit–Comb Ware settlement of Pogorelovka, in the forest zone of the Dnieper left bank; and Neolithic settlements with a developing agricultural economy, such as those of the Bug–Dniester and Linear Pottery cultures (Anthony 2007).

The Mariupol culture, appearing ca. 5400–5200 BC, refers to the most distinctive regional aspect of the Neolithic Dnieper–Donets culture, around the Dnieper rapids, with the appearance of deep ossuary pits filled with multiple layers of skeletons, accompanied by copper objects, ceramic vessels, polished axes, and other unusual grave gifts, and correlates with the appearance of domesticated animals in the Dnieper valley (Anthony 2006).

In the Late Neolithic Don–Volga–Ural area there was the Orlovka culture in the steppe Volga basin, the Lower Don culture to the south-west, the Samara culture in the east, and the Voronezh-Don culture between the Volga and Don sites. These four cultures were related, and some researchers combine them into the Mariupol cultural-historical area, with similarities between their material culture suggesting a human migration ca. 5200–5150 BC to the north from the steppe Don region into the valleys of the Don, Medveditsa and Volga, reaching the Voronezh basin (Kotova 2008).

The first settlements of the Samara culture are found at Syezzhe during the mid–6th millennium BC, showing rituals and decorations made of shells and the fang of a wild boar, stone axes and other goods, similar to those found at Mariupol in the north Pontic area (Figure 10). Differences in pottery in the

Volga–Ural region seems to indicate the arrival of migrants distinct from the previous Elshanian territory (Morgunova 2015):

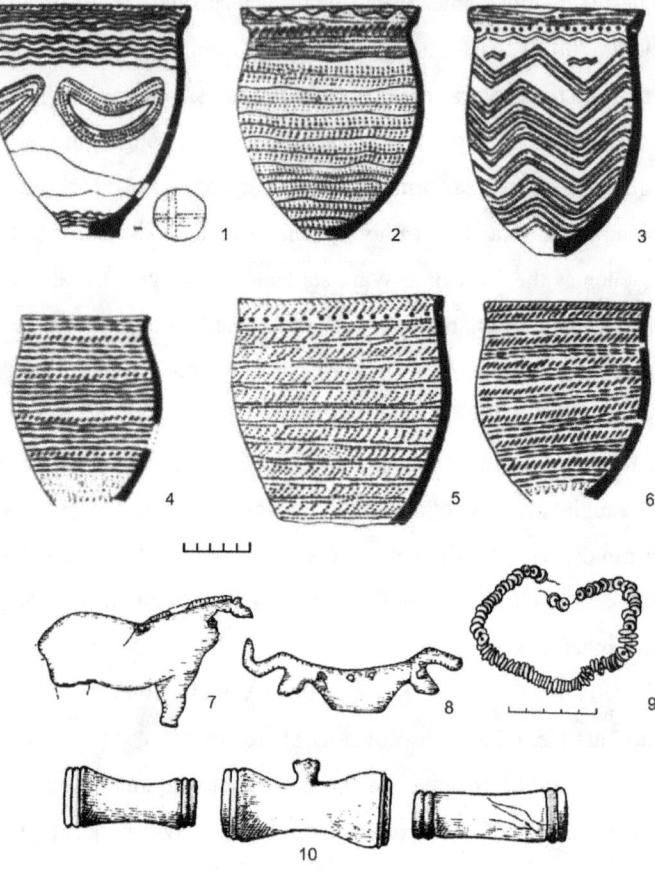

Figure 10. Materials from the Syezzhe burial ground: 1–3 pottery I group; 4–6 pottery II group; 7–8 bone amulets; 9 shell beads; 10 ornament from wild boar fang. From Morgunova (2015).

- The first type of pottery includes high vessels with small flat bottom, pronounced rim-like collars, made of clay containing silt with an admixture of shell, and shaped using plastic moulds. The surfaces were painted with ochre, and motifs are complex with meander patterns and zigzags, made with incisions and comb stamps.

- Pottery of the second type differs from the first typologically and technologically: necks are prominently made with the help of rows of deep pits and grooves; bottoms are large and flat; and not all vessels have collars. They are made from silt produced in water basins, and surfaces are covered with motifs made with comb stamps, which point to a connection with local Neolithic cultures.

Based on the evolution of pottery, those of the first type seem to have been made by a small group of foreigners that became assimilated by the second group. Due to the similarity with Mariupol pottery, foreigners apparently came from the west, and their influence is seen ca. 5300–4800 BC, corresponding to the dating of the Azov–Dnieper culture and Trypillia A1 (Morgunova 2015).

Their burial ritual consists in the extended position on their backs, similar to the Dnieper–Donets culture, with a ritual deposit of red ochre, broken pottery, shell beds, a bone harpoon, and bones of two horses. These are funeral feast deposits similar to the above-grade deposits found later at Khvalynsk (Anthony 2007).

The important social and economic significance of the horse – in terms of horse remains and symbolism – in the Pontic–Caspian steppe clearly points to the eastern area of the steppes between the Tisza and the Urals as the most likely zone of earliest domestication, as well as (later) use for transport and horseback riding, although the precise dates of each event are not known.

The first quantitative leap is seen in the Middle Volga region (in a close relation to the Volga–Ural region) during the Neolithic. The Samara culture reveals itself as a specialised horse-breeding group based on the increase in the number of horse remains, and in their wide inclusion in cults, rituals, and funerary ceremonies, evidenced by horse pendants, ornamented metacarpi, horse bones, and sacrificial altars. The domesticated horse expanded only later to the western agricultural zones (Dergachev 2007).

A third expansion of hunter-gatherer pottery spread from the Volga–Kama region to the east ca. 5000 BC, connected to influences from beyond the Urals, showing a more elaborately decorated ware (with bands of pits and impressions made from comb stamps), spreading north and west in the Sperrings and Säräisniemi 1 cultures (Piezonka 2015).

iii.5. Early Indo-Europeans and Uralians

The appearance of I2a1b1a2-CTS10057 lineages (formed ca. 8800, TMRCA 8800 BC) in Baltic hunter-gatherers in the 7th–6th millennium BC and slightly later in Ukraine Neolithic cultures—presumably all of hg. I2a1b1a2a2-Y5606 (formed ca. 7900 BC, TMRCA ca. 6800 BC)—points probably to continuous connections of the north Pontic region with the Baltic area, either as remnant populations of the post-Swiderian expansion, or—more likely, since this lineage is also found in hunter-gatherers from the Iron Gates and in later Balkan populations—through the late spread of hunter-gatherer pottery to the north. This is supported by the difference in ancestry between Mesolithic and Neolithic samples, with a clear shift towards WHG (Mathieson et al. 2018).

This conceivable expansion of I2a1b1a2-CTS10057 lineages with hunter-gatherer pottery in mainly east–west and south–north migration waves, coupled with the equally inferable expansion of R1a1a-M198 lineages through the forest zone, suggests that these stepped cultural expansions in eastern Europe were to some extent driven by acculturation of certain groups or alternatively by the resurgence of hunter-gatherer groups among migrating settlers, evidenced by their different Y-chromosome bottlenecks.

Haplogroup I-M170 is found in sixteen samples of the Dnieper–Donets culture from Volynsky and Deriïvka sites spanning the whole 6th millennium, and one sample of the Azov–Dnieper culture from Vovnihy (Mathieson et al. 2018). Eleven individuals of R1b1b-V88 lineages found in Deriïvka during the second half of the 6th millennium BC, at the same time as other I-M170 lineages, may represent a resurgence of this haplogroup from the previous Mesolithic population of the area, rather than an actual population movement from the west Pontic area.

Two samples of the 5th millennium BC, presumably from early Sredni Stog cultures, show genetic continuity with ancestry similar to the previous samples of the north Pontic area: one from Deriïvka (ca. 4630 BC), and one from Vovnihy (ca. 4430 BC), of hg. I2a1b1a2-CTS10057 (Mathieson et al. 2018). Another sample from Deriïvka (ca. 4870 BC) is a clear outlier with fully Anatolian farmer-like ancestry, clustering closely with Balkan Neolithic and Chalcolithic samples, but shows haplogroup I2a1b1-M223, and thus probably continuity of the male population of the west and north Pontic area.

Most likely, the population of the Don–Volga–Ural region eventually represented the Early Proto-Indo-European community by the time of the coexistence of western (Mariupol) and eastern (Samara–Orlovka) communities. The presence of R1b1a1-P297 subclades in a hunter-gatherer from Samara ca. 5600 BC and later in Khvalynsk ca. 4700 BC (see below *§iv.2. Indo-Anatolians*) supports the continuity of these lineages in the Don–Volga–Ural region and the North Caspian steppes, at least since the expansion of Elshanian pottery, in a population possibly already admixed with other lineages typical of Pontic–Caspian steppe and forest-steppe cultures—like R1b1a-L388, R1b1b-V88, R1a1-M459, or I-M170. This haplogroup variability supports the nature of the region during the Mesolithic as the sink of multiple migration waves, supporting thus the described interaction of cultures.

The likely presence of West Siberian hunter-gatherer ancestry (see *§v.8. Palaeosiberians*) in Khvalynsk seems to support this influence from the east. Ancient contacts between Pontic–Caspian and Kazakh steppe populations are also suggested by the cline formed between Eneolithic samples from the Northern Caucasus Piedmont (see below *§iv.2. Indo-Anatolians*) and West Siberian HG ancestry. This cline can be inferred from the ancestry of Maikop samples from the North Caucasus steppe (see below *§V.2.2. Maikop*), which can be roughly described as of North Caucasus steppe and West Siberian HG ancestry.

The expansion of material culture from the west may alternatively suggest an expansion of certain lineages from a Lower Volga region into the Volga–

Ural forest-steppes, to form the Orlovka and Samara cultures. A less likely possibility is the integration of lineages from North Caspian (Kairshak and related) groups in the area during the formation of the Samara culture. The presence of materials related to Khvalynsk—probably representing a late expansion—and Mariupol groups—possibly through imports—near the Moksha, Sura and Oka rivers (Artemova, Ikonnikov, and Prikazchikova 2018) attests to the continued strong connection of the Volga–Don forest-steppe and forest regions during the Neolithic.

Indo-European has been considered a branch of Indo-Uralic that was transformed under the influence of a Caucasian substratum (Kortlandt 2002; Bomhard 2017), which would imply the absorption of certain Caucasian traits by an Indo-Uralic branch already separated from Uralic. This event can be inferred in genetics from the admixture in the Volga–Ural region, evident in the diversity of haplogroups—including the likely local "resurgence" of R1b1a2-V1636, R1a1b-YP1272, and Q1a2-M25 lineages (see below *§iv.2. Indo-Anatolians*)—and in the eventual admixture with local steppe populations of elevated ANE/CHG ancestry.

On the other hand, haplogroup R1a1a1-M417 (formed ca. 6600 BC) was probably expanding with other R1a1a-M198 lineages, within the north-east European forest zone, likely resurging over an Indo-Uralic-speaking community, and expanding over likely Eurasiatic-speaking groups—which may account for the lack of the strong Caucasian-like features proper of Indo-European—perhaps already close to the north Pontic area. Whether these lineages expanded Early Proto-Uralic through north-eastern Europe, or they acquired the language when infiltrating the north Pontic post-Swiderian cultures, remains unclear.

IV. Early Æneolithic

IV.1. Central Europe

IV.1.1. Post-LBK

About 4900 BC, central Europe continued to be occupied by peoples descended from LBK communities, such as the Rössen culture in central and southern Germany, the Stroke-Ornamented Pottery culture of eastern Germany and Bohemia, and the Lengyel culture of Poland, Slovakia, and Hungary. These groups followed the same way of life as LBK farmers through most of the 5th millennium BC. An important development of this period is the exchange with Mesolithic foragers of southern Scandinavia, represented mainly by stone axes (Midgley 2004).

The Mesolithic Ertebølle culture is found ca. 5400–3950 BC in the western Baltic area, including southern Scandinavia and northern Germany between the Elbe and the Oder rivers, roughly contemporary with LBK communities to the south. It comprises an older, aceramic phase (ca. 5400–4600 BC), and a younger phase with pottery, T-shaped antler axes, and imported axes. T-shaped antler axes had a wide Neolithic distribution, appearing first in southern regions, like agriculture and cattle. Flint technology was based on blades used to produce arrowheads with a transverse edge, end scrapers with a convex edge, and tanged scrapers with a concave edge (Midgley 2004).

Contacts with other Mesolithic groups of the Atlantic façade and Comb-Ceramic groups in the eastern Baltic are evidenced by their similar pottery, featuring shallow, oval bowls (presumably used as lamps), and pointed-bottom vessels including small beakers and medium and large pots used for drinking, cooking, and perhaps storage. This pottery is only rarely found in groups to the south. Settlements appear in coastal and riverine zones apt for fishing, and comprise large central sites occupied more or less continuously, as well as small, seasonal sites. Burials can be found far from the coast and on high elevations, and consist mostly of inhumations, usually in the extended supine position, with men given knives, daggers and axes, and women ornaments (Midgley 2004).

IV.1.2. Megalithic culture

Megalithic tombs appear in western Europe among post-Neolithic societies along the western Mediterranean but principally expands along the extensive Atlantic façade, reaching from western and northern Iberia, France, Sardinia, mainland Italy, and the British Isles to the Shetland Islands, the Northern European Lowlands and Scandinavia. Tens of thousands of large stone burial monuments were built, using large boulders to form chambers and passages covered with mounds of earth or cairns of stones (Müller 2014).

The first grave chambers start ca. 4700 BC in north-western France, from pre-megalithic monumental sequence and transitional structures, as a response to different influences: the Blique/Epi-Rössen tradition of non-megalithic long barrows of the Passy type were erected in the Paris Basin; the eastern concept of trapezoidal-rectangular long barrows reached the northwest coast of France; and stone cists and early passage graves, similar to Epicardial and Cerny grave traditions further to the south, are found in these and other mounds (Schulz Paulsson 2019).

The culture appears in Corsica and Sardinia as simple dolmens ca. 4200 BC, while on the British Isles it emerges ca. 4000 BC, and in north-central Europe and southern Scandinavia they start ca. 3650 BC within the Funnel Beaker culture, starting on the western coasts of Oland and Gotland. The *allées couvertes* and gallery graves both developed as of ca. 3600 BC in inland settings from Brittany over to the Paris Basin and to central Germany (Müller 2014). The distribution emphasises the maritime linkage of these societies and a diffusion of the passage grave tradition along the seaway, accompanied by economic and social changes in Europe (Schulz Paulsson 2019).

Most tombs were collective burials, opened repeatedly to bury bones of newer generations, pushing bones of earlier generations aside to make space. They clearly represent ceremonial monuments, where the deceased members of the same community or clan were buried together (Midgley 2004). The high mobility of Neolithic society probably facilitated the diffusion and convergence of the new architectural developments and underlying ideology, but there is also evidence of demic diffusion (Müller 2014).

iv.1. European Neolithic farmers

Samples from Late Mesolithic/Neolithic sites show continuity of I-M170 lineages and previous ancestry; so e.g. in a sample from central Norway (ca. 3940 BC) of hg. I2a1a2-M423 (Günther et al. 2017), in subsequent samples of the Nordic Neolithic, and in samples from the Zedmar culture (ca. 4500–3000 BC), related to Mesolithic Ertebølle, showing typical hunter-gatherer mtDNA U5b1 (Bramanti et al. 2009).

In Neolithic individuals from southern France and Britain there is a greater affinity to the Iberian Early Neolithic farmers than to central European ones. This is confirmed by haplotype matching, consistent with the same ancestral populations bringing the Neolithic to Britain and Ireland. Chronological modelling suggests that NWAN-related ancestry arrived in Britain ca. 3975–3722 BC, marginally earlier in the west and rapidly dispersing to other regions (Olalde et al. 2018; Brace et al. 2019).

The Middle Neolithic resurgence of hunter-gatherer-related ancestry in central Europe and Iberia was driven more by males than by females, evidenced by the expansion of hunter-gatherer-associated Y-chromosome haplogroups I2-M438, R1-M173, or C1-F3393 in seven out of nine male individuals in Iberian Neolithic and Copper Age, and nine out of ten individuals in Middle–Late Neolithic central Europe, including the Globular Amphorae culture (Lipson et al. 2017).

Similarly, there is continuity of Mesolithic I2a2-M436 lineages in the British Isles, with fifteen out of nineteen samples reported in Great Britain (ca. 8750–2500 BC) corresponding to this subclade, at least during the Neolithic probably corresponding to I2a1b1a1a1-L1195 (formed ca. 5100 BC, TMRCA ca. 3600 BC) at least three of them from the south-west (ca. 4000–3500 BC) of subclade I2a1b1a1a1b-L1193 (TMRCA ca. 3500 BC). At least one sample from the centre (ca. 3600 BC) is of hg. I2a1a2a-L161.1, which appears more often among megalithic builders.

Individuals buried in megaliths from the British Isles (ca. 3800–3100 BC) and Ansarve in Scandinavia (ca. 3500–3200 BC) show an ancestry similar to other contemporaneous farmer groups, with a majority of their ancestry related to Early Neolithic farmers and a partial admixture component related to European Mesolithic hunter-gatherers. The genetic connection found between western European Neolithic groups from the British Isles to Scandinavia is driven by NWAN-related ancestry, rather than hunter-gatherer-related ancestry, and seems not to include central European farmers, which suggests a migration along the Atlantic coast (Sánchez-Quinto et al. 2019).

Kinship of buried individuals is reflected in Y-chromosome haplogroups, with only I-M170 lineages reported: at least four I2a1b1a1a1-L1195 out of nine in Scotland, one of them of subclade I2a1b1a1a1b-L1193; at least five out of eight I2a1a2a-L161.1 in Scotland and Sweden, with three I2a1a2a1a1-Y3749/FGC7126 (formed ca. 4900 BC, TMRCA ca. 3400 BC) reported in

Ansarve (Sánchez-Quinto et al. 2019). Both subclades may also be related to the Neolithic expansion of Atlantic farmers through the Mediterranean route.

Individuals from megalithic burials in the north (with over-representation of males) do not show systematic differences with geographically proximate non-Megalithic burials. Societal complexity during the Neolithic contrasts with the identification of a highly inbred elite individual in Ireland, strongly suggesting that the elaboration and expansion of megalithic monuments was associated (at least in some regions) with dynastic hierarchies (Cassidy 2018).

IV.2. Khvalynsk–Novodanilovka

A climatic improvement is seen peaking in the mid–5th millennium BC, with mild summers and winters, less precipitation, and an increase in steppe grass-cover with more varied vegetation (Binney et al. 2017). These changes favoured the importance of a subsistence economy based on animal husbandry, which had spread into the Pontic–Caspian area only centuries earlier. This specialisation was coupled with fundamental innovations (Parzinger 2013):

- Kurgan burials substituting flat graves.
- Dead in the crouched position, substituting the previously standard supine position.
- Rich grave goods revealing social stratification.
- Animal husbandry including widespread horse imagery and likely horse domestication.

IV.2.1. Genesis and expansion of Khvalynsk

Early Khvalynsk (ca. 5300–3900 BC) probably began from an autochthonous group of the Middle Volga region—near the site that gives the culture its name—closely related to the previous Neolithic Volga–Ural groups (Samara and Orlovka), expanding rapidly into neighbouring regions. Changes during the second stage of sites showing continuity with Samara material culture, at the beginning of the 5th millennium, are marked by strong influences from the Khvalynsk culture, with pottery showing technological changes and more variety, as well as typical Khvalynsk features, such as clays containing silt, wicker elements in ornamentation, etc. Pottery also partly continues the previous Syezzhe tradition. (Vybornov et al. 2016).

The Khvalynsk culture started to settle in the south of the Volga valley, reaching about 4900 BC wormwood deserts to the north-west of the Caspian area, and the Mangyshlak Peninsula in the east of the Caspian area, witnessed by many finds of ceramics with comb decoration, where the Khvalynsk population partly assimilated the native inhabitants of the North Caspian culture in the Lower Volga. A part of the North Caspian culture probably migrated to the Saratov Trans-Volga region, where it was assimilated to the Orlovka culture (Kotova 2008).

The Khvalynsk culture expanded to the south and west along the Lower Danube into the north Caucasian region from ca. 4800 BC, with the Nalchik cemetery in the northern Caucasus steppe being synchronous with this early stage (Vybornov et al. 2018). At the same time, Khvalynsk expanded to the west into the Don–Kalmius interfluve, developing a significant area in the north Pontic region with the so-called Novodanilovka group, including synchronous findings reaching the lower Danube region and beyond with the so-called Suvorovo group (Kotova 2008).

The sudden expansion of Khvalynsk settlers from the Volga–Ural in all directions (Figure 11) marks the development of a Khvalynsk–Novodanilovka cultural-historical area, dominating over the steppes from the lower Danube to the Middle Volga, the Caucasus, and the Trans-Caspian area showing common funerary sacrifices of domesticated sheep, goats, and cattle, connected with an increase in the number and diversity of new types of body ornaments in graves, some made of exotic materials, including copper, and polished stone maces (Anthony 2016).

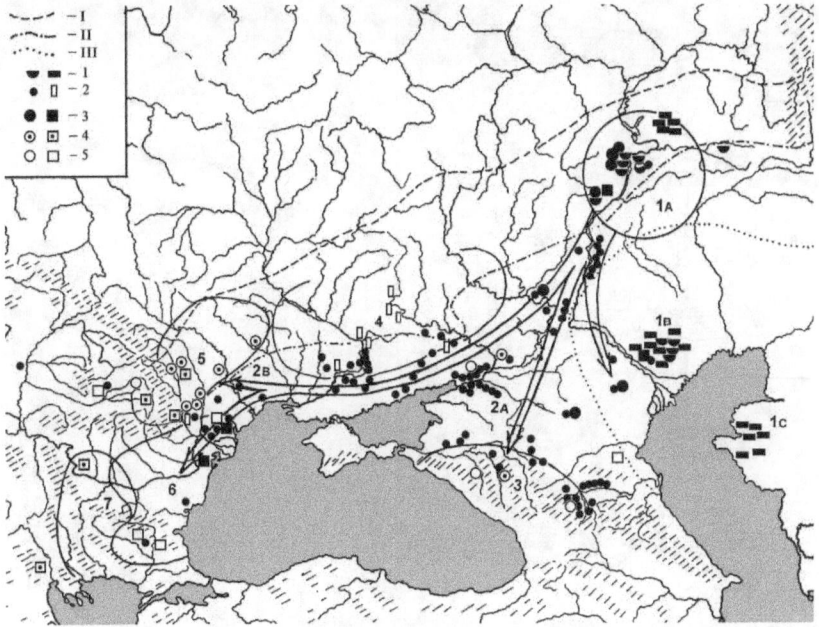

Figure 11. Map with likely route of migration of Khvalynsk settlers into the northern Caucasus and the Balkans. I) southern border of forest-steppe; II) southern border of steppe; III) border of semidesert. Cultures: 1) Burial complexes and settlements of Khvalynsk (1A – Middle Don; 1B – Northern Caspian; 1C – Western Caspian); 2) Burial complexes and remains of the Novodanilovka type (2A – Eastern; 2B – Western); 3) Pre-Maikop (Steppe Eneolithic); 4) Sredni Stog; 5) Cucuteni–Trypillia; 6) Bolgrad– Alden' – Gumelniţa–Karanovo VI; 7) Krivodol–Sălcuţa. From Dergachev (2007).

IV.2.2. Horses

The Khvalynsk culture, genetically related to Samara, preserves traditions of the ritual, cultural meaning, the treatment of the horse imagery in funeral contexts, including altars, horse bones, and funerary rites. At the same time, it is in this precise culture that the image of the horse—included in the social symbolism, such as horse-head pommel-sceptres—acquires for the first time a special, maximum social significance. The appearance and subsequent widespread distribution of similar social symbols in the whole Khvalynsk–Novodanilovka cultural-historical area, through the expansion of Novodanilovka-type objects, could be considered as another qualitative leap in the social significance of the horse.

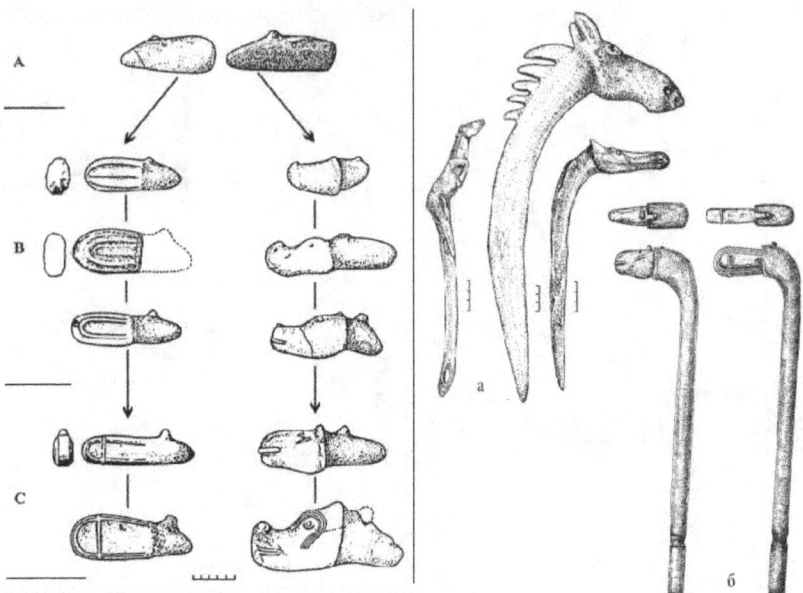

Figure 12. Left: Development of sceptres after Dergachev (2007), with more abstract (left) and more realistic (right) horse-head motif. Right: Restoration of sceptres: (a) Models for restoration – Late Mesolithic and Neolithic elk-head staffs from the east European forest zone (b) supposed appearance of Eneolithic sceptres on the hundle with the clutch. From Dergachev (2007).

Especially relevant is the position of stone-carved horse-head pommel-sceptres (Figure 12), whose almost synchronous appearance during the first half of the 5th millennium bc in Khvalynsk–Novodanilovka sites can be followed with certain precision into different subgroups, according to subsequent developments in their shape (Figure 13). Sceptres are an important cultural phenomenon, with strong symbolic functions as a divine object, used in times of peace, in times of war, and in a system of ritual power. The account of the divine and human genealogy of Agamemnon's sceptre in the Iliad bears testimony to their likely increased relevance for earlier warrior-priest chiefs (Dergachev 2007):

"Then among them lord Agamemnon uprose, bearing in his hands the sceptre which Hephaestus had wrought with toil. Hephaestus gave it to king Zeus, son of Cronos, and Zeus gave it to the messenger Argeïphontes; and Hermes, the lord, gave it to Pelops, driver of horses, and Pelops in turn gave

it to Atreus, shepherd of the host; and Atreus at his death left it to Thyestes, rich in flocks, and Thyestes again left it to Agamemnon to bear, that so he might be lord of many isles and of all Argos."[7]

The explosion of horse symbolism has been interpreted as a signal of the start of horse-riding technique in the eastern cultural area, with the horse becoming a necessary instrument for long-distance travel, for transport, for raids, and for war (as a means for quicker movements rather than their use for mounted war *per se*), facilitating thus the culture's rapid expansion. While bone remains show a similar proportion in the contemporary early Sredni Stog groups of the north Pontic steppe and forest-steppe areas during the 5th millennium BC, supporting the relevance of the domesticated horse in their subsistence economy, horse remains are strictly limited to an economic context, without social or symbolic meaning (Dergachev 2007).

Clear archaeological evidence for the development of horseback riding is found in the early–to–mid–4th millennium BC in the Botai-Tersek culture of central Asia, in the expansion of Repin herders, in Maikop and Transcaucasian cultures, and in Armenia. Archaeology points thus a likely expansion of the technique from a single source at the turn of the 5th–4th millennium BC, which is compatible with an earlier, autolimited expansion of incompletely tamed horses (not adapted for many relevant tasks seen in later cultures) with Khvalynsk chieftains in the mid–5th millennium BC (see *§V.7.2. Late Repin*).

Horse palaeontological and archaeological data suggests that the Urals were a potential constraint for the dispersal of horses between Europe and north-central Asia, and that suitability for the species steadily improved in western and south-eastern Europe from ca. 6,000 to 3,000 years ago. The Caspian Sea was possibly the westernmost boundary of Asian horses, which probably became adapted to the new areas in the Pontic–Caspian steppes and to the west from the 5th millennium BC on (Leonardi, Boschin, et al. 2018).

---

[7] *Homer. The Iliad with an English Translation* by A.T. Murray, Ph.D. in two volumes. Cambridge, MA., Harvard University Press; London, William Heinemann, Ltd. 1924. Retrieved (2018) from <http://www.perseus.tufts.edu/>.

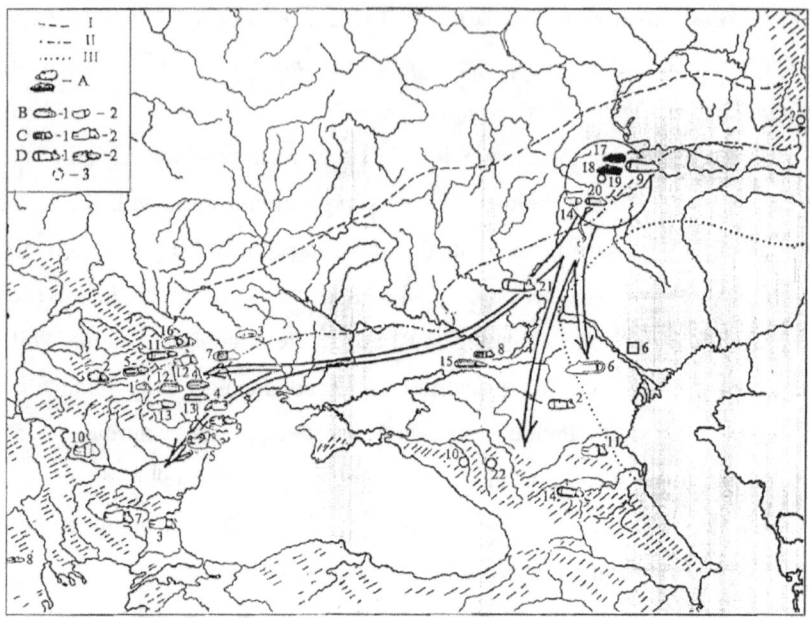

Figure 13. Map of regional–chronological morphological developments of horse-head sceptres. A) Original (Khvalynsk) variant. B) More complex forms including schematic Byrlelesht' and realistic Khlopkovo-Ariushd variants. C) Transitional forms, including schematic Konstantinovka-Velen' and realistic Drama-Fedeleshen' variants. D) Extremely complex forms, including schematic Arkharinskiy and realistic Kasimcha-Suvorovo variants. From Dergachev (2007).

IV.2.3. Kurgans

The use of barrow burials or *tumuli* has been argued to appear as a local north Caucasian feature, with origin in the 5th millennium BC, and also in the southern Caucasus, with the Leilatepe culture, from where it spread north. Other barrows are later found up to north-west Iran. Although evidence is too scarce to select a precise origin, the early Khvalynsk–Novodanilovka burials, on the Pontic–Caspian steppes, are the first to feature rich, ochre-sprinkled graves under kurgan-like structures (Korenevskiy 2012).

The addition of the tradition of ochre staining (originally from the steppes) to the emerging proto-kurgans supports that these structures emerged with the contacts of steppe cultures with the Caucasus. The standard posture on the back with knees raised, with their heads to the north and east (Anthony 2007), characteristic of Khvalynsk-type burials, point to the expansion of the

Khvalynsk–Novodanilovka cultural-historical area as the starting point of this tradition in the steppes.

All members of society are considered represented in the earliest Khvalynsk cemeteries, although there is a clear emerging trend during their expansion for elite male burials to predominate. Rich grave assemblages include stone clubs and axes, animal-head sceptres, long flint blades, and ornaments for clothing, many made of copper. These rare copper objects like rings and beads, most likely from western industries, are more common in elite male graves, as are animal sacrifices and red ochre (Murphy and Khokhlov 2016).

The emerging kurgan structures were probably not simple pits filled with earth. There was a belief that the funerary structure was the place where the buried moved to another world, and in that sense similar funerary structures reflect certain egalitarian ideas, so the evolution from collective necropolis to the rich grave assemblages reflect the meaning of prestige objects as symbols that emphasise social status, and thus an evolution to a kinship-based, elite-dominated organisation into small families, as well as a potential function in the transition to the afterlife (Korenevskiy 2012).

The new social elites identified themselves through grave goods and grave construction, marked the status with clothing (copper jewellery) and symbols of power (mace and sceptre). The existence of similar children's graves supports the membership to social groups being acquired by birth. This common evolution in the whole Khvalynsk–Novodanilovka cultural-historical area supports the emergence of tightly structured elite social groups expanding from the east (Parzinger 2013).

Ceremonial skull-scraping of the parietal bone, consisting of one to seven gouges about 2–3 cm in length in the parietal bone surface, may appear mainly in mature adults (Khokhlov 2016), with some cases clearly associated with elite burials. Zoomorphic sceptres represented probably a ritual source of power for Khvalynsk chieftains, political and/or religious leaders, as evidenced by the unique zoomorphic carving found in Ekaterinovka (a riverine settlement) in the second half of the 5th millennium, resembling a toothed fish or reptile, rather than the most common horse-related motifs expanding with Novodanilovka–Suvorovo settlers. The finding of similar elk-head staffs in Mesolithic–Neolithic cultures of northern and eastern Europe and the Trans-Urals region may suggest an ancient cultural connection of this tradition through northern Eurasia.

Early kurgan-like or *proto-kurgan* constructions in the Pontic–Caspian steppes are found thus associated with the expansion of Khvalynsk–Novodanilovka chiefs, featuring similar constructions to mark elite graves: rooves made from separate slabs with cairns are known in the Dnieper and Volga regions (17% of burials in early Khvalynsk were superimposed with stone cairns or had a single stone marker); cists with cairns are known from the northern Donets and Azov areas; and a unique cromlech is found in the Dniester–Danube area, among Suvorovo graves. Apart from these stone constructions, in the Volga and northern Caucasus region sometimes natural hills or small earthen or wooden constructions are used as burial markers (Rassamakin 1999).

IV.2.4. Khvalynsk economy

Stable isotopes in the bones of Khvalynsk individuals show that their diet depended to a large extent on fish (Schulting and Richards 2016), although domesticated animals also feature prominently in the Khvalynsk culture, accompanying thus a food-producing economy (Vybornov, Kosintsev, and Kulkova 2015).

Sheep and goats were sacrificed more than any other species in ochre-stained ritual deposits in Khvalynsk; cattle predominated on the lower Dnieper; and horse bones dominated in between. Given the data from the predominant diet, domesticated animals may have been reserved for their use as a ritual and feasting currency associated with the political competition between a new social rank of elites (Anthony 2016; Anthony and Brown 2011).

There is an evident causal relationship between the emergence of a warrior class of social leaders and the spread of cattle herds through widening grazing lands, and between their rapid westward spread and imagery evidencing the domestication of the horse (and its use as a riding mount). Deposits from Ukraine, the Caucasus or the Urals were not used at this moment, so metal imports came probably from Balkan and Carpathian mines, which supports a dense network of extensive trading links through the north Pontic steppes up to the Volga (Parzinger 2013).

The Khvalynsk expansion represents thus the start of the Eneolithic era, as the time of development of a prestigious economy that marked social elites through different valuable objects, many of them obtained through exchange networks, reflecting the direct or indirect involvement of the owners. Among them were items requiring high skills or complex manufacturing techniques (different woollen tools, sceptres, stone bracelets); tools that occupy an important role in labour, war and industry (stone flat axes, arrowheads, knife-like plates, and chips of flint); iconic objects (bone plates from canine fang, pins, bone sticks with a hole); beads (from bone, stone, shell, and bead washers that could be collected in whole garlands, acquiring a special value); copper jewellery (beads, rings, bracelets) (Korenevskiy 2012).

IV.2.5. Suvorovo

Since the early 6th millennium BC objects and ideas flew in a single direction, from central (Linearbandkeramik) and south-eastern (Old European) cultures to the Pontic–Caspian steppe. This changed from the mid–5th millennium on, when movement is seen also from the steppe to the west, driven by the newly found human mobility (Heyd 2016).

The Suvorovo–Novodanilovka group appear in the north-west Pontic area, the Lower Danube, and Dobruja, reaching to the south the Upper Thracian Plain and northern Greece, and to the east the east and central Carpathian Basin, up to central Europe. They are recognised by their rich individual primary graves displaying ostentatious prestige goods, whose inventory included jewellery (shell necklace, copper goods, rarely gold), tools (flint and copper) and weaponry. The most conspicuous objects are the high technology flint inventory, with axes, long blades, and triangular silex spearheads sticking out (Heyd 2016).

Individuals were laid on their back, extended or crouched, with slightly bent knees on the side, and oval to square graves (Figure 14). Ochre staining of the entire grave is the norm, including whole ochre pieces. A short, still quite shallow mound can be seen over the grave, sometimes with circular stone structures either around the grave (as in Suvorovo) or around the mound. Zoomorphic (usually horse-head) sceptres were particularly common in the south, while in the north-west and west Pontic areas (and in the east Carpathian Basin) they represent mostly isolated finds, and comparable pieces are abstract stone sceptres and stone mace heads with knob decoration. The farthest south that these materials are found is Suvodol–Šuplevec, northern Macedonia, in the south-east Balkans (Heyd 2016). To the west, the farthest finds are in the Csongrád–Kettőshalom site, dated ca. 4370–4240 BC (Horváth et al. 2013).

The most recent radiocarbon dates show that these findings appear in south-east Europe from about 4600 BC—contemporaneous with the rich graves of the Necropolis from Varna I—to ca. 4000 BC. Steppe imports are found in Gumelniţa, in the Lower Danube region, from about 4400 BC, pointing to an established trade network (Reingruber and Rassamakin 2016). A the end of the Early Eneolithic, the Varna necropolis ceases to function, and the Suvorovo elites disappear, although Cernavodă I, continues to show burials similar to north Pontic findings ca. 4000–3750 BC (Heyd 2016).

While the Suvorovo expansion marked the beginning of long-distance exchange of prestige goods with the steppe, and they did not represent a massive migration, the "infiltration" of settlers was enough to cause the abandonment of settlements by the Gumelniţa population of the left bank of the Danube when the steppe tribes appeared. There is little evidence of armed conflict, so it is possible that the crisis of local agricultural economy may have caused this abandonment. There may have been a gradual, peaceful process of assimilation of the Suvorovo settlers by the local Gumelniţa population, favoured by the crisis of the local agricultural economy, which is seen by many scholars as the process by which the Cernavodă I culture came into existence (Heyd 2016).

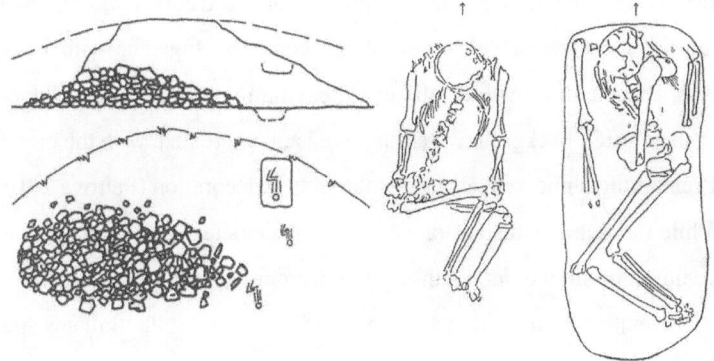

Figure 14. Suvorovo/Novodanilovka graves Kjulevca, and Targovište-Gonova Mogila, grave 1 (Bulgaria) after Govedarica (2004) and Heyd (2016).

These kurgan findings show funerary rites and technology, but there are no associated settlements. They appear in parallel to complex organised settlements like those of Cucuteni–Trypillia, Bolgrad–Aldeni, Varna, Kodžadermen–Gumelniţa–Karanovo VI and related cultures, with which they participate in economic exchange (Suppl. Fig. 7). Based on the number of graves among these groups, we can say that the size of the infiltration must have been rather small (Heyd 2016).

IV.2.6. Varna

It was probably the arrival of Suvorovo migrants what triggered the idea of lavish grave furniture and the display of wealth, prestige, power, and social position in the graves of Copper Age sedentary farming communities of southeastern Europe. The Varna I cemetery is the clearest representative of the expansion of the new mentality to the Balkans, and has been recently dated more exactly to ca. 4590–4340 BC (Krauß et al. 2017).

Characteristic ceramics are based on features known from the late Chalcolithic ceramic complexes in Durankulak (tell and necropolis), Devnia, the Varna lake settlements and necropolis. Fine ceramic is thin-walled, with a light grey turning into black, burnished, smoothed surface, with pots having upper cylindrical and rounded parts or being bi-conical forms with vertical handles. The composite S-profile and strongly outward curved mouth rim were one of the most typical elements of the complex, together with ring-like bottoms, and the most emblematic ornamentation was the "Ezerovo" type, in which the motif's background was engraved and encrusted, with the ornament itself remaining embossed, as well as the flutted decoration (Petrova 2016).

While the richness displayed by the Varna cemetery and its accumulation of wealth are unique in south-eastern Europe, similar accumulations of material wealth are encountered in isolated finds all over the Balkans and the Carpathian Basin, reaching Greece and Anatolia. Metallurgy requires material and skills which are not readily available, which means that elites kept control of them by limiting people's ability to access and produce metals themselves. In fact, except for the distinct material culture, the rich Varna burials and the Novodanilovka burials are essentially equivalent (Heyd and Walker 2004).

Graves and hoards demonstrate thus sharp inequality over wide parts of south-east Europe in the 5th and 4th millennium BC, showing thus social stratification, also displayed in the form of house sizes and pottery inventories (in quantity and quality) within settlements. There is thus a pattern of robust social institutions and enhanced complexity, of lineages and powerful chieftains, of networks and bonds persistent in time and space, reflected in Varna, in mega-villages of middle and late Trypillia, and in many other sites in south-eastern Europe (Heyd and Walker 2004).

iv.2. Indo-Anatolians

Three samples dated after ca. 4700 BC have been analysed from the Khvalynsk cemetery, described as hosting typically southern and northern individuals. One high-status burial, buried supine with raised knees and an assemblage of 293 copper artefacts (this grave alone accounting for ca. 80% of copper objects in the Khvalynsk cemetery) represents thus a high-status individual, member of an elite group of patrilineally-related families that was probably successful during the Indo-Anatolian expansion, reported as of haplogroup R1b1-L754[8] and mtDNA H2a1, unique in the region (Mathieson et al. 2015).

The individual of haplogroup R1a1-M459 (xR1a1a-M198), mtDNA U5a1i, also buried on his back with raised knees, probably represents a commoner, remnant of a local population, showing more EHG-like ancestry. An old male of Q1a-F903 lineage (probably Q1a2-M25, see above *§ii.5. Caucasus hunter-gatherers*), mtDNA U4, and higher CHG/ANE component related to steppe eneolithic samples, who died from blows to his skull, suggests that the origin of this extra ancestral component found in Khvalynsk (and much elevated later in sampled Yamna) individuals comes from the admixture of Samara hunter-gatherer-like elites from the Don–Volga–Ural region with northern Caucasian or northern Caspian steppe populations, or both, during their expansion.

Two individuals from Progress in the Northern Caucasus Piedmont (dated ca. 4600 BC and 4150 BC), of haplogroup R1b1a2-V1636+, and one from Vonyuchka (ca. 4300 BC) show elevated ANE ancestry[9], which confirms the presence of this component in regions of the northern Caucasus with early pit

---

[8] Tentative analysis with Yleaf does not yield a subclade beyond R1b1-L754 in Wang et al. (2019) supplementary materials. It is probably safe to assume, as did the authors of the paper that published the sample (given the relevance of its assemblage, and the likely influence of his family in the Khvalynsk expansion) that it may have belonged to haplogroup R1b1a1b-M269.

[9] Both populations can be modelled as of EHG (ca. 40–60%) and the remainder from a CHG-related basal ancestry, with Vonyuchka slightly shifted to Iran Neolithic, whereas Maykop receives CHG-related ancestry (ca. 86%) apart from Anatolian Neolithic (ca. 10%) and EHG (ca. 4%), according to Wang et al. (2019) supplementary materials.

grave burials—related to the Don–Caspian steppes—and support its expansion in the Don–Volga–Ural region linked to steppe elites, likely through exogamy of expanding Khvalynsk settlers (Suppl. Graph. 4). Both Eneolithic Samara and north Caucasus steppe populations analysed to date show no gene flow from Anatolian farmers, unlike contemporary samples from the north Pontic region and later samples from the Yamna culture (Wang et al. 2019).

The finding of R1a1b-YP1272$^+$ as a Maikop outlier in the same kurgan in the late 4th millennium (see below *§v.2. Early Caucasians*), and of R1b1a2-V1636$^+$ among Yamna individuals of the Caucasus in the early 3rd millennium BC (see below *vi.1. Disintegrating Indo-Europeans*) further supports the possible relevance of these lineages in the Indo-Anatolian expansions associated with Khvalynsk, or else their widespread presence among Pontic–Caspian steppe populations before the Khvalynsk expansion.

Based on continuity of ancestral components in later samples from Afanasevo, the early Khvalynsk community eventually stabilised (probably after ca. 4500 BC) in its admixture, remaining close to the analysed samples from the north Caucasian steppes, deriving more than 60% of ancestry from EHG, and the remainder from a CHG-related basal ancestry (Wang et al. 2019), in what can be taken a model of the so-called "Steppe ancestry" in later samples. This homogenisation of the Don–Volga–Ural, Kuban, and north Pontic areas suggests continued exogamy of Khvalynsk clans dominated by elite males with other related groups from the steppe.

Other analysed elite Khvalynsk individuals (ca. 4250–4000 BC), from the riverine Ekaterinovka settlement, whose material culture is interpreted as chronologically intermediate between late Samara and early Ivanovska–Khvalynsk materials (Korolev, Kochkina, and Stashenkov 2019), have been reported by Khokhlov (2018) as within the R1b1a1b-M269$^+$ tree (formed ca. 11300 BC, TMRCA ca. 4400 BC), which confirms the continuous presence of R1b1a1-P297 subclades in the region. The estimated split and successful spread of R1b1a1b1-L23 (formed ca. 4300 BC, TMRCA ca. 4200 BC), subclade of R1b1a1b-M269, further supports its association with patrilineally-related clans that expanded with early Khvalynsk around the mid–5th millennium BC.

The finding of a rare R1b1a1b-M269 subclade R1b1a1b2-PF7562 (formed ca. 4400 BC, TMRCA ca. 3400 BC) in the Balkans, Central Europe, Anatolia, and the Caucasus (Myres et al. 2011; Herrera et al. 2012) may support their association with the early spread of Indo-Anatolian speakers, although their late TMRCA points to a recent expansion linked to the spread of Yamna migrants (see below *§vi.1. Disintegrating Indo-Europeans*).

The earlier, Epipalaeolithic–Early Mesolithic origin of haplogroups R1b1a2-V1636, R1a1b-YP1272, and R1b1a1b-M269 compared to their late estimated successful expansion around the mid–5th millennium BC supports their emergence among local populations of diverse haplogroups around this time of population movements in the region. Since R1b1a1-P297 lineages were probably the latest to successfully spread into the Volga–Ural area, the presence of other lineages among Khvalynsk males suggests the resurgence of indigenous haplogroups, probably prevalent among certain local Pontic–Caspian groups before the expansion of the North-Eastern Technocomplex and hunter-gatherer pottery.

Steppe ancestry has been found in one female child from the Varna I cemetery (ca. 4711–4450 BC), from the earliest burials of the first phase, richly furnished; in a young male from Smyadovo (ca. 4550–4450 BC), of a Balkan Copper Age culture (Mathieson et al. 2018), of hg. R-M207[10]; and in a Greece Neolithic sample, probably also from the middle to late 5th millennium BC (Wang et al. 2018). All these samples prove the expansion of Suvorovo settlers to the south into the southern Danube regions and beyond, up to northern Greece (Suppl. Graph. 5). However, the presence of few individuals with

---

[10] The sample has been officially and repeatedly reported as of hg. R-M207. It is also positive for SNPs CTS9018, CTS9018, PF6452 (defining the R1b1a1b-M269 tree), but shows negative SNPs for P equivalent, which makes its actual haplogroup unclear. Additional information by Richard Rocca.

Steppe ancestry among two dozen Copper Age Balkan samples from the region (ca. 5000–4000 BC) further supports the nature of the Suvorovo expansion as a rapid infiltration of few steppe chieftains among local east Balkan populations.

The presence of R1b1-L754 samples in the north Pontic region and in the Balkans during the 5th millennium does not prove they belong to the known lineages associated with Khvalynsk, and the presence of confirmed R1b1b-V88 subclades in both regions (Mathieson et al. 2018) frustrates their interpretation as belonging to any specific subclade. On the other hand, the presence of I2a1b1a2a2-Y5606 in Neolithic samples from the north Pontic area, and of I2a1b1a2a2a-L699 lineages expanding with Yamna (see *§vii.5. Palaeo-Balkan peoples*), suggests that some of these lineages were also integrated in the Khvalynsk society, and may appear associated with Suvorovo–Novodanilovka settlers. Other Pontic–Caspian steppe lineages, like R1a1-M459 or Q1a-F903, could have also accompanied these early Indo-European elites, before the further Y-chromosome bottlenecks seen in Repin and early Yamna.

IV.3. Eastern Europe

IV.3.1. Metalworking

The Copper Age began in Bulgaria ca. 5200–5000 BC, but Old European copper-trade network included the Pontic–Caspian steppe groups only after ca. 4600 BC. In the period ca. 4800–4000 BC, Trypillia ceramic imports appear at the Neolithic Dnieper sites. In the forest-steppe region, they occur on a number of sites belonging to the Kyiv–Cherkassy variant of the Dnieper–Donets community, and later imports reach into the forest zone, into the territory of the Pit–Comb Ware culture. Prestige objects begin to appear at this time on the north Pontic region, too, marking the beginning of the prestige exchange (Rassamakin 1999).

There is no gap between the Neolithic cemeteries of Nikolskoe, Lysogorskoe or Mariupol and the emergence of the first Sredni Stog burials, which mark the advent of the Eneolithic. In fact, certain prestige objects appear in Neolithic cemeteries before their demise, and flint workshops on the Donets—which become quite relevant during the beginning of prestige exchange in the region—can be traced back to the late Neolithic (Mariupol) industry (Rassamakin 1999).

The expansion of Khvalynsk–Novodanilovka connected Early Eneolithic sites, from the Lower Danube (Suvorovo, Cernavodă I) to the Kuban region, bordering on the pre-Caucasus region (with pre-Maikop Trans-Kuban culture) to the steppe and forest-steppe Volga region of the Khvalynsk culture. The expansion of Suvorovo to the Lower Danube, with its contact with rich local agricultural settlements, sets into motion a long-distance prestige exchange system, and the tradition of rich burial assemblages that expands through cultures of the north-west and north Pontic region (Rassamakin 1999).

The economy of the region included sheep–goat, cattle, and horse bones, and it seems that sedentarism was the rule, with hunting playing a significant part in the diet. Trade in this period was based around copper and copper artefacts, from two main extraction regions: the Middle Danube area and

Thrace. Finds from the steppe up to Khvalynsk show that Novodanilovka was associated not only with the distribution of the first copper artefacts in the steppe, but also with the establishment of an independent metalworking focus in the Black Sea region, which used Thracian–Lower Danubian and Middle Danubian ore, as well as Trypillian, Varna, and Gumelniţa technology (Rassamakin 1999).

The lack of complex copperworking in early Khvalynsk suggests that all the copper finds in the Volga and pre-Caucasus region were imports from the west, and rich copper assemblages in the Dnieper and Donets regions seem to occur at regular intervals or suitable stopping places along the main route (Figure 15), which—together with the flint processing remains—points to north Pontic groups as intermediaries (Rassamakin 1999).

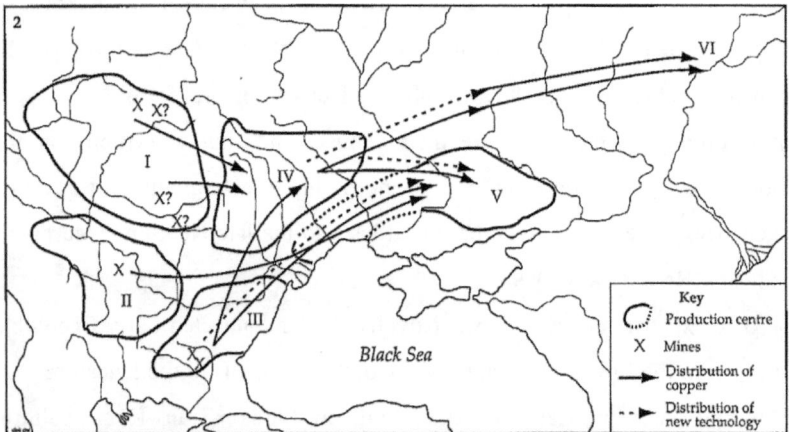

Figure 15. Distribution of copper and new technology in the steppe area from the different production centres (after Ryndina 1993, fig. 1). I-VI are production regions (metalworking centres): I) Tisza – Transylvania; II) Middle Danube; III) Thracian – Lower Danube; IV) Carpathian-Dnieper; V) Northern Pontic; VI) Middle Volga. From Rassamakin (1999).

Metal hoards of the initial Chalcolithic (ca. 4800–4200 BC) coexist with the first great wave of the Alpine jadeites. The Danubian axe and adze hoards phenomenon flourishes in central Europe, covering the area between the Meuse and the Vistula rivers. By 4600/4500 BC, a 'Europe of hoards' extends from Brittany to the Carpathian Mountains, in non-metalworking Neolithic societies. Isolated finds of giant ('elite') mounds, the Carnac mounds, built for a single individual, are found in this copperless western Europe at the same time as those in Varna I. A vast distribution network of Alpine axeheads and its corresponding hoard phenomenon in the west is thus comparable to the contemporaneous copper hammer-axe horizon in the initial east European Chalcolithic (Jeunesse 2017).

Late LBK groups, including the Lengyel, Tiszapolgàr and Bodrogkeresztúr cultures, as well as contemporaneous cultures of northern and western France, like the Cerny culture of long barrows (ca. 4800–4300 BC), show some burials with stereotypical grave goods (weapons for men, ornaments for women) which stand out from other burials, showing thus the elite status of certain individuals (usually males), although without much difference with other graves (Jeunesse 2017).

IV.3.2. Early Sredni Stog

The Bug–Dniester culture follows a division being proposed between an aceramic phase dated to ca. 5500–4900 BC and a ceramic phase ca. 4900–4400 BC, with apparent similarities in fabric, form, and decoration of pottery with the parallel developments in the Dnieper–Donets culture. This change marks the transition from Mesolithic hunter-gatherer societies to Neolithic ones, featuring domesticated animals and arable agriculture (wheat and barley, as well as millet, oats, vetch and rye) to differing degrees (Telegin et al. 2015).

The early Sredni Stog culture is characterised by a distinctive incised line and dot decoration, that spans from the Lower Don in the east to the Cucuteni–Trypillia settlements. Similar pottery decoration connected these cultures to earlier north Pontic Neolithic decorative features. Typical assemblages of these sites include typologically distinctive flint and stone artefacts, such as long knife-like blades, triangular flint spears and arrowheads, and flat axe-adzes, as well as distinctive perforated antler artefacts (Rassamakin 1999).

Early Sredni Stog settlements in the north Pontic area include the earliest burials of Stril'cha Skelya, Oleksandriia on the Oskol (a tributary of the

Donets), Aleksandrovsk on the Donets, Igren VIIII, Razdolnoe on the Kalmius, as well as Vasylivka, Deriïvka 2 on the Dnieper, the island of Vinogradny, and Sredni Stog II; possibly some burials of the Lower Don, a border with the eastern region, may have part of the culture, too. Contacts with neighbouring steppe cultures are evident in imported Sredni Stog materials in the Khvalynsk settlement of Kara-Khuduk in the Caspian region, and in pre-Maikop Svobodnoe in the Kuban region (Rassamakin 1999; Manzura 2005).

Chronologically, the culture corresponds to the Cucuteni–Trypillia A3–4 and B1 agricultural settlements (ca. 4800–4000 BC), which show the same type of pottery in terms of technique and decoration, also found in Gumelniţa A2 (ca. 4500–3950 BC). The first copper and gold objects in the north Pontic region are associated with this period. Flint extraction and flint-working loci including mines appear in the Donets zone, the products of which correspond to artefacts from prestige burial assemblages. Oleksandriia was one such flint-processing locus, where a large quantity of both finished projectile tips, axes, long blades, semi-finished products, and also production waste was found (Rassamakin 1999; Manzura 2005).

Early contacts of north Pontic populations with the Hamangia culture from the Dobruja region (ca. 5250–4500 BC) is seen in imports including adornments from copper, cornelian, marine shells and pots in steppe sites, and plates from bone and nacre, pendants from teeth of red deer in Hamangia sites. The Hamangia influence was especially important in the burial rites of the steppe population, and may have caused the use of stone in graves and above them, pits with alcove, and new adornments of burial clothes. The strongest impact is seen to the north of the Sea of Azov in the early Sredni Stog culture, with the adoption of the new religious element potentially connected with the formation of the centre of steppe metal working (Kotova 2016).

IV.3.3. Cucuteni–Trypillia

The Cucuteni–Trypillia agrarian culture sites show complicated rules and networks of social organisation at different levels, which may be reconstructed as follows (Müller et al. 2015):

- the household level shows open communication between neighbours and the whole settlement, linking together neighbouring households but not separating them from others (peaceful neighbourhood principle);

- specialisation between households at an economic level in respect to their integration in processing primary (e.g. cereal) and secondary (e.g. weaving) products, with smaller houses showing more primary activity, and larger houses showing fewer activities of primary subsistence production;

- household clusters of ca. 5 houses link households spatially (house-ring principle), potentially based on generational contracts / lineages;

- the economic and political linkage of households to quarters, represented by mega-structures as the focal point and by supra-household economic specialisations (mega-structure principle);

- the overall settlement, which needs a political institution to direct the spatial planning of the site and combined economic activities (mega-site principle).

During the 5th millennium, a strong, long-lasting, east–west orientated exchange network can be observed in the north Pontic area between the Cucuteni–Trypillian culture in the forest-steppe and north Pontic groups of the coastal steppe, including the site at Deriïvka (Reingruber and Rassamakin 2016). These close interactions were also maintained between the north Pontic and Khvalynsk populations, and it seems to have been driven by an interest in metal objects. In fact, the Khvalynsk centre of metalworking was formed under western influences, among which the Early Trypillian centre dominated (Kotova 2008)

The exchange systems in the north Pontic area during the Eneolithic included intercommunal exchange within likely related ethnolinguistic groups (such as the Sredni Stog internal exchange of natural resources, like flint materials); exchange with the nearest neighbours (such as prestigious exchange between Sredni Stog and the Trypillian, Kyiv–Cherkassy, Donets and Middle Don cultures, as well as northern Caucasus and Crimea); and long distance exchange, made far from friendly villages, probably created under the Khvalynsk–Novodanilovka network (Kotova 2008).

The revolution of herding, travel, and raiding—and thus the change in the steppe—had come with horseback riding, appearing ca. 4800 BC in early Khvalynsk, and spreading south- and eastward with Suvorovo–Novodanilovka elites. They came from the eastern steppe, and they were probably involved in raiding and trading with the north and west Pontic areas during the Trypillian B1 period, before and during the collapse of Old Europe (Anthony 2007).

At the end of this period, the system of interrelationships disappears: there are no late Sredni Stog pottery in Cucuteni–Trypillia; the pottery changes; there is no evidence for the production and distribution of flint artefacts of the old type; and new types of arrow- and spearheads with distinctive notched bases substitute the old projectile points in the region (Rassamakin 1999).

IV.3.4. Forest Zone

The first pottery appeared around the ancient Lake Saimaa basin ca. 5100 BC, followed by the Early Asbestos Ware (EAW) culture ca. 4700 BC, which used asbestos as a tempering material. This culture was the prevailing type of archaeological assemblage for several hundred years, but declined and disappeared around the early 4th millennium BC.

In mixed forest regions of the central Russian lowland plain to the Volga area and the Kama valley, Mesolithic–Neolithic hunter-gatherer groups like Lyalovo (ca. 5000–3650 BC) and Volosovo (ca. 3650–2300 BC) show late pricked and comb–stamped ceramic. The characteristic settlement shows partially sunken earth-houses or dugouts (*poluzemlyanki*), and vessels are simply formed, with a round or pointed base and imprint-decorated outer surfaces. Tools include harpoons made of flint, bone, and horn. Copper objects are rare findings, but bone or stone animal figures are typical and associated with forest fauna, such as bears, fish, beavers, etc. (Parzinger 2013).

In the southern area, near the north Pontic forest zone up to the Don River, this stage of Rudnyaya culture shows continuity in relation to the previous Late Neolithic period, and cultural interaction is observed with the eastern Baltic area and through the Western Dvina (Mazurkevich et al. 2009).

iv.3. Early Uralians

Two individuals of the 5th millennium BC, presumably from early Sredni Stog, show continuity with ancestry similar to the previous samples of the north Pontic area: one from Deriïvka (ca. 4630 BC), and one from Vovnihy (ca. 4430 BC) of hg. I2a1b1a2-CTS10057 (Mathieson et al. 2018). Another sample from Deriïvka (ca. 4870 BC) is a clear outlier, with fully Anatolian farmer-like ancestry—clustering closely with Balkan Neolithic and Chalcolithic samples—but shows haplogroup I2a1b1-M223, and thus probably continuity of the male population.

In the Balkans, Copper Age populations contain significantly more hunter-gatherer-related ancestry, contemporary with the 'resurgence' of hunter-gatherer ancestry in central Europe and Iberia, and consistent with changes in funerary rites ca. 4500 BC to extended supine burial, in contrast to the Early Neolithic tradition of flexed burials. An important population replacement is supported by the presence of mtDNA haplogroups H, HV, W, K, and T in twenty-eight Trypillian samples from the Verteba Cave, contrasting with typical pre-Eneolithic lineages (Wakabayashi et al. 2017). Trypillian samples of the Middle Eneolithic (see *§v.6. Late Uralians*) show mostly Anatolian-related ancestry (ca. 80%) with contribution of hunter-gatherer ancestry (ca. 20%) intermediate between WHG and EHG, consistent with the local population (Mathieson et al. 2018).

Based on the lack of individuals of R1a1a-M198 lineages during the 6th and 5th millennium BC—the most likely Early Uralic speakers—in the sampled populations from the Pontic–Caspian area, and their presence there and among Baltic hunter-gatherers during the Mesolithic, it is possible that communities of this lineage were at this time mainly part of the forest or forest-steppe regions from the Middle and Upper Dnieper basin, and spread to the south ("resurging" in the area) only after the Khvalynsk–Novodanilovka expansion.

This wide distribution of haplogroup R1a1-M459 in the eastern European forests is supported by an individual from Yuzhnyy Oleni Ostrov (ca. 6300 BC), of hg. R1a1-M459, reported as both outside and (tentatively) within the R1a1a-M198 tree (Haak et al. 2015); the individual from Khvalynsk (ca. 4600 BC), of hg. R1a1-M459 (xR1a1a1-M417), probably R1a1b-YP1272, like a later Maykop sample from the Northern Caucasus Piedmont (ca. 3230 BC); an individual from Serteya VIII (ca. 4000 BC), of hg. R1a-M420 (Chekunova et al. 2014); and a sample from Kudrukūla, Estonia (ca. 3000 BC), of hg. R1a1b-YP1272 (Saag et al. 2017). Contacts between the Upper Dnieper–Upper Don forest cultures and those of the forest-steppe continued during the Neolithic (Mazurkevich et al. 2009), which justifies their eventual infiltration down the Dnieper during the so-called steppe 'hiatus'.

IV.4. Fertile Crescent

IV.4.1. Anatolia and the Levant

There is no perceptible break in cultural continuity between the beginning and the end of the Anatolian Chalcolithic. However, certain gaps and discontinuities observed in many regions and periods, usually attributed to lack of adequate research, coupled with relevant changes in local cultures, point to a shift of the previous east–west influence to (at least partially) a west–east direction of innovations in certain Anatolian sites (Schoop 2011).

In general, this period of the 6th millennium shows thus the existence of wide communication systems, with continuity of previous traditions but with the introduction of foreign decoration techniques. Shortly before 5500 BC, a number of changes can already be seen in the Fikirtepe groups around the Bosporus (mainly settled on its eastern part), which point to a connection with the Vinča culture in the Southern Balkan region. In the Lake District, 'vinčoid' pottery is observed postdating the Fikirtepe tradition: it belongs to the dark-faced monochrome group, but there is some decoration with motifs in the stab-and-drag technique. Similar material is found in neighbouring regions (Schoop 2011).

During the 5th millennium, in the Middle Chalcolithic, a period of significant cultural development emerges. Near the western coast, Ubaid influence is noticed in urban plans and in pottery typical of the Halaf/Ubaid transitional period. In the Cappadocian margin of the Anatolian Plateau, which showed monochrome pottery decorated with different techniques in the Early Chalcolithic, the site of Gelvery-Güzelyurt shows pottery with swirling designs, executed in a stab-and-drag technique, which represents Balkan influences in the 4th millennium BC (Schoop 2011).

To the north, Late Chalcolithic İkiztepe (ca. 4500–4000) shows striking parallels with early to middle 4th millennium BC assemblages from the southern Balkans. This culture shows increasingly strong typological connections with materials further inland. While pottery traits point to continuity of traditions, notable innovations in shapes and decoration point to a *koiné* that encompasses most of Anatolia, the northern Aegean, and the southern Balkans. This period, since the early 5th millennium BC, coincides with the evidence of the production and consumption of metals, either simple metal artefacts (flat axes, pins, awls) or as crucibles or slag (Schoop 2011).

In south-east Anatolia, the Halafian 'heartland' developed since the 6th millennium BC, from its previous small or very small communities to large settlements which represented regional centres in a two- or three-tiered settlement hierarchies, with sedentary farming as the main subsistence economy, although cattle maintained its relevance for this originally semi-nomadic culture based on pastoral herding (Özbal 2011).

Close contacts and interaction between Halaf and Ubaid from Southern Mesopotamia had already been ongoing for a millennium, and possibly a crisis caused by its demographic and geographical expansion led the culture to a different organisation system. From about 4700 BC, though, Ubaid influence is increased in northern Mesopotamia, across a broad east–west arc (Özbal 2011). Southern Mesopotamian communities seem to have moved northwards, given the sudden social and cultural change in certain sites, first in the northern border, then in the Upper Euphrates.

A transformation began which eventually led to the disappearance of the way of life of the Halaf communities: new material culture, with new types of domestic architecture, village arrangements, public buildings, pottery, and other daily life objects; new economy, with less varied and more agriculturally-orientated production system; and a new social structure with sedentary population, a society that ceased to be egalitarian, family and not clan as the basic social unit, and emerging elites. The hybridisation of the two cultures produced innovations that spread in a southern direction, too (Frangipane 2015).

Farther south, the Levant Late Chalcolithic shows burial customs, artefacts and motifs with an origin in earlier Neolithic traditions in Anatolia and northern Mesopotamia. Characteristic of this culture are the secondary burials in ossuaries with iconographic and geometric designs. Artistic expressions have been related to northern regions related to finds, ideas, and later religious concepts, such as the gods Inanna and Dumuzu. The knowledge and resources required to produce metallurgical artefacts in the Levant have also been hypothesised to come from the north (Harney et al. 2018).

IV.4.2. Caucasus and Mesopotamia

The Chalcolithic in the Caucasus begins with foreign contacts from eastern Anatolia and Mesopotamia through the Taurus Mountains, giving rise to a new social and economic network ranging from the south-eastern Caucasus to the Kuban region in the steppe. The Maikop culture is thus the dominant northern Caucasian tradition, known from its extremely wealthy tomb assemblages, and probably born out of an indigenous group with distant economic connections to the south. The pre-Maikop phase appears in sites like Nal'chik and Meshoko in the late 5th and early 4th millennium BC (Sagona 2017).

Characteristic features of the Maikop culture include the adoption of barrow burials, shifting settlements on elevated positions—on foothills overlooking a river valley, but avoiding rugged highlands—with short occupations, abundance of metalwork, and widespread connections with the Near East and Europe. The greatest concentration of settlements occurs in the north-west, around the Kuban River system. The eastern half of the northern Caucasus, judging by the hundreds of Pit-Grave burials, belonged to the steppe cultures. The spread of the Pit-Grave building tradition in pre-Maikop is likely related to the expansion of Khvalynsk settlers into the neighbouring region (see *§IV.2.3. Kurgans*), but the southern burials—including small, mud-brick burial chambers, possibly reflecting an idealised house—have also been linked to central Asian and northern Iranian influence, which would have been added to the exotic imports of turquoise, silver, gold, carnelian, lapis lazuli, and cotton (Sagona 2017).

The southern Caucasus Chalcolithic groups are distinguished from Neolithic cultures by a more flexible lifestyle, reflected in varying modes of occupation (from permanent villages to seasonal camps, from open plains to caves); a capacity to benefit from resources across a wide range of environmental zones, including at higher altitudes; diverse subsistence strategies, incorporating wine-making; external networks, based on a flow of commodities; and advancement of metallurgy (Sagona 2017).

The Chaff-Faced Ware horizon forms part of a tradition that reached from the north Syrian and Mesopotamian plains through the middle part of the Araxes Valley and Azerbaijan to north-western Iran, known in the Fertile Crescent as Amuq F. It is found in the first half of the 4th millennium, with Azerbaijan showing slightly earlier dates. This is a homogeneous culture that reflects standardisation and technological simplification. In the later periods of the culture (as well as in north-west Iran), the influence of the Ubaid tradition of Upper Mesopotamia can be seen in ornamentation (Sagona 2017).

Connections with the Neolithic, evident in the earlier period with circular dwellings furnished with a central hearth, disappear later on (after ca. 4300 BC) as small, multi-roomed rectangular structures appear, with an evolving social structure, heavy exploitation of tree fruits, and more complex wine production industry. Single or multiple pit–graves with barrow burials are the standard, with the deceased in a flexed position with no preference as to side, showing the start of the 'sacrificial' metals in assemblages, possibly to strengthen the kinship-related social status (Sagona 2017).

The Sioni horizon is a local, imprecisely defined culture based on ceramics found in south-eastern Caucasus and on the Iranian side of the middle Araxes Valley, as well as in easternmost Anatolia. Its early phase is dated ca. 4800–4000 BC, and its late stage ca. 4000–3200 BC. Sites are characterised by flat settlements with variable building tradition. It probably emerged as local communities moved away from the alluvial plain into the foothills, as they were able to exploit a wider range of resources and pastures. Pottery has relatively few forms and a limited range of ornamentation, and their lithic technology is difficult to reconstruct (Sagona 2017).

iv.4. Late Middle Easterners

Chalcolithic peoples from Hajji Firuz in north-western Iran (ca. 6000–5700 BC) and from Seh Gabi in eastern Iran (ca. 4800–3800 BC) can be modelled as a mixture of western Iran Neolithic with significant contributions from a CHG-like population (ca. 63%) and the Levant (ca. 20%), becoming thus more 'western', consistent with their shift in the PCA. In Anatolia, the low genetic diversity of early Middle Eastern farmers during the early Neolithic was broken by a wave of 'eastern' ancestry from Iran Chalcolithic (ca. 33%), which eventually reached south-eastern Europe before at least ca. 3800 BC. These migrants brought also J-M304 lineages—typical of Caucasus and eastern Iranian populations—to the late Neolithic central and western Anatolia (Lazaridis et al. 2016; Kilinc et al. 2016).

This 'eastern' ancestry may have been caused by interactions between central Anatolia and the Fertile Crescent in the late Pre-Pottery Neolithic B (Özdoğan 2008), a migration related to other interregional exchanges, or admixture among local populations. The Tepecik-Çiftlik site's presumed role as an obsidian hub, and its cultural links with the Levant, might have started already before the Pottery Neolithic (Kilinc et al. 2016).

Although traditionally associated with an east–west movement of peoples, it could well represent the opposite direction, thus including expanding Anatolian-speaking peoples through northern Anatolia, from the west to the central part. Later samples from Bronze Age south-western Anatolia (ca. 2800–1800 BC) show this 'eastern' contribution of CHG-related ancestry, but lacking steppe-related EHG and WHG ancestry (Lazaridis et al. 2016).

The Chalcolithic population from Areni in modern Armenia (ca. 4350–3500 BC) also shows similar components to neighbouring Anatolian and Iranian Chalcolithic samples, but with a different distribution: Anatolia Neolithic (ca. 52%), Iran Neolithic (ca. 30%) and EHG (c. 18%). This, coupled with the different haplogroup found, L1a1-M27 (formed ca. 15000 BC, TMRCA ca. 6100 BC), points to a different population in the southern Caucasus Piedmont (Lazaridis et al. 2016). The appearance of mtDNA hg. H2a1 and U4a (more typical of the Pontic–Caspian steppes) among these samples, as well as their position closer to steppe populations, speaks in favour of female exogamy.

Before the emergence of the classical Maikop culture, the three sampled Caucasus Eneolithic individuals of Darkveti-Meshoko from Unakozovskaya, in the north-west Caucasus Piedmont (ca. 4600–4300 BC), present a genetic profile similar to Iranian Chalcolithic samples, with predominant haplogroup J2a-M410, possibly both J2a1a1a2b2a3b1a-Y11200 (formed ca. 5900 BC, TMRCA ca. 5800 BC). This increased assimilation of Chalcolithic individuals from Iran, Anatolia, and Armenia is in accordance with the Neolithisation of the Caucasus, which started in the floodplains of the great rivers of the southern Caucasus in the 6th millennium BC, from where it spread to the western and north-western Caucasus during the 5th millennium BC (Wang et al. 2019).

Haplogroup J2a2-L581[+](formed ca. 14100 BC, TMRCA 13100) also appears in one sample from Seh Gabi (ca. 4700 BC), and hg. J2b-M12 in two samples from Hajji Firuz (ca. 6050–5850 BC), with hg. G2a1a-Z6553 (ca. 5750) and G1a1b-GG313[+] (ca. 3900 BC) in Seh Gabi pointing to a mixture of these haplogroups since the sampled Iran Neolithic individuals, compatible with a migration wave of J2a2-L581 lineages connecting the northern and southern Caucasus regions ca. 5500–4500 BC (Wang et al. 2019). This haplogroup is also found later in Anatolian Bronze Age samples and in Old Hittites.

Samples of the Late Chalcolithic in the southern Levant, from the Peqi'in Cave (ca. 4500–3900 BC), attributed to the Ghassulian period (Figure 16), can be modelled as deriving ancestry from local Levant Neolithic peoples (ca. 57%), Iran Chalcolithic (ca. 26%), and Anatolian Neolithic (ca. 26%), suggesting the spread of Iranian agriculturalists into the Levant. They overlap in the PCA with a cluster containing Neolithic Levantine samples, shifted slightly toward Levant Bronze Age samples. Their prevalent Y-DNA haplogroup, probably in twelve of thirteen samples reported, is T1a1a1b2-CTS2214 (formed ca. 6700 BC, TMRCA ca. 6700 BC), with only one sample of E1b1b1b2-Z830 subclade, also suggesting an important population replacement in the region.

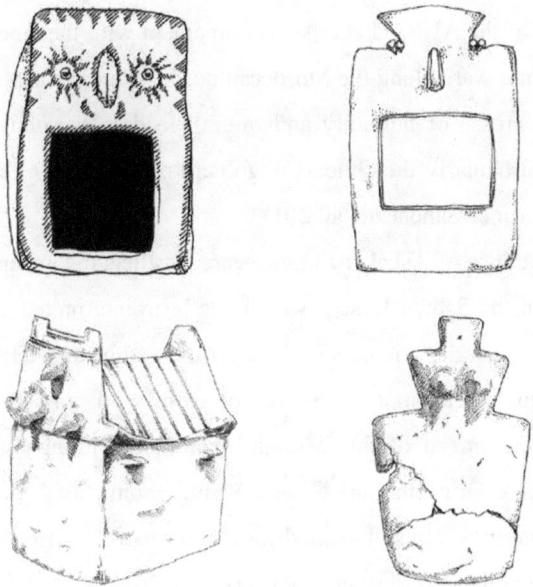

Figure 16. Ossuaries from the Peqi'in Cave. Image modified from Jaruf (2017), using figures adapted from Yannai and Porath 2011, and Gal et al. 2011.

IV.5. Africa

Towards the end of the African Humid Period there were some dry peaks due to fluctuations in rainfall and increasing dryness, with serious impacts on human settlements. Demographic modelling suggests a relative human population maximum in the central and western Sahara near the end of the AHP (between 4700–4300 BC), followed by a population collapse coinciding with the end of the AHP. This crisis is coincident with the emergence of the Ashakar-Skhirat ware along the Moroccan north-Atlantic coast, and also with the abrupt decrease of humidity and rainfall leading to sudden changes in vegetation, particularly the Gineo-Congolian taxa, in the western Sahara ca. 4500 BC (Martínez Sánchez et al. 2018).

The appearance of Ashakar-Skhirat ware parallels the expansion of cattle pastoralism in the Sahara Desert, which has been interpreted as an adaptive mechanism against the arid conditions after the end of the AHP. Pastoralism required frequent relocations in search of fresh wild grasses for livestock, resulting in the spread of this ceramic technology throughout central and northern Sahara during the late 6th and 5th millennium BC. Similar types of distribution have been linked to southward and westward dispersal of pastoral groups from the Sahara (Martínez Sánchez et al. 2018).

Megalithic stone monuments with different architectural features and various ceremonial purposes, which possess characteristics of social relevance and temporality as to be considered 'places to be remembered', also appear and diffuse at the end of the AHP (6th-5th millennium BC) and spread afterwards (di Lernia 2013). After ca. 4500 BC, herding spread south and east, and pastoralists built megalithic structures at Wadi Khashab in the Red Sea Hills, and established large cemeteries at Kadreo, Kadruka, and R12 in the Nile Valley. Elaborate mortuary traditions continued for more than a thousand years at sites like Jebel Moya, as agricultural lifeways and early states developed.

iv.5. Late Afrasians

Before 5500 BC, population curves for the eastern Sahara, the Atlas & Hoggar, and central Sahara follow broadly synchronous variation. After the population decline period ca. 5500–4500 BC, there are divergent responses: the eastern Sahara, extremely arid today, underwent a rapid population decline, with occupation shifted towards the Nile Valley, probably giving rise to the Hamito-Semitic isolation. The finding of hg. R1b1b2a2a-V1589/Y7771 and subclade R1b1b2a2a1-V69 in modern populations of the Arabian Peninsula may indicate its wide distribution in eastern Africa, just before the Semitic and Hamitic expansions.

To the north and west, the Atlas & Hoggar mountain region declined equally rapidly, possibly isolating the pre-Berber-speaking population (Brierley, Manning, and Maslin 2018). An important population replacement happened in north-west Africa between the Early Neolithic and Late Neolithic samples from Kelif el Boroud (ca. 3700 BC), with ancestry shifting (up to 50%) from Iberomaurusian to European Early Neolithic. Reported mtDNA haplogroups include K1, T2, and X2, proper of Anatolian and European Neolithic populations; and Y-DNA T-M184, observed in European Neolithic individuals. This is probably related to contacts between both sides of the Gibraltar strait at this time, before the Bell Beaker expansion (Fregel et al. 2018).

Central Sahara, on the other hand, had a much more gradual decline in population, attesting flexible and adaptive strategies that co-evolved with the drying environment, living in balance with the available pasture (Brierley, Manning, and Maslin 2018). The persistence and expansion of haplogroup R1b1b-V88 in this region dominated by modern Chadic speakers is thus most likely the result of less population pressure and continued expansion of Pre-Chadic peoples, contrasting with the harsh environment, faster life history, and haplogroup replacement experienced to the north and east.

Two Neolithic individuals from Takarkori in the central Green Sahara (radiocarbon-dated to the early 5th and mid–5th millennium BC, respectively) show a basal mtDNA haplogroup N that branched off immediately after the Palaeolithic sample Oase 1, and before all present-day N-derived mtDNAs. This finding could be explained as from a local subclade that branched off just after the differentiation from L3 within Africa, or as a back-migration after its expansion out of Africa (Vai et al. 2019).

V. Middle and Late Æneolithic

V.1. Africa and the Levant

At the end of the AHP, with increasingly arid conditions, mountain ranges like Tibesti, Tassili-n-Ajjer, and Ahaggar—forming a major topographic feature spanning more than 2,500 km. from southern Argelia to northern Chad—acted as important water towers in contrast to the surrounding plains, providing populations settled on the windward side with more persistent rain runoff. Because of that, some of the earliest direct evidence for the exploitation of domestic livestock, use of milk products, and the construction of cattle tumuli come from the heart of the central Sahara (Brierley, Manning, and Maslin 2018).

The emergence of rock art depicting livestock scenes and stone monuments with associated domestic animal remains in the middle Holocene attest to a highly formalised expression of a wider Saharan "cattle cult", with isotopic analysis of animal bones from the region demonstrating seasonal transhumance similar to strategies followed by modern traditional pastoralists (Brierley, Manning, and Maslin 2018).

In the Levant, discontinuity in archaeology with the previous period is marked by dramatic changes in settlement patterns, large-scale abandonment of sites, many fewer items with symbolic meaning, and shifts in burial practices, such as secondary burial in ossuaries, which disappear completely. This supports the view of profound cultural upheaval leading to the extinction of whole populations, associated with the collapse of the Chalcolithic culture in the region.

v.1. Early Semites

Population replacement is supported in genetics by the genetic discontinuity between the Chalcolithic and the Early Bronze Age period (Harney et al. 2018). Proto-Semites, who probably lived at the beginning of the 4th millennium still on the savannahs of the then still pale-green Sahara, migrated north after ca. 3900 BC, once the region reverted back to a desert climate (Lipiński 2001). Whether they migrated through the horn of Africa to Arabia or through the Nile to the Sinai before reaching the Levant is unclear, although the apparently closer relationship of Proto-Semitic to Proto-Berber, and the appearance of Semitic languages quite early in both the Syrian desert and the Levant, seem to support its initial diffusion through east Africa rather than Arabia.

The presence of haplogroup J1a2a1a2d2-Z1865 (formed ca. 5500 BC, TMRCA ca. 5200 BC) and its subclade J1a2a1a2d2b-Z1853 (TMRCA ca. 4900 BC) in modern south Arabian populations suggests an expansion of this subclade potentially from southern or northern Arabia in ancient times. Subclade J1a2a1a2d2b2-Z2331 (TMRCA ca. 3800 BC) is associated with modern Semitic populations widely distributed through the Middle East, with its oldest subclade J1a2a1a2d2b2a-Y15152 (TMRCA ca. 3800 BC) present today in modern Jewish populations. The presence of J1a2a1a2d2b2-Z2331 subclades later in Canaanites from Sidon (ca. 1750 BC) and in Levant BA from 'Ain Ghazal (ca. 2100 BC) further supports the connection of this haplogroup with the expansion of Proto-Semitic.

The presence of haplogroup J1a2a1a2d2b2b2c-Z2329 (formed ca. 3600 BC, TMRCA ca. 3600 BC) in a Pre-Ptolemaic Egyptian individual (ca. 670

BC) (Schuenemann et al. 2017) suggests the widespread distribution of this haplogroup J1a2a-L620/Z2356 (formed ca. 16000 BC, TMRCA ca. 12700 BC) along the Levant, from the north to the Sinai Peninsula and beyond, as evidenced by its widespread presence in modern Arabic tribes.

The arrival of Proto-Semitic migrants might have caused the collapse of the related cultures Amratian (Naqada I, in Egypt) and Ghassulian (in the Levant) ca. 3500–3350 BC, both in turn possibly related to Minoans. The expansion of Semites from the Levant may have in turn caused the first split into a western and an eastern dialect, and the further expansion of the latter to the north and east across the Syrian steppes, into south-east Anatolia and southern Mesopotamia.

In northern Arabia, crop cultivation and other features traditionally used to define the Neolithic do not seem to have been practised until the arrival of the Bronze Age (ca. 3000–1200 BC), a situation that contrasts strongly with the Fertile Crescent, where sedentary communities were present since the Epipalaeolithic (Scerri et al. 2018). This significant cultural change also suggests a potential demic diffusion to the region.

V.2. The Caucasus

V.2.1. Chaff tempered ceramics

In the late 6th millennium BC, eastern Anatolia, the Upper Euphrates Valley, Syria, and northern Mesopotamia were involved in a system of interactions. It seems that this network of expanding influence in a south–north axis is repeated in the final phases of the Ubaid period, during the Chalcolithic, from ca. 4500 BC onward, in a process of transformation of the role, function, and meaning of the ceramics, with extreme simplification of decoration and formal standardisation (Sagona 2017).

The diffusion of chaff tempered ceramics in eastern Anatolia and the Caucasus has been thus linked by researchers to the presence (ca. 4250–3500 BC) of northern Mesopotamian groups involved in such economic activities as pastoralism and commerce (trade in metal ores or raw materials). The presence of "indigenous" sites distinguished by their continuation of local pottery suggests a complex system of complementary interactions between groups of differing origins and different cultures (Mesopotamian and Transcaucasian) that occupied different areas depending on their different economic activities (Sagona 2017).

The adoption of funerary customs such as elite tombs built with mudbricks but under funerary tumuli (from the north Caucasian tradition), and the presence of fortifications in the area, strengthen the increased Syro-Mesopotamian influence overlapping the cultural substratum of Late Chalcolithic communities of the Caucasus. The emergence of a regional centres and a stratified society with elite groups in northern Mesopotamian communities probably triggered the structural and organisational changes in the South Caucasian—and eventually eastern Anatolian—communities, to adapt themselves to the growing demand from the south. They show an increasing territorial mobility, pastoral specialisation, and the capacity to exploit ecologically different resources. Some findings point to the formation of small local elites imitating the Mesopotamian structure (Sagona 2017).

Before ca. 3500 BC, scarce Pre-Kura–Araxes settlements are found in northern areas. The culture shows little continuity with previous Chalcolithic cultures, and the Red-Black Burnished Ware displayed by the culture shows technological and cultural links to certain settlements of eastern Anatolia and the Upper Euphrates. The synchronous appearance of these sites suggests a common network of information, trade, and culture. On the other hand, the strong typological, functional, and ornamental similarities with the southern Caucasus suggests a connection with the southern Caucasian domestic model (Sagona 2017).

V.2.2. Maikop

The turn of the 4th millennium BC saw the development of various cultural traditions in south-east Anatolia, north-east Syria and north-west Iran; on the northern fringe, these traditions manifested themselves in the Maikop culture. In fact, the first high-status burials containing gold and gemstone jewellery (including carnelian, turquoise and lapis lazuli) appeared in the northern, rather than southern, centres ca. 4000–3750 BC. With regard to funeral rites and stylistic characteristics of jewellery pieces, these graves have many parallels with early Maikop burials (Sagona 2017).

Few settlements are known from the classical Maikop stage (ca. 3800–3000 BC), with few fortified central places and a majority of open areas composed by groups of small and ephemeral villages of ca. 1–2 ha, with house plans of varied shapes, not articulated through foundations or postholes. Hearths played an important role. Subsistence economy was most likely based on cattle breeding, probably including transhumance, as well as on other animal husbandry (mainly pigs) and probably agricultural means (Sagona 2017).

In the classic phase, Maikop circular pit–grave burials became larger, and showed symbolic features like a flat top (probably a cultic platform), had a stone gridle delineating its circumference, and a trend to seal smaller barrows under a 'roof', so as to create a cemetery-like structure (Figure 17). Red ochre was ceremonially sprinkeled on the deceased, placed in a flexed position on their right side, head pointing south (Sagona 2017).

Most Maikop burial assemblages and constructions do not share the magnificence of the wealthy barrows, and are simple, rectangular earthen pits beneath a shallow tumulus, although they share the same principles. Despite the abundant metalwork, there is little evidence of extractive mining or metallurgical craftsmanship. The society appears divided thus sharply in two levels, with few individuals being regarded as the 'chieftains' and buried with luxurious assemblages. They were probably a sign of the emergent elite ideology in the Caucasus, absent in the southern territories, as well as monuments affirming territoriality (due to their visibility) and veneration of ancestors (Sagona 2017).

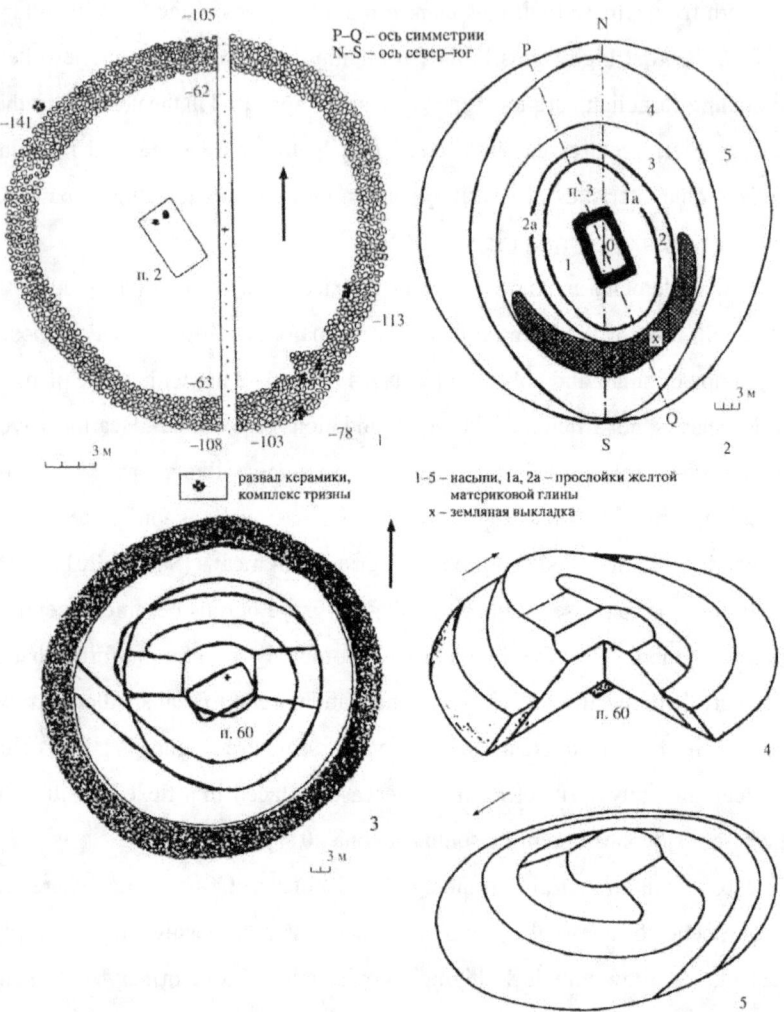

Figure 17. Kurgans of the Maikop type. 1 – kurgan II of the Sunzhenski cemetery; 2 – kurgan III of Brut, 3; 3 – Zamankul (plan and reconstruction of the mound). Modified from Korenevskiy (2012).

Four types of tomb chamber are distinguished (Sagona 2017):

- A rectangular, earthen pit with rounded corners, edged around the base with stones, and with a roof of timber logs.

- Rectangular or circular burial on the ground surface made of wooden planks or field-stones.

- Rectangular, one- or two-chambered tombs, built above ground with slabs of stone, with access through a porthole entrance. Ornamentation is rare. This type is typical of the megalithic tradition.

- Stone cist tombs built with slabs set into a pit (identical to the previous), with access through the roof.

In its late phase, Maikop metalwork diversified, with metalsmiths working gold, silver, and copper. The source of copper-nickel-based objects seem to lie in metal ores to the south of the Caucasus, while arsenical-copper objects—concentrated in the Kuban region—probably had a local origin. Typical Maikop pottery shows a limited range that emphasises rounded and simple profiles, like globular pots and jars, and hemispherical bowls and cups (Sagona 2017).

The Uruk expansion in Mesopotamia after about 3700 BC intensified during the late Uruk period (ca. 3350–3100 BC), and its expansion reached toward the gold, silver, and copper sources in the Caucasus Mountains. The Maikop culture of rich chieftains' graves with Mesopotamian ornaments probably developed from this trade network in the North Caucasus Piedmont. A western and probably also a later eastern southern trade routes have been proposed, through the shores of the Black and Caspian seas respectively (Anthony 2007).

Connections with the Near East are evident in the occasional cylindrical seals (Rollsiegel) in Maikop assemblages. It seems that a distinctive technique of making thin-walled jointless beads from gold was a regional technological development of Maikop culture goldsmiths. This was deeply rooted in the Near Eastern tradition of ritualisation of the production and use of jewellery pieces made of gold, silver and gemstones. The jewellery traditions of the Maikop culture had no successors in the Caucasus or the adjacent steppes. In the third millennium BC, the goldsmiths of Europe and Asia had to reinvent the technique of making thin-walled jointless gold beads from scratch (Trifonov et al. 2018).

To the north, the existence of steppe–Caucasian trade is supported by Maikop imports found in the north Pontic steppe from the Dniester to the Lower Volga in the east, but no Caucasian imports have been found in the Volga–Ural region. Late Maikop peoples, most likely speaking languages ancestral to modern Caucasian languages, probably interacted with individuals from Repin and late Khvalynsk cultures, and the contact was most direct on the lower Don. Late Maikop graves incorporated carved stone stelae like those of western Yamna. The trading of drugs, wool, and horses has been proposed as main steppe imports into Maikop (Anthony 2007).

v.2. Early Caucasians

The two Maikop samples from this period in the Northern Caucasus Piedmont show largely continuity with Caucasus Eneolithic samples, but with a clear additional contribution of Anatolian Neolithic-related (possibly AME) ancestry (ca. 15%) compared to them. Five Maikop outlier samples from the steppe (ca. 3600–3100 BC) represent a likely expansion of Maikop peoples to the area and their admixture with the previous Khvalynsk and local settlers, suggesting their acculturation in the region, evidenced by their admixture closest to ANE.

In terms of haplogroups, one sample from Baksanenok (ca. 3350 BC) is reported as within the K-M9 trunk, possibly L-M20. The acculturation of the North Caucasus region may also be inferred from haplogroups of outliers, which show one Q1b2b1b2-L933[+] (formed ca. 13600 BC, TMRCA ca. 6600 BC) and another R1a1b-YP1272[+], in contrast to previous Eneolithic (J-M304) and later (L-M20) haplogroups (Wang et al. 2019). Both individuals were buried in the same kurgan in Sharakhalsun and with similar radiocarbon dates (ca. 3350-3105 BC), and a later individual attributed to the Yamna culture in the same site (ca. 2780 BC) also shows a typical Indo-Anatolian lineage R1b1a2-V1636. Another outlier shows hg. T1-L206.

Horse trade, including wheels, carts, and the possibility of a quicker transport of metals into Uruk, is proof of an indirect contact between steppe herders and Mesopotamia. The need of exported domesticated horses to be accompanied by experienced breeders and riders from the lower Don offers a solid framework to support the hypothesis of the presence of Late-Indo-European-speaking peoples in Mesopotamia, and thus allow for Indo-European borrowings in Sumerian (Sahala 2009-2013).

Nevertheless, the scarcity of proofs for wooden vehicles in the region before the first attested one in Sharakhalsun, as well as bioarchaeological investigations of common representations which point to an emphasis on cattle as driving force—instead of highlighting the means of transportation, as in the Yamna culture—seriously challenge the hypothesis of large-scale mobility in the piedmont and the Caucasus (Reinhold et al. 2017). The condition of Pre-North-West Indo-European (likely spoken by the late Repin culture expanding westward) as an Euphratic superstratum of Sumerian (Whittaker 2008, 2012) would require a more detailed explanation of internal and external cultural influence, and reasons for potential language replacement and expansion in Mesopotamia.

V.3. Anatolia

In the Late Chalcolithic (ca. 4250–3000 BC), the western cultural koiné characteristic of the previous period continues, at least in south-west Anatolia and along the southern and middle Aegean coast. Further to the east, cultural developments show a different trend, with influence from the Late Uruk pottery reaching ca. 3500 BC eastern Anatolian regions, including the Plateau. This influence coincides with the start of Transcaucasian contacts, too. Regional fragmentation is the rule, and the Black Sea coast shows no eastern influence. On the contrary, ring-shaped figurines—flat objects of stylised female shape made of silver, lead, or gold—connect the Plateau to the southern Balkans, which is seen as evidence of close contacts with the west Pontic area, particulary Bulgaria (Palumbi 2011)..

In Syrian and northern Mesopotamian regions, south-eastern Anatolia, and the Upper Euphrates, the period ca. 3500–3000 BC was associated thus with the formation of large and important regional centres, a general reorganisation of the craft production – with growing specialisation –, and the emergence of stratified societies and elite groups. Late Chalcolithic architecture evolves into settlements of huge dimension. Increasing social and economic complexity lead gradually into the Early Bronze Age (EBA), although nothing suggests the emergence in the area of proto-urban structures typical of the Upper Euphrates or northern Syria (Palumbi 2011).

The Ubaid or Uruk expansion is supposed to have affected wide regions in the Middle East, although the precise regional mechanisms are still unknown. For example, whereas the Upper Euphrates shows indigenous social complexity, the Upper Tigris Valley shows a resistance to foreign influence. South-eastern Anatolian and northern Mesopotamian populations interacted ca. 3500–3000 BC with settlements in the Upper Euphrates, Upper Tigris, and beyond in their quest to obtain raw materials. Uruk colonists seem to have expanded to the north. Although it had been presumed that early state systems were restricted to the southern Mesopotamian alluvium or Uruk-influenced sites to the north, different sites show that indigenous societies which whom Urukians interacted had independently evolved into complex administrative centres in south-eastern Anatolia, like Hamoukar, Tell Brak, or Arslantepe (Palumbi 2011).

V.3.1. Arslantepe

Two periods reveal the centrality of the Arslantepe site, located in the Malatya plain, as a place of cultural boundary in the network of interregional relations of the Middle East. The first is the Late Chalcolithic period, over the entire 4th millennium BC, when a very early hierarchical and politically centralised society developed on the site; and the second period refers to the 2nd and early 1st millennium BC, when Arslantepe was affected by the eastward expansion of the Hittite state (Frangipane, Manuelli, and Vignola 2017).

During the first half of the 4th millennium, Arslantepe had already developed a political system in which the elites had gained some control over staples, and developed well-established circuits for the centralisation and redistribution of foodstuffs, carried out in public ceremonial areas. The material culture of the site was local, although it seems to have been a powerful regional centre, whose leaders interacted with their Mesopotamian neighbours. The elites had their residence separated from the common houses, on top of a mound, close to temples, which indicates their social importance and the symbolic emphasis on their prestige and growing power. Their role as central authority included control over food and its redistribution in ceremonies and feasts (Frangipane, Manuelli, and Vignola 2017).

Around 3500–3400 BC, the main temples were abruptly abandoned, and a radical change occurred in the power system, which led to an extraordinary development of the Arslantepe society towards a stronger and more centralised structure. A monumental, imposing, tall building with very thick walls— although much smaller than the previous buildings—was built, without any cultic or religious features. This building and the courtyard, whose entrance was decorated with stamped lozenge motifs and wall paintings, constituted the core of the new public area. There seems to have been a throne room opened towards the courtyard, a place for audience, and also a private section for authorised persons (Frangipane, Manuelli, and Vignola 2017).

The public area may have been conceived for the leader to address the public and held audiences in a ceremonial environment, now without any cultic or religious connotations. On a corner, a temple with a floor plan identical to the audience building shows that cultic and religious rites were of restricted access, probably for people of high status. Authority was thus excercised without any religious mediation, and elites preserved the religious authority, detaching themselves still more from the rest of the population. Economic and administrative rooms were added, evidenced by intensive sealing and sophisticated accounting system. Interesting are certain scenes, like those of bulls pulling a cart or plough driven by a coachman (depicting a ploughing scene), and a transport of an eminent person on a threshing sledge car found on a cylinder seal (Frangipane, Manuelli, and Vignola 2017).

v.3. Early Anatolians

Anatolian has long been considered the first language to branch off of the Proto-Indo-European trunk, due to its peculiar archaisms (Trager and Smith 1950), even before the proposal of a Late Indo-European community from which all other known Indo-European languages branched out (Meid 1975; Kortlandt 1990; Lehmann 1992; Dunkel 1997; Melchert 1998; Adrados 1998; Ringe 2006; Mallory and Adams 2007; Beekes 2011). Based on the known Khvalynsk migrations of the previous period, and on the presence of a prehistoric geographic and genetic barrier in the Caucasus Mountains (Wang et al. 2019), the most likely route of expansion of Proto-Anatolians lies in the Balkans (Anthony 2007), which is supported by the presence of Balkan outliers with Steppe ancestry (see §iv.2. Indo-Anatolians).

The main question has turned thus to the when and how of the migration into Anatolia of Proto-Anatolian speakers. One important cue, based on its relevance for Suvorovo–Novodanilovka chieftains (see §IV.2.2. Horses), is horse domestication: it is found at Çadir in north-central Anatolia already in the early 4th millennium BC, continuing into the 3rd millennium (Arbuckle 2009), representing thus the earliest evidence of its presence in Anatolia, comfortably earlier than Late Chalcolithic remains of eastern Anatolia or the earliest representations of a wheeled vehicle by Sumerians ca. 3100 BC, probably pulled along by oxen (Sagona 2011).

Similarities between the Varna culture (lasting until ca. 4200 BC) and that of İkiztepe on the central coastal region of the Black Sea strongly imply close ties between the eastern Balkans and central Anatolia (see §IV.4.1. Anatolia and the Levant), with this population having been proposed as cultural predecessors of the Hittites (Bilgi 2001, 2005) based on its connection with Balkan Early Eneolithic pit grave cultures, including extended, supine inhumations with the use of ochre, as well as the use of ring-shaped idols

(Zimmermann 2007), and also craniometric features proper of south-eastern Europeans (Welton 2010). The lack of similar remains in western Anatolia may suggest an ancient maritime connection to continental Europe through the coasts of the Black Sea rather than by way of a land route (Özdoğan 2011). This is compatible with the Anatolian Chalcolithic sample of Barcõn, Marmara Region, north-west Anatolia (ca. 3800 BC) showing "eastern" contribution (see *§iv.4. Late Middle Easterners*), but no Steppe ancestry (Lazaridis et al. 2016).

The lack of relevant cultural or genetic connections in north-west Anatolia may also suggest the infiltration of small groups of Proto-Anatolian speakers who have not left much traces in other intermediate Balkan regions, either. In any case, the Sea of Marmara had become a true cultural barrier during the Chalcolithic, separating south-eastern Europe from Anatolia, as evidenced by the split of the "Balkano-Anatolian Culture Complex" by the turn of the 5th to the 4th millennium BC, at the end of the Vinça Period (Özdoğan 2011).

Based on the likely presence of Anatolian speakers ca. 2500 BC in south-eastern Anatolia, it is tempting to locate the arrival of pioneer Proto-Anatolian speakers in İkiztepe, north-central Anatolia, via the south-eastern Balkans— whether by land or sea—and their expansion southward into central Anatolia with the sociopolitical change at Arslantepe ca. 3500–3400 BC. The lack of genetic traces from the steppe on south-western and central Bronze Age samples (see *§vii.4. Aegeans and Anatolians* and *§viii.13. Assyrians and Hittites*) may suggest a low genetic impact of the Anatolian migration, or the replacement of this early population with eastern migrants, or both. Among the investigated eighteen ancient individuals from the Late Chalcolithic to the Early Bronze Age in Arslantepe, there is no evidence of a major genetic shift, although there is high heterogeneity compared to other Anatolians, and more Iran Neolithic-related ancestry (Skourtanioti et al. 2018).

V.4. Steppe package

V.4.1. Kurgan cultures

After the expansion of Suvorovo–Novodanilovka chieftains through south-eastern Europe, and the use of kurgan burials by Cernavodă I and related groups (first half of 4th millennium), there is a process of coexistence and acculturation in the north-west and west Pontic areas at the end of the Eneolithic, ca. 3600–2900 BC, where the first burial mounds indicate a lack of standardisation (Figure 18). This process is simultaneous with the evolution of Late Copper Age communities north of the Black Sea, such as Lower Mikhailovka, Trypillia C (including Usatovo), Late Kvityana, Late Deriïvka, Late Sredni Stog, post-Mariupol, and eastern cultures like Repin, Maikop, etc (Frînculeasa, Preda, and Heyd 2015).

In these late Eneolithic 'kurgan cultures', primary graves consisted usually of small mounds (only later became enlarged), were orientated to various directions, and individuals lied in a contracted position to the side or (continuing earlier periods) in an extended position. Grave pits were more oval than rectangular, and ochre was sparsely used (if at all). Both males and females were buried, and only rarely had they assemblages. The most prominent burials with inventories are those of Trypillia C2, Horodiştea–Folteşti and particularly Baden–Coţofeni (later evolving into Usatovo) at the Lower Danube. These are mostly local developments, although there might have been some infiltration of local steppe peoples from the Lower Mikhailovka and Kvityana into the Lower Danube (Frînculeasa, Preda, and Heyd 2015).

Unlike Marija Gimbutas' claim of succeeding 'kurgan population waves' into south-eastern Europe, the Eneolithic period shows merely a long-term, low-level population interaction between similar steppe environments north and west of the Black Sea, continuing some of the cultural patterns left by Suvorovo–Novodanilovka chiefs ca. 4600–4000 BC, representing therefore local populations integrating 'eastern' burial customs in their own rituals

(Frînculeasa, Preda, and Heyd 2015). Which of these populations might have been direct cultural heirs of the Suvorovo migrants, and which showed mere remains of their earlier influence, is unclear.

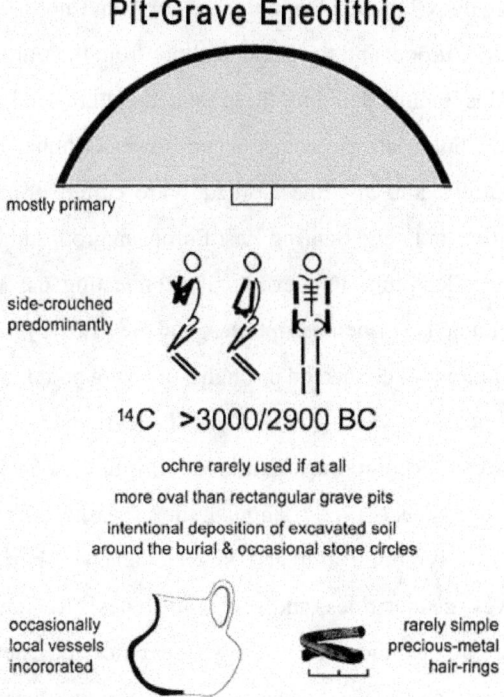

Pit-Grave Eneolithic

Figure 18. Burial schemes of pit graves found in the Lower Danube region during the Eneolithic. Modified from Frînculeasa, Preda, and Heyd (2015).

Precisely in these steppe areas north-west and west of the Black Sea are few pit–grave kurgans found ca. 3700–3000 BC, of variable shape and rituals, from the Prut–Siret–Plain to the south-west Balkans, including Horodiştea II, Gordineşti-Cernavodă II, Folteşti, Horodiştea-Folteşti, and to the south into the Dobruja and the Eastern Thracian Plain in Cernavodă III and Ezero A1. Such simple, 'steppe-related graves' are also to be found to the west in cultures like Boleráz and Baden in the Carpathian Basin (Frînculeasa, Preda, and Heyd 2015).

There are thus similar cultural findings all over Europe since the mid–4th millennium, unifying regions that were previously separated in material culture, as well as in social, economic and ritual aspects: so the predecessor of the Baden complex, Cernavodă III–Boleráz, spread from the Lower Danube to the Bodensee in the northern Alps (beginning ca. 3700/3600 BC); and later, ca. 3350–2700 BC, the Globular Amphorae in north and north-east Carpathian Basin and central Europe, and the Baden culture from the Carpathian Basin to the Northern Alps. Included among these related cultures and ceramic groups are those of the whole Balkan Peninsula and Lower Danube, almost reaching north-west Anatolia; and also the Corded Ware culture (Single Grave and Battle Axe culture, and neighbouring East European groups in the first half up to the mid–3rd millennium BC), eventually connecting the Volga with the Rhein and Scandinavia (Frînculeasa, Preda, and Heyd 2015).

All these cultures are connected through a unifying pottery, fine ceramic— often drinking and eating ware—with identic shape (round to tappered bottom) and emblematic cord decoration (with mixed forms like the Cucuteni C-Ware), apart from prestige objects (viz. triangular silex spearhead and the European dagger idea), and symbolic aim and key elements of burial rituals (like individual graves, gender roles, and social attributes). Regional and cultural differences lie in technique/technology, specific subsistence economy, settlement patterns, and social organisation. It seems that these cultures were therefore united in certain essential social, spiritual, and religious aspects (Frînculeasa, Preda, and Heyd 2015).

Apart from this, it is also apparent that the expansion of Suvorovo chiefs must have set in motion the start of the "Secondary Products Revolution", which becomes full-fledged in eastern Europe ca. 3600 BC, and includes traction, dairy farming, horse riding, and wool production (Figure 19). This revolution brought about changes in economic and social complexity, population growth, density pressure, expansion to secondary environments, deforestation and increasing pasture, and easier transport, greater mobility, regional specification or territorial competition. The new emphasis is thus on cattle, with a marked rise in its numbers, and a diminishing number of pigs (with a later, gradually rising number of goats and sheep), and it eventually affects all aspects of life, including social and spiritual beliefs (Sherratt 1981).

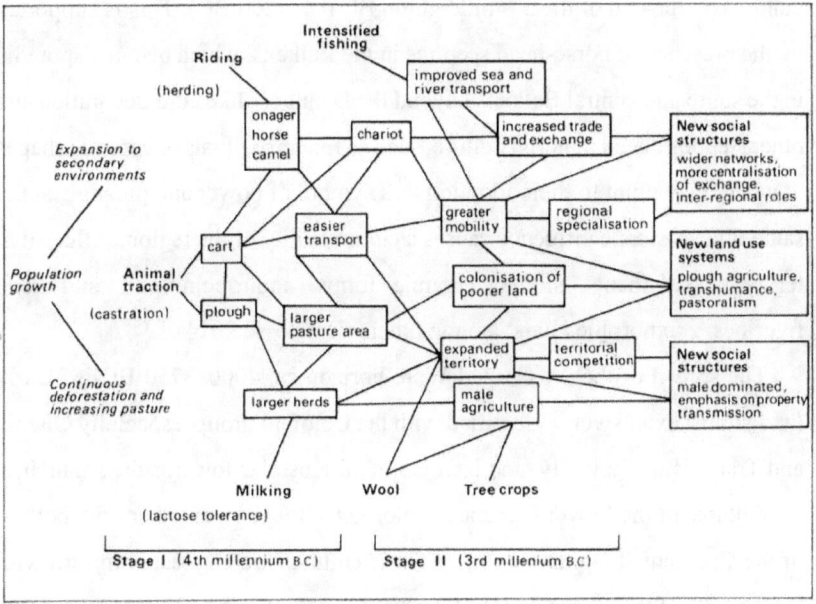

Figure 19. Original graph of the 'Interaction of the components of the secondary products complex through time', by Andrew (Sherratt 1981).

V.4.2. Corded ware

Corded ware refers to corded decoration of pottery assemblages, made with a cord, which has been proposed to be originally derived from twined hemp (the rope of which is used to control herds), hence related to cattle-herding cultures; or from wool cords, hence related to sheep and wool processing. Whichever the case, its spread over a great part of central Europe was mediated by the Globular Amphorae culture, which popularised the drinking vessels and their corded ornamentation (Bulatović 2014).

In the first horizon of Corded Ware culture, a cord is twisted, or wrapped around a stick, and then pressed directly onto the fresh surface of a vessel leaving a characteristic decoration (*Wickelschnur*). This technique appears as a non-native trait in the Early Eneolithic Bubanj–Salcuţa–Krivodol cultural complex, with no correlation in autochthonous Neolithic traditions, possibly from influence of the Suvorovo–Novodanilovka and north-west Pontic cultures, at the end of the 5th millennium BC (ca. 4200 BC). This is supported by the presence of horse-head sceptres in the Balkans, which become sporadic in the south and central Balkans beyond the Danube—like cord decoration and other steppe-related material culture, such as funerary rituals or ceramic shapes —which may point to their adoption as a symbol of power and prestige, at the same time as steppe influence causes a cultural unity of the region, reflected in tell-type settlements, similar ceramic forms, anthropomorphic and bone figurines, zoomorphic altars, among others (Bulatović 2014).

The spread of the 2nd Corded Ware horizon ca. 4000–3750 BC is clearly (and almost exclusively) identified with the Coţofeni group, especially Oltenia and Transylvania initially, and later Coţofeni-Kostolac towards the south, into the cultures of the Lower Danube and northern Bulgaria, together with pottery of the Cucuteni–Trypillian culture. These cultures are connected in turn with movements of the steppe-related Cernavodă I community in the Danube delta, to the south into Ezerovo (Bulatović 2014). The decoration is applied with a real cord, but ornaments are shorter and do not cover the whole girth, which is found only later in the classic central European corded ware.

The corded ware decoration was adopted widely during in the middle to second half of the 3rd millennium BC in the north-west and west Pontic areas, the Balkans, and the north and north-east Carpathian Basin. Only in the central Balkans were new steppe elements noticed during this period, which may point to a closer cultural relationship of this area with the Pontic–Caspian steppe. The Usatovo culture, settled in the territory of the Trypillian culture, eventually replaced the Coţofeni culture at the time of the expansion of the third horizon of the Corded Ware culture into Central Europe.

The third horizon of the Corded Ware culture appeared in the late Eneolithic / Early Bronze Age in areas to were Coţofeni and related cultures had expanded during the second horizon, including central (Vučedol, Bell

Beaker) and northern (Classical Corded Ware culture), as well as central and southern Balkans, Greece and the Peloponnese, including the Eastern Thracian Plain, and also a western area in the Adriatic coast (through a southern route from Bell Beaker or Vučedol). These ornamentations are often local innovations connected to previous regional Eneolithic cultures (Bulatović 2014).

There was a long-term connection between the north-west Pontic steppe area and the border of the forest zone up to the eastern Baltic area, centred on the Dniester–Buh limes (encompassing the Dniester, Dnieper, and Buh rivers). It also included the areas between the Vistula and the Dnieper (with the Lesser Poland area) – which topographically form a natural *continuum*.

V.5. Northern Europe

V.5.1. Funnel Beaker culture

The Funnel Beaker, Funnelbeaker or *Trichterbecher* (TRB) culture (ca. 4000–2800 BC) was spread through the North European Plains, to the north of the groups that formed the late LBK groups. Its characteristic pottery includes funnel-necked beakers, two-handled or four-handled amphorae, flasks, bowls, and flat clay disks, with limited ornamentation consisting of a series of stabs below the rim. Later, decoration of the vessel body with vertical incisions is common. Flint tools include (pointed or thin, butt-end, flat-trimmed) daggers, axes, round scrapers, transverse arrowheads, and knives. Flat hammer axes and club heads are made of ground stone, and amber heads and pendants are used as ornaments (Midgley 2004).

During the early period (ca. 4000–3500 BC), only small plots are cultivated, with cereal appearing late. Livestock—cattle, pig, and few sheep–goat—are more important than agriculture. Settlements with earthen long barrows are small and mobile, in the vicinity of lakes and streams or on the coast. Extensive swidden agriculture and ploughing appear ca. 3600 BC and replace wild resources in the diet. In this period (ca. 3600–3200 BC), a three-tier settlement pattern expands, organised around regional causewayed enclosured centres, surrounded by small communities (each with a settlement), a cluster of megalithic tombs, and bog deposits. During the final period (ca. 3200–2800 BC), habitation becomes concentrated in still larger settlements (Midgley 2004).

Initially, the burial rite consisted of non-monumental burials, including flat graves, a simple inhumation in the extended supine position, without a mound, which probably continued the Ertebølle tradition and expanded it through the Northern European Plains. The introduction of the new settlement pattern (ca. 3600 BC) coincides with the period of intense construction of megalithic tombs, which included only selected bones—with bodies skeletonised elsewhere—and elaborate sacrifices in the bogs. This period also coincides with the appearance of enclosures and changes in the landscape compatible with the deliberate creation of open areas, for both cereal cultivation and grazing (Midgley 2004).

Ditch systems, the so-called *causewayed enclosures*, are among the most impressive monumental constructions of Neolithic Europe. They comprise a succession of elongated single pits, possibly dug collectively by individual small groups, such as families. In the TRB North, causewayed enclosures are always accompanied by megalithic tombs, and their function may vary, possibly including a defensive role. Approximately every second generation, a burying and renewed digging phase is carried out, which—based on the settlement patterns and small size of farmyards—suggests that inhabitants from different farmyards convened at intervals in order to cooperate at particular celebrations (Müller 2014).

These structures appear scattered all over Europe, developing first in the Paris Basin connected with late Chaséen and the early Michelsberg cultures, which did not initially have grave fields as south-eastern Europe nor megalithic tombs as western Europe. The central German region (Baalberg) began their construction ca. 3800 BC at the latest, and they appear after a temporal and spatial gap—excluding west, south-east, and east TRB groups—in the TRB North group ca. 3600 BC (Müller 2014).

Megalithic tombs appear ca. 3650 BC and spread through the TRB area. They consist of the collection of matching boulders to form a corridor covered with a capstone and, in contrast to earlier stone cists, to leave an open access and thus form a chamber that may be re-entered recurrently. The construction needs a supportive mound, which may also serve as a ramp during the building process. A variable number of objects may be deposited in or in front of the grave chamber (Müller 2014).

V.5.2. Lublin–Volhynia

The late phase of the Malice culture showed intense contacts during the second part of the 5th millennium BC with Trypillia: on a moderate scale, Trypillian men seem to have sought their wives in the area of the Malice culture, and women moved to the Trypillian settlements. These women were probably responsible for the unidirectional transfer of material culture, i.e. the numerous imitations of the Malice ceramics, and the long-lasting (selective) traditions of Malice pottery passed down in their new environment (Kadrow 2016).

The Lublin–Volhynia Painted Ceramic Ware culture, which lasted a good part of the 4th millennium BC, is closely related to the Wyciąże–Złotniki group (starting ca. 4200/4100 BC), with which it coexisted until the mid–4th millennium BC, both in turn coexisting between the late 38th and early 35th centuries BC with early Funnel Beaker and early Baden influences. The later Wyciąże/Niedźwiedź materials, beginning in the mid–4th millennium BC, coexisted partially with the Lublin–Volhynian culture, and with the Funnelbeaker culture until ca. 3100 BC (Novak 2017).

These east-central European groups show some key elements in common (Wilk 2018):

- Concentration of graves in separate cemeteries;
- differentiation of burials with regard to sex (the principle of the 'left–right' side, different burial goods for males and females);

- stratification of graves with regard to the richness of their inventories (this mainly applied to copper artefacts);

- occurrence of indicators of the richest male burials (a copper dagger in Wyciąże, a copper battle axe, a small axe and a chisel in Książnice);

- allocation of a separate area for elite burials (the eastern burial area in Książnice, and the south-eastern and north-central part of the necropolis in Wyciąże), as well as one for egalitarian burials (the western area in Książnice, and the south-central and western part of the cemetery in Wyciąże).

These patterns of social and religious behaviours, partly due to the expansion of the steppe package, stemed probably from areas lying to the south, beyond the Carpathian Mountains. Similarities with the late Polgár groups and areas of the Tisza river include especially the different treatment of the deceased depending on their sex, age, and social rank. In the Lublin–Volhynia culture, there is an opposition male–female observed particularly in the consistent positioning of males on the right, and females on the left side, a ritual norm that divided the deceased from early childhood (Zakościelna 2010).

Nevertheless, differences in the size of cemeteries and orientation of burials, as well as in the details of the burial goods (smaller frequency of copper artefacts, particularly in prestigious, heavy items like battle axes, axes, and daggers) and local pottery used support that these influences were not caused by migrations from the south, but were rather due to processes of selective cultural transmission (Nowak 2014).

During the 4th millennium BC, the Danube (youngest Malice culture and classic Lublin–Volhynia culture) and Trypillian settlement complexes came into contact on the upper Dniester, and in the Styr and the Horyn rivers in Volhynia. This contact helped continue the previous forms of marital exchange, resulting in the further popularisation of the Danube culture in Trypillian settlements. It seems that the demand from Lublin–Volhynian groups favoured the expansion of flint working technology (such as trough-like retouch, or long flint blades) from Trypillian sources (Kadrow 2016).

These interactions continued in the mid–4th millennium, until the Funnel Beaker culture quickly drove out and replaced the Danube population in western Volhynia and the upper Dniester basin during its pre-classic and early classic phase. During this stage, even as an agrarian community, it has been described as closer to the sub-Neolithic groups than its predecessor, the LBK culture, with an observed archaeological "confusion" concerning wider areas beyond the Polish Lowlands. This is exemplified by the emergence of the Zedmar culture (Zedmar-type materials), the impact of Comb Ware culture or other cultural impacts on the Narva culture, changes in the Neman culture (or Pripyat–Neman culture and rise of the Neman culture), and in Poland the Linin-type materials (Adamczak, Kukawka, and Małecka-Kukawka 2016-2017).

During its classic phase, in the second half of the 4th millennium BC, Funnel Beaker migrants settled more intensively the upper Dniester basin up to the Hnyla Lypa river, and western Volhynia up to the Styr river, coexisting and interacting with Trypillian settlements for many generations (Kadrow 2016).

V.5.3. Forest Zone

After the deglaciation of the Scandinavian Ice Sheet, the Great Lake of Central Finland formed around 6500 BC, dominating the whole region at the time of the Holocene climatic optimum. Its waters dispersed into two basins: one of them formed Lake Saimaa, which continued to tilt in the southeast direction. Around 3845–2795 BC, Lake Saimaa eventually burst through its southern boundary, burrowing a 23–km-long valley and flooding towards Lake Ladoga along the River Vuoksi, the new outlet of Lake Saimaa. The Vuoksi breakthrough is considered a natural disaster on a massive scale (Oinonen et al. 2014).

Drastic changes to the shoreline occurred, with dried shallow lakes and rivers, lakes isolated from the main body of water, and thousands of square kilometres of emerged new land, creating a patchy habitat soon populated by pioneer flora (e.g. *Picea abies*, and spruce and Scots pine replacing mixed conifer-deciduous forests) and fauna adapted to the gradually cooling climate that increased the local biodiversity over the following 100–400 years. Lake Ladoga rose 1–2 m, altering the shoreline ecosystem and burying several settlements (Oinonen et al. 2014).

These environmental changes were coupled with cultural transitions of hunter-gatherer populations that lived in the area. The introduction of a completely new archaeological assemblage, the Typical Comb Ware culture (ca. 3800-3450 BC) heralds the appearance of Neolithic traits in the forest zone. It was a relatively uniform culture that covered a vast area ranging from the Urals to the Baltic Sea, and from Northern Ukraine to the Arctic Ocean, although in southern Finland and Karelia variants of the older types remained still in use. The rapid spread of the Typical Comb Ware culture was almost contemporaneous with the disappearance with the Early Asbestos Ware culture, and has been considered the most influential and innovative culture of eastern Fennoscandian prehistory, introducing pit houses with rectangular timber-frames, red-ochre graves, and exotic materials such as amber, flint and copper, rare in earlier periods. It coincided with the population maximum of Neolithic Fennoscandia, and with the increased salinity of the Baltic Sea and Holocene climatic optimum, which is linked to high productivity of terrestrial, lacustrine and marine ecosystems (Nordqvist and Mökkönen 2016).

The moose population explosion in southern Finland was due to the creation of huge areas rich in grazing lands for large ruminants, and came to an end one or two centuries after the Vuoksi breakthrough, when wetlands developed into old forests and spruce-dominated forests. The Typical Comb Ware culture, predominantly of maritime hunter-gatherers concentrated on seal hunting and fishing in the coastal areas, adapted to the new resources.

Change in subsistence strategies include the increase of moose remains from ca. 3% in the Early Asbestos to ca. 24% in the Typical Comb Ware period (especially significant in the Lake Saimaa region), falling back to ca. 5% in later periods. All this led to a population maximum in the region that lasted for approximately two centuries (ca. 3800–3600 BC), declining after 3600 BC with the change from open grassland into old forests (Nordqvist and Mökkönen 2016).

The disintegration of the Comb Ware phase began ca. 3500 BC, coinciding with the influence of the Volga–Kama region and the birth of several variants of Asbestos- and Organic-tempered Wares, although no break has been observed in cultural development (Nordqvist et al. 2012). These groups also maintained vast and varying intra- and interregional contact networks. During this period of 3500–3000 BC a shift to drier and cooler conditions is found in the steppes, with steppes expanding, and therefore also Yamna pastoralists and their cattle following them. The emergence of the poor Volosovo and Garino-Vor metallurgy in the 4th millennium BC (see *§VIII.15.1. Balanovo*) has been attributed to external influences from Yamna.

Between 3500–2000 BC an interruption in cultural continuity is found in the forest zone, coinciding with a major change in the environment, with selective felling and subsequent regeneration of forests in the Pit–Comb Ware area (Mazurkevich et al. 2009; Poska and Saarse 2002). This could have been caused by the complex movement of peoples in this period, as reflected by the interaction or "checkerboard of regional cultures covering the rolling hills and valleys of the forest steppe zone" (Anthony 2007), and a complex set of cultures is found in the east European forest zone, different from central European cultures (Czebreszuk and Szmyt 2004).

v.5. Northern Europeans

TRB samples from central Europe include: from the old Baalberge group, one individual from Esperstedt (ca. 3850 BC), of hg. I2-M438, and two from Quedlinburg (ca. 3650–3500 BC), one of hg. R1b1b-V88; one from the later

Bernburg culture (ca. 3200 BC) in Esperstedt, of hg. I2a1a2a1a1-Y3749; and one from a west TRB group in Sorsum (ca. 3200 BC). All of them show a typical ancestry composed of NWAN ancestry plus contributions of WHG ancestry, forming a close cluster with other Early Neolithic farmers from Europe (Haak et al. 2015), as well as with contemporary Iberian Middle Neolithic and Hungary Chalcolithic samples, although with lesser WHG contribution than the samples of the Michelsberg culture (ca. 4600–3000 BC) from Blätterhöhle (Gamba et al. 2014; Krause-Kyora et al. 2018).

An Early Neolithic TRB sample from Kvärlöv in the Skåne region also shows similar ancestry, with hunter-gatherer contribution either from WHG or Baltic hunter-gatherers, rather than SHG (Mittnik, Wang, et al. 2018). On the other hand, a female from Syltholm in Denmark (ca. 3700 BC), before the transition to the Neolithic, shows entirely WHG ancestry, without any significant trace of EHG or NWAN ancestry, suggesting that EHG did not reach southern Denmark in Prehistory, and that NWAN ancestry had not still reached this region (Jensen et al. 2018).

TRB communities responsible for the spread of agriculture to Poland have been proposed to be formed by indigenous northern European Mesolithic peoples who adopted farming locally rather than by incoming exogenous Danubian farmers from central Europe. However, post-LBK samples of the Lengyel culture from the Brześć Kujawski group (ca. 4500–4000 BC) and of the TRB culture from Kuyavia (ca. 3500 BC) cluster together with Early/Middle Neolithic European farmers, with one Brześć Kujawski outlier showing an intermediate position with WHG, and another clustering together with WHG. One sample from the Brześć Kujawski group shows hg. G2a2b2a1a1a-U1, and one TRB sample shows hg. C1a2b-Z38888, while mtDNA shows a mixture of farmer and hunter-gatherer lineages (Fernandes et al. 2018).

At the transition to the northern Middle Neolithic (ca. 3300 BC) there was an intensification of agriculture in Denmark and in western and central Sweden, accompanied by the erection of megaliths, with Middle Neolithic (MN) TRB samples from western Sweden being directly derived from Early Neolithic TRB. In eastern central Sweden, settlements became concentrated along the coast, shifting towards marine resources. This early Pitted Ware culture (PWC), contemporaneous with MN TRB, shows an admixture and position in the PCA intermediate between SHG and MN TRB. Both MN TRB and PWC groups show continuity with hg. I2-M438 lineages (Skoglund et al. 2012; Raghavan et al. 2014; Skoglund et al. 2014; Mittnik, Wang, et al. 2018).

In the eastern Baltic, samples from the Mesolithic Kunda and Early Neolithic Narva cultures in Latvia and Estonia had an ancestry intermediate between WHG (ca. 70%) and EHG (ca. 30%) show a dramatic shift with the introduction of the Middle Neolithic Comb Pit Ware culture, with more EHG-related ancestry: from 65–99% EHG (and 1–32% WHG), with two individuals showing 100% EHG (Mathieson et al. 2018; Mittnik, Wang, et al. 2018). This suggests that a westward migration of peoples accompanied cultural changes in the region.

Individuals from the Forest Zone were not found to have received genetic influx from Anatolian-farmer-related genes during the Mesolithic or Neolithic, and therefore an inner cultural diffusion of pottery, farming and metallurgy is assumed for the population of the Baltic and Dnieper Rapids (Jones et al. 2017). The presence of a R1b1-L754 (xR1b1a2-M269) lineage in a Middle Neolithic sample from the Baltic may support both the continuity of (a part of) male lineages in the region, and the arrival of these lineages (probably R1b1a1-P297) from the west.

V.6. North Pontic area

V.6.1. Late Sredni Stog

At the end of the 5th millennium BC, Trypillia ceramic imports appear at the Neolithic Dnieper sites. In the forest-steppe region, they occur on a number of sites belonging to the Kyiv–Cherkassy variant of the Dnieper–Donets community, and later imports reach into the forest zone, into the territory of the Pit–Comb Ware culture. Prestige objects begin to appear at this time on the north Pontic region, too, marking the beginning of the prestige exchange (Rassamakin 1999).

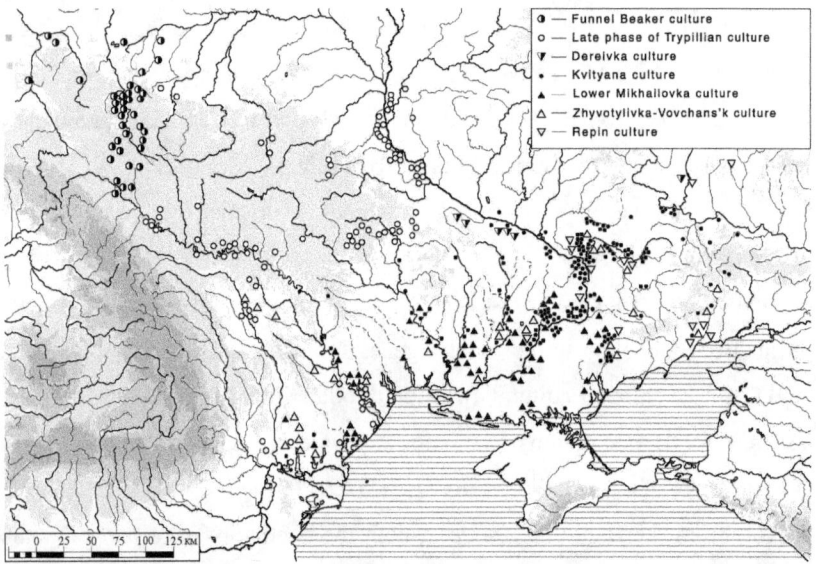

Figure 20. Late Eneolithic cultures of the north Pontic region. Image modified from Tolochko (1997).

After 4000 BC, different groups were formed in the steppes (Figure 20). In the west, distinct late Sredni Stog cultures appear, remaining in contact with Trypillian villagers, and some cultural assimilation seems to have happened east of the Dnieper ca. 3700–3500 BC. The Kvityana (also "post-Mariupol") culture is characterised by supine burials and specific pottery, and encompasses the Dnieper steppe and forest-steppe region, the Azov region, and the Donets. Its emergence was probably related to the development of the

Lower Mikhailovka culture, and the culture is conservative and archaic in appearance, manifested in the burial rite involving a supine position, and in the pottery with no corded or caterpillar track decoration. Eventually, the Kvityana culture would expand to the south-west, with typical assemblages found in Usatovo territory (Rassamakin 1999).

Sredni Stog settlements appeared in the Middle and Lower Dnieper and Lower Don valleys and surrounds, with lifestyle based on restricted mobility, and orientated to valley resources. Their subsistence was varied, according to its distribution between the forest-steppes and the steppes, but a hunter-fisher-gatherer economy remained prevalent: people fished, hunted small and large game including wild horse, and kept some sheep-goats, cattle and pigs (as well as dogs). They also probably herded horses and there is evidence for control of the horse by bridles (Whittle 1996).

An important event in the history of population movements in this period was the appearance of Neolithic tribes of the Pit–Comb Ware culture in north-east Ukraine. Analogues of this culture have been found in the region of the Volga–Oka Rivers. The appearance of new ethnocultural groups in Ukraine resulted in an increasingly heterogeneous regional population, which differed in their cultural, religious, and anthropological traits (Telegin et al. 2015).

The Deriïvka culture, known from settlement materials in the Dnieper, at the sites of Deriïvka and Molyukhov Bugor, and distinctive pottery in Oleksandriia on the Oskol, is found in different forest-steppe regions in the Dnieper and the Donets basins, limiting to the south with Kvityana, and to the north with Pit–Comb ware cultures of the forest zone. Its pottery shows consistent features, such as a weak profile and slightly elongated proportions, with high, straight mouths, evenly cut off at the rim, and conical bases (Rassamakin 1999).

Deriïvka (ca. 4000–3000 BC) is on a promontory of the Omelnik river, a tributary of the Middle Dnieper, and represents a settlement of 60 by 40 m, including hearths, pits and two or more large rectangular structures with slightly earth-sunk floors (Figure 21). Other areas seem to have been given over to specific tasks, connected with pot use, bone tool manufacture and preparation of fishing gear and fish processing. Ducks and several species of fish show the importance of riverine resources. The abundance of horse remains could have come from both wild and managed animals, but scarce hints at the proportion of males may indicate they were domesticated.

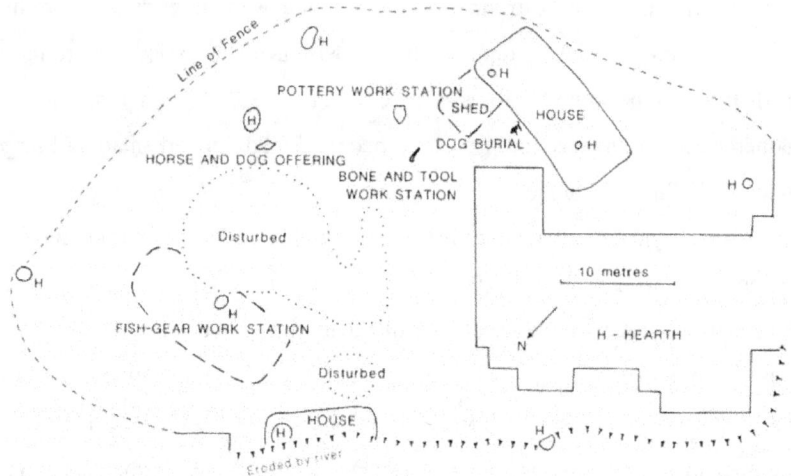

Figure 21. Simplified plan of the Sredni Stog occupation at Deriïvka. After Telegin and Mallory. Image modified from Whittle (1996).

Deriïvka shows a wide range of ceramics and anthropomorphic and zoomorphic figurines associated with Trypillia CI-CII, and ornamental corded compositions of Deriïvka ceramics are just like those from late Trypillia Usatovo, Gorodsk, and Tsviklovtsy (Rassamakin 1999).

The Molyukhov Bugor and Deriïvka sites from the Deriïvka culture show a clear reliance on equine products, mainly horse exploitation (ca. 15–50%), hunting (ca. 15–50%, mainly of deer and elk), fishing, and cattle breeding (ca. 15–40%). Nevertheless, the absence of ruminant dairy product residues suggests a relatively unsophisticated knowledge of ruminant domestication, which implies the presence of wild rather than domesticated horses. Agricultural tools and cereal impressions in pots suggest plant exploitation as a complementary activity (Mileto et al. 2017).

This primitive agriculture probably emerged as an imitation of Trypillian agriculture, based on the finding of numerous hoes, querns, and flat-bottomed vessels and flint tools similar to sickle blade elements. The fish and wild boar representations of the zoomorphic plastic art, apart from the high percentage of hunted fauna, support the high reliance on hunting and fishing as the main subsistence economy, depending on the ecological niche (Rassamakin 1999).

The Deriïvka culture emerged as the transformation of some part of the Neolithic forest-steppe tribes, since pointed base pottery is a characteristic feature of its assemblages. From the mid–5th millennium BC on, contacts with newcomers from Neolithic communities of pit–comb pottery is seen in the Dnieper–Don interfluve. This interaction is continued by Eneolithic communities, up to late materials (ca. 4000–3750 BC) with stroke and pit-comb complexes, material of Ksizovo type, and rhomb–pit pottery (Smolyaninov, Skorobogatov, and Surkov 2017).

There was local production of flat-bottomed basins (particularly at Deriïvka) and jugs (particularly at Molyukhov Bugor) within the specific Trypillian tradition. Corded decoration appeared first on cultures with a closer link to Trypillian cultures, and was thus prevalent among pottery from Molyukhov Bugor, in contrast to the scarce findings in Oleksandriia and Deriïvka, where they appear during later stages of their occupation (Rassamakin 1999).

Further up the Dnieper (mainly on its left bank) was the Pivikha culture, with its northern border reaching the Kyiv region, and with its imports reaching to the south into Mikhailovka. North of the Pivikha culture, Neolithic sites of the Dnieper–Donets culture still remain active in this period, with the forests on the left bank of the Dnieper showing sites of the Pit–Comb Ware culture, whose imports can be seen on the Deriïvka and Oleksandriia settlements (Rassamakin 1999).

The Lower Mikhailovka culture covers a wide area and period, showing a characteristic internal unity and consistent manifestations: analogous pottery in settlements and burials, interment ritual, and burial structure, with ditched cromlechs, an entry orientated to the southwest, mounds all made the same way, and a compound construction with clay overlying the black earth. It was located on the Dnieper–Danube steppe region, and chronologically it begins ca. 4000–3500 BC, with some overlapping of territory and culture with Kvityana (Rassamakin 1999).

The Mikhailovka I site reflects parallel adaptations with the north Pontic forest-steppe areas consisting of permanent settlements in river valleys with fortifications, and occasional farming west of the Don, evidenced also by occasional grain impressions in their vessels. Among domestic animals, mainly sheep and goat bones are found (and occasionally pigs), with only up to 7% of horse remains (used for secondary products rather than meat), and utilisation of wild resources, consistent with a sedentary way of life (Parzinger 2013; Mileto 2018).

V.6.2. Zhyvotylivka–Vovchans'k and Gordineşti

The Zhyvotylivka–Vovchans'k-type sties appear as burials cut into existing kurgans between the Prut and the Don. Their pottery belongs to forest-steppe (Kasperovo-Gordineşti) type, not to the steppe (Usatovo) variant. It appears to be a late, eastwards expanding culture, which was replaced eventually by Neolithic sites of the emergent Pit–Comb Ware culture, e.g. in the Samara tributary on the east bank of the Dnieper (Rassamakin 1999).

The period ca. 3500–3000 BC is characterised by a cultural break-up in the north Pontic area. Cultures from the previous period see their territories much reduced, or divided into smaller, more localised units. The Trypillia world continues to lead this area, but the settlement pattern is noticeably altered, the overall density of sites decrease dramatically, and the culture breaks down into individual groups with different burial traditions: in the Dniester region, Vykhvatinsk; in the steppe zone, Usatovo, which absorbed some features of the Lower Mikhailovka culture; in the Prut and Middle Dniester regions, Gordineşti (corresponding to Horodiştea on the Prut–Siret interfluve, and on the Lower Danube to Cernavodă III); Sofievka in the forest-steppes of the Middle Dnieper (Rassamakin 1999).

Trypillia increases its influence over Deriïvka, where corded decoration ("pre-Corded Ware"), plastic art, and bowls appear. The fate of Pivikha is not clear. To the south, Lower Mikhailovka remains intact on the Azov region and the Crimean steppes. To the north, the Kvityana culture survives in its initial core zone. The Dnieper–Buh group of sites emerges with mixed features between Trypillia, Lower Mikhailovka and Kvityana (Rassamakin 1999). All these terminal Eneolithic units developed gradually, probably as an adaptive response to climatic conditions over the course of the 4th millennium, and adopted a way of life similar to their EBA successors in the area (Harper et al. 2019).

A true Trypillian colonisation wave happened in what seems a mass exodus of Trypillian communities to the steppe (Manzura 2005). Gordineşti tribes expand to the south, into the zone of the Usatovo sites, and to the east and southeast, towards the Dnieper. The Zhyvotylivka–Vovchans'k burial assemblages are linked to this culture, and connected the forest-steppe Buh, Dniester, and Prut regions with the Lower Don and the northern Caucasus, where the late stage of the Maikop culture (the Novosvobodnaya sites) continued (Figure 22). Maikop cultural elements became more widespread in the steppe zone, and also Konstantinovka vessels appeared in Gorodsk settlements, probably with Zhyvotylivka–Vovchans'k as intermediaries (Rassamakin 1999).

The connection between Pre-Caucasian (Maikop) and Late Trypillian cultures that had moved to the left bank of the Dnieper points not only to Caucasian imports, but to a likely Caucasian immigration in a series of small shifts or 'shuttle' movements, possibly with the aim of exchange, trade, spoils of war, borrowing of technological devices, etc. This migration is linked to the creation of "bridge" communities, like the Zhyvotylivka-Vovchans'k cultural group, and the Late Trypillian Gordineşti group (Ivanova and Toschev 2015).

The expansion of Zhyvotylivka graves across the Pontic steppes, from the Carpathians to the Lower Don and the Kuban Basin clearly signals a rapid dissolution of former cultural borders, and the beginning of active movements of peoples, things and ideas over vast territories (Manzura 2016).

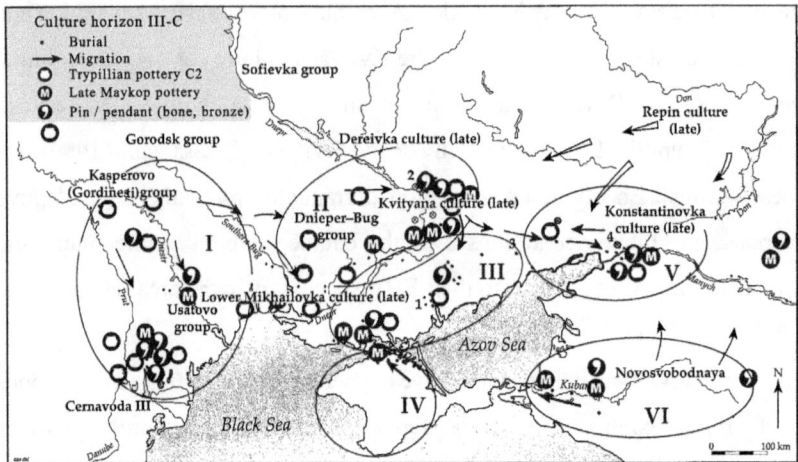

Figure 22. Disintegration, migration, and imports of the Azov–Black Sea region. First migration event (solid arrows): Gordineşti–Maikop expansion (groups: I – Bursuchensk; II – Zhyvotylivka; III – Vovchans'k; IV – Crimean; V – Lower Don; VI – pre-Kuban. Second migration event (hollow arrows): Repin expansion. After Rassamakin (1999), Demchenko (2016).

V.6.3. Globular Amphorae and Proto-Corded Ware

The easternmost area of the Funnel Beaker culture had become more Baden-like with the expansion of the Baden culture in its western area ca. 3300–2900 BC (with findings up to 2600 BC). On the east, the influence of the neighbouring Trypillian culture is seen from ca. 3000 BC, either from earlier (cf. Troyaniv, Koshilivtsy, Brînzeni, Zhvaniets, or Vychvatintsy) or later groups (cf. Gorodsk, Kasperivtsy, Sofievka, Horodiştea-Folteşti, Usatovo). In this period of reduction or concentration of settlements, vertical hierarchical relationships (central sites) were replaced by heterogeneous horizontal network links, eliminating the previous cultural boundaries, thus promoting the spread of foreign (Baden and Trypillian) stylistic influences (Kadrow 2018).

The later periods of shorter oscillations of more wet and drier sub-periods ca. 3700–3000 BC, and especially 2800–2200 BC, may have caused some of the population movements seen in Baden and younger Funnel Beaker culture phases, and older Corded Ware culture (CWC). Slash and burn techniques of agriculture—especially those practised by Trypillian and Funnel Beaker populations—must have intensified effects of natural growth of humidity (ca. 3400–2400), incrementing fluvial activities in west Ukrainian river valleys, and increasing deforestation processes (Kadrow 2016).

There is a trend during in the late 4th millennium to intensified expansion towards maximum inhabitation and agricultural use of all ecological zones, associated with favourable environmental conditions. The traditionally densely inhabited areas of Lesser Poland, the Carpathian foothills, Sandomierz Basin, Lublin Upland and Volhynian Upland, as well as those on the northern plains, are maintained, but there is a clear expansion towards less favourable areas, like the Carpathian Mountains and the Sudetes, with colonisation of the forest zone, especially Mazovia and central and north-eastern Poland (Kadrow 2016).

This reinforced pastoral tendencies in economy and caused changes in settlement patterns, with a reduction of great central settlements and the appearance of fortified settlement centres concentrated in some regions. Other parts of the population became increasingly specialised in stock breeding, leading a mobile way of life, favouring smaller mobile groups tied by kinship links instead of village-like communities, which may have favoured the initial expansion of a Proto-Corded Ware population in the area, among other groups. The brief phase of significant cooling ca. 2900–2850 BC may have favoured the initial, swift and synchronous migration of the group through northern Europe (Rassamakin 1999).

In the forest-steppe zone, herding and hunting activities intensified, while agricultural traditions were preserved, as shown by the Sofievka, Kasperivtsy, and Gorodsk groups. From the end of the 4th millennium BC, mobile parts of the late Trypillian populations moved to the steppe zone, absorbing more and more steppe elements: among others, cord ornamentation (in Vykhvatintsy, Troyaniv, and Gorodsk groups), pottery forms (Vykhvatintsy, which served as prototype for the Thuringian Apmphorae, dispersed along the Dniester river, too), flat burials with bodies in contracted position on the left or right side (Vykhvatintsy, reminding of Polgár culture different male-female position, and later Corded Ware burials, and also Lower Mikhailovka, under a mound without stone constructions). (Rassamakin 1999).

At the end of the Trypillia culture, its agricultural system collapsed completely. The cattle-centred economy of the agricultural Trypillia culture was probably carried into the steppes in the 4th millennium BC (Rassamakin 1999), and this is probably behind its adoption as main subsistence economy in Proto-Corded Ware groups, at the same time as west Yamna further specialised the trend that expanded with Repin.

The Globular Amphora culture (GAC), emerging in the same area as the Funnel Beaker culture but with a more mobile character, based on the scarcity of settlements found, shows a similar reliance on livestock (with decreasing relevance of cattle and increasing relevance of pigs), and characteristic pottery, with globular-shaped pots with two or four handles (see *§VI.3. Classical Corded Ware culture*).

GAC settlers on the north Pontic area, identified as the culture's eastern group, formed a strong long-distance structure for the circulation of cultural patterns, in which three subsystems or route foundations can be distinguished: the Volhynia, the Podolia, and the Siret (Moldavian) subsystems. These systems were used for about 500 years, from the 30th to the 25/24th centuries BC, and left Pontic elements assimilated by societies inhabiting the Lowlands on the Vistula river, such as adaptations of the funerary rite to the horse, and potentially the early reception of niche grave structures by the Złota culture (Klochko and Kośko 2009).

The expansion of Globular Amphora culture groups to the southeast, through the supposed cradle territories of the Corded Ware culture (Volhynia,

Podolia, and upper Dniester river basin) likely destroyed the oldest, primary structures of the Proto-Corded Ware communities. This Proto-Corded Ware population found refuge and conditions for further development in south-eastern margin zone of the Funnel Beaker culture territories, penetrating at first the upper parts of the loess uplands like typical Funnel Beaker sites, but on the margins of their range, and also on areas avoided by Funnel Beaker settlement agglomerations (Kadrow 2016).

These close contacts between the Carpathians and the Dniester or Buh rivers with GAC included steppe cultural patterns traditionally identified with Yamna, but there might have been multiple sources of inspirations, such as CWC groups from the Vistula drainage basin and also CWC groups from the Carpathians, both of which are known to have settled in neighbouring regions for a long time. In turn, the origin of 'steppe' elements in the CWC may have been much more complex (see *§V.5.2. Lublin–Volhynia*).

Corded Ware settlements showed thus a continuation of subsistence strategies of Funnel Beaker, with cattle and sheep herding and cereal farming playing an important role, but also (in the first occupied sites covered with primeval forests) showing all possible subsistence strategies, like hunting and foraging. The deforestation caused by the Funnel Beaker culture in the eastern part of the Carpathian Foreland, creating opened local landscapes, allowed Corded Ware herdsmen to enter these territories and practise their way of pastoral economic activities without problem (Kadrow 2016).

v.6. Late Uralians

The finding of *Yersinia pestis* in two Funnel Beaker individuals from the Frälsegården passage grave in Sweden (ca. 2900 BC), basal to other Bronze Age strains, and not found in nearby Pitted Ware culture, suggests that Neolithic farming villages suffered epidemics before the arrival of steppe-derived populations. Their higher human and animal densities may have helped spread the disease, which seems to have expanded ca. 3700 BC in its basal strain, and then ca. 3300 BC in its typical Bronze Age strain associated with

Corded Ware and Yamna. This supports the spread of the disease prior to steppe migrations into Europe, but geographically associated with the steppe. Trypillian mega-settlements are thus best candidates for the emergence of the ancestors of plague lineages (Rascovan et al. 2018).

The connection of Trypillia with TRB in Sweden is probably to be found in the expansion of the Globular Amphorae culture and the mobility of its population, which likely helped spread the disease throughout northern Europe. The major genetic turnover that happened during the Neolithic demographic collapse, and the associated change in settlements and economic structures, particularly in Late Trypillian groups (abandoning of large villages and adoption of mobile herding), may be therefore explained by the spread of this deadly pandemic that benefits precisely from the agglomeration of big settlements (Rascovan et al. 2018). This population displacement probably set in motion a complex expansion of peoples in the north Pontic area, which ended with the migration of some groups—already associated with cattle-herding economy—to the north-west through the Volhynia–Podolia region, *pushed* by Late Trypillian groups.

The individual from Serteya VIII, in western Russia at the border with Belarus (ca. 4000 BC), of hg. R1a-M420 (Chekunova et al. 2014), and an EHG-like individual from Kudruküla, Estonia (ca. 3000 BC), of hg. R1a1b-YP1272 sample (Saag et al. 2017), both from Combed Ceramic-related groups, support the existence of R1a–dominated Neolithic groups in the eastern European forest zone during the 4th millennium BC, including the north Pontic forested areas.

Individuals from the north Pontic forest-steppe spanning the 4th millennium BC (classified as *Ukraine Eneolithic* samples) form a cline spanning from the Mesolithic/Neolithic cluster to the Northern Caucasus in the PCA, having a mixture of hunter-gatherer-, Steppe- and NWAN-related ancestry (Mathieson et al. 2018). This is compatible with the expansion of forest groups from the north (or resurgence of local hunter-gatherer populations of the forest-steppe) into areas previously occupied by Novodanilovka settlers of Steppe-like ancestry, and admixture with them through exogamy (Suppl. Graph. 6). The earliest one, from Oleksandriia (ca. 4000 BC) shows the highest contribution of Steppe ancestry—connected to north Pontic populations rather than the Don–Volga–Ural area[11]—while later samples from Deriïvka (ca. 3500 BC and 3100 BC) show more hunter-gatherer-related ancestry. This Steppe-related ancestry also found later in Corded Ware individuals may be more properly referred to as Forest-Steppe ancestry.

The sample from Oleksandriia is reported as of haplogroup R1a1a1-M417[12] (formed ca. 6600 BC, TMRCA ca. 3500 BC), which has an estimated expansion date ca. 3800 BC based on modern populations (Underhill et al. 2015), roughly coincident with the split of haplogroup R1a1a1b-Z645 (formed ca. 3500 BC, TMRCA ca. 3000 BC). Its isolated finding in the Middle Dnieper forest-steppe region, together with the known interaction of this area with forest cultures to the north, suggest a replacement of the previous Novodanilovka settlers with male migrants from northern forested areas, spreading Uralic languages with them into the north Pontic forest-steppe.

Copper Age Trypillian samples from the Verteba cave have been reported as having approximately 80% NWAN-related ancestry, with ca. 20% of hunter-gatherer-related ancestry (intermediate between WHG and EHG), consistent with their origin in early European Neolithic farmers admixing with hunter-gatherers from the region. Their prevalent Y-DNA haplogroup G2a2b2a-P303 (formed ca. 12400 BC, TMRCA ca. 9700 BC) further confirms their direct evolution from the first farmers from Anatolia (Mathieson et al. 2018). There is also a sample of haplogroup E-M96, also found in expanding

---

[11] The ancestry of the Oleksandriia individual can be modelled best as a mixture of Progress Eneolithic with local populations, which may include a combination of Varna-, Ukraine Neolithic-, or Comb Ware-related samples. The lack of Samara/Khvalynsk-related ancestry supports the origin of his ancestry as an admixture of locals from the North Pontic forest-steppe region with Suvorovo–Novodanilovka-like populations of Steppe ancestry.

[12] Tentative SNP calls with Yleaf from Wang et al. (2019) supplementary materials yield hg. R1a1a1-M417, without further subclade.

Neolithic farmers (see *§iii.2. Early European farmers*). At the mtDNA level, Trypillia shows typical Neolithic farmer haplogroups, with a closer connection with Funnel Beaker samples (Nikitin et al. 2017), which supports the described long-term exogamy practice with groups of the northern Carpathian area, taking Funnel Beaker as a proxy for them.

A late Trypillian outlier from the Verteba Cave (mean date ca. 3230 BC, compared to the others ca. 3700 BC), of hg. G2a2b2a1a1b1a1a1-L43 (formed ca. 2600 BC, TMRCA ca. 1600 BC), shows contribution of Steppe ancestry similar to the individual of Oleksandriia (Mathieson et al. 2017), clustering closely with the previous Suvorovo-related outliers (Wang et al. 2018). This sample, likely predating the expansion of late Repin/Yamna settlers in the area, supports the presence of Steppe ancestry in the north Pontic area driven by previous Novodanilovka settlers (see *§iv.2. Indo-Anatolians*). Novodanilovka-related peoples were eventually assimilated after their cultural demise by other groups expanding into the north Pontic forest-steppe and steppe areas, as evidenced by this male of a clear Neolithic Y-chromosome haplogroup but elevated Steppe ancestry. The same origin of Steppe ancestry is thus to be expected for Proto-Corded Ware populations of hg. R1a1a1-M417 stemming from the north Pontic forest-steppe area.

Samples of the Globular Amphora culture from sites in Kuyavia and Podolia (ca. 3400–2800 BC) form a tight cluster, showing thus high similarity over a large distance. Both groups have more hunter-gatherer-related ancestry than did Middle Neolithic groups from central Europe, representing thus a resurgence of this ancestry in central European farmers. They harbour mainly NWAN ancestry with WHG contributions (ca. 25%, similar to the level seen in Chalcolithic Iberian individuals), but no significant Steppe-related ancestry, representing thus a barrier to gene flow. This barrier to gene flow is also related to previous groups of the north Pontic steppe, which harbour hunter-gatherer-ancestry composed mainly of an EHG-related component (Mathieson et al. 2018). The greater similarity of GAC peoples to Middle Neolithic individuals from Hungary, Iberia, and Sweden, rather than geographically closer populations, supports their origin in north-central Europe rather than the east (Tassi et al. 2017).

The prevalent haplogroup in GAC samples from Kierzkowo (3100–2900 BC), Ilyatka (ca. 2900–2700 BC), and Koszyce (ca. 2880–2776 BC) I2a1b1a2b-Z161 (formed ca. 8500 BC, TMRCA ca. 7800 BC), found widespread across Europe, from Iberia Middle Neolithic and Chalcolithic to hunter-gatherers from the Iron Gates (ca. 6500–5700 BC). Its parent haplogroup I2a1b1a2-CTS10057 and sister clades are also found widely distributed between the Baltic and west Ukraine since the Mesolithic (see *§iii.5. Early Indo-Europeans and Uralians*).

The finding of subclade I2a1b1a2b1-L801 (formed ca. 7800 BC, TMRCA ca. 2000 BC) in most samples from Koszyce, Sandomierz, Mierzanowice, Wilczyce, and in samples of the Złota group (Schroeder et al. 2019), apart from late Corded Ware samples from Poland (Fernandes et al. 2018), supports precisely this subclade as the main lineage expanding with GAC settlers from east-central Europe, and the potential connection of the GAC population with the earliest Corded Ware groups (see *§VI.3.1. Genesis of the Corded Ware culture*). Their varied mtDNA lineages, H, J, K, U and W, support a mixture of hunter-gatherer and Neolithic populations (Tassi et al. 2017).

V.7. Don–Volga–Ural region

V.7.1. Late Khvalynsk–early Repin

To the east of the north Pontic area, the evolution of the early Khvalynsk–Novodanilovka cultural-historical area gave eventually way to different interconnected local cultures after ca. 4000 BC, whose main demographic base were the previous expanded patrilineally related clans, developing in close contact with each other, but also in contact with neighbouring emerging cultures. Their language is to be associated with an evolving Late Proto-Indo-European (Anthony 2007).

The Caspian steppes linked to the Lower Volga, the Lower Don, and the northern Caucasus region were characterised by the evolution of kurgan burials; the spread of catapult-shaped bone pins (later modernised as hammer-headed pins ca. 3000 BC in the early Yamna culture on the north Pontic area), developed in parallel in Repin and northern Caucasus areas; pottery assemblages similar to those expanded previously under the Khvalynsk–Novodanilovka period (i.e. similar to the one continued by Kvityana, but distinct e.g. from the innovative Deriïvka); etc. (Anthony 2007).

On the Lower Don, the Konstantinovka culture appeared as a continuation of the previous kurgan cultures with rich assemblages, distinct from the flat cemeteries of the Dnieper region. Characteristic of this culture is its orientation towards the Maikop culture, whose influence remains initially restricted to the Lower Don region (Rassamakin 1999).

The change of early to late Khvalynsk on the Lower Volga is defined by the Kara-Khuduk site in the Caspian region, and Razdorskoe on the Lower Don, and thus dated to ca. 3900–3800 BC, but it must have lasted at least until the expansion of the Repin culture. Pottery from this site includes incised ornament and rim form which mark its difference with the early Khvalynsk culture (although a similar rim fragment comes from the Khvalynsk cemetery), and connects this region to the Middle Eneolithic Pontic–Caspian steppes. The third stage of the Samara culture (marked by the Totsky-type grave goods, from Ivanovka), attested until the mid–4th millennium BC, can be included in the late Khvalynsk culture (Rassamakin 1999).

The early Repin culture (beginning ca. 3900/3800 BC) appeared in the region west of Khvalynsk, north of the Konstantinovka culture, and east of the Deriïvka and Kvityana cultures. It emerged on the territory of the previous Neolithic Lower Don culture, and continued the local tradition of pitted linear decoration, and clay bosses applied at the base of the neck. Settlements of the Repin type were few and short-lived, whereas the ritual of individual burials under kurgans became more widespread, replacing big soil burial grounds.

Settlements and burials point to a subsistence economy based on stockbreeding and nomadic or semi-nomadic way of life. The economic changes in Repin brought about a transformation in the social and spiritual sphere, too (Rassamakin 1999).

Its pottery is original, combining characteristic features of Eneolithic pottery, showing thus continuity in methods, technology, and morphology with neighbouring Volga–Ural (Khvalynsk – Samara) and Near Don (now inherited by Kvityana) areas, with demonstrated technical and technological continuity between Khvalynsk and Repin traditions (Vasilyeva 2002; Salugina 2005). Typologically, it comprises high vessels with profiled necks and spherical or flat bootms. Technologically, it included silt or clay containing silt, with an admixture of ground and shells and some organic solutions; vessels were made with the help of moulds, and their surface was smoothed and then decorated with comb stamps in different motifs (Rassamakin 1999).

V.7.2. Late Repin

V.7.2.1. Cattle-breeding and horseback riding

At the end of the 5th and during the 4th millennium, the steppe region was characterised by dry climatic conditions worsening gradually, with short-term but violent floods in the Volga–Ural region, and the peak of aridity happening during the late Repin / early Yamna period in the mid–4th millennium BC (Khokhlova et al. 2018). Forest in the river alleys receded markedly, and semi-arid landscapes appeared in the areas with the lowest rainfall to the south of the Pontic lowlands (see below Figure 25). Steppe grass-cover changed and pastoral productivity fell by ca. 50–60% in the whole region, which must have affected all cultures of the area, benefitting more specialised and mobile types of animal husbandry (Parzinger 2013).

In the fourth millennium, sheep–goat still dominated the domesticated animals of the north Pontic area (e.g. Mikhailovka I, Sredni Stog II, or Usatovo, all showing up to 60% of sheep–goat remains), and it probably also composed the majority of the diet (together with cattle) in the Don–Volga–Ural region.

Until 3500 BC, steppe populations were still largely hunters, gatherers and fishers who had herding as an adjunct to their foraging-centred economy (Anthony 2016). Unlike settlements to the west in the north Pontic area, cultivated cereals do not appear during the Eneolithic in the Don–Volga–Ural steppes, though.

The Repin culture, characterised by its cattle-breeding subsistence economy and semi-nomadic way of life, with much less reliance on hunting and fishing, must have emerged and spread benefitting from the expanding grasslands and retreating settlements of neighbouring cultures. Extensive use of broad, unsettled (or abandoned) steppe areas with little access to water was facilitated by highly mobile groups, no doubt thanks to horse-aided herding and wagons, without which the rapid adaptation and improved economic performance of this regional group would be unimaginable (Anthony 2016).

Wagons revolutionised the pastoral economy by providing bulk transport for tents, water, and supplies in combination with horseback riding, for which there is clear contemporary evidence in the Botai culture. Riding increased the number of animals a single herder could watch and control, and also improved a wagon-based mobile way of life (Anthony 2016). It is disputed whether wagons were also pulled by horses apart from bovines, though, because there is no direct evidence of the use of draft horses.

The widespread remains of burials and the rare finding of settlements, represented by seasonal camps of herders, starts in this period and continues into the late Repin / early Yamna stages. The eventual appearance of wagon parts in burials show a transition of a herding tool to a generalised symbol of a home in everyday life: by putting wheels (made of poplar especially for the occasion) in the corners of a burial, it turns into the last home on wheels for the dead. This also offers a potential explanation of how covered wagons— necessary for long travels—might have spread (Morgunova and Turetskij 2016).

All modern domestic horses investigated to date only show ca. 2.7% of Botai-related ancestry, with Przewalski's horses being feral descendants of Botai domesticates. This supports the existence of a different centre of domestication becoming the source of all modern domesticated horses, which incorporated minute amounts of Botai ancestry during their expansion. Ancient specimens from Russia, Romania and Georgia show this ancestry, suggesting its expansion to the east and south before ca. 2000 BC (Gaunitz et al. 2018). The most likely candidates for the expansion of the domesticated horse into Europe and Central Asia are therefore Yamna settlers from the Pontic–Caspian steppe.

The closest specimen found comes from Dunaujvarus, an East Bell Beaker site in Hungary, at the end of the 3rd millennium BC, which shows a contribution of ca. 39% of the branch ancestral to all horses, DOM2, apart from contributions form a source close to Iberian specimens. The Dunaujvarus branch is more archaic than the branch found in two roughly contemporary Sintashta sites, which suggests that the origin of both branches were domesticates spread with Yamna (Fages et al.).

The Dunaujvarus specimen also shows an archaic Y-chromosome lineage of horse domesticates, of haplotype Y-HT-3, shared with one of the Botai–Borly lineages, apart from specimens from Aldy Bel, Iron Age Estonia, Xiongnu, and Iron Age France, before further intense founder effects under a closely related lineage Y-HT-1 during succeeding periods; and it also shows an mtDNA line shared with Botai samples, with a sample from Lebyazhinka IV, and with different Eurasian domesticates. All this supports an origin of the expansion of DOM2 in late Repin–Yamna, and thus the presence of the ancestor of modern domesticated horses in late Khvalynsk–Repin (Wutke et al. 2018; Fages et al.).

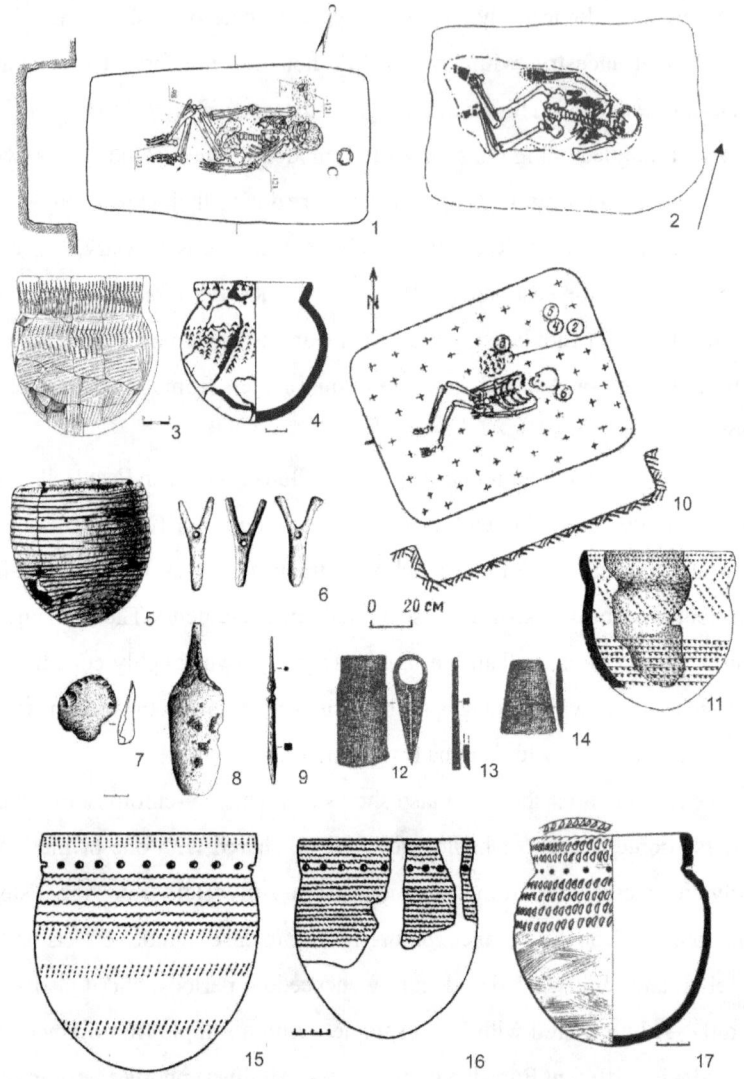

Figure 23. Materials of the Repin type: 1, 2, 10 burials under kurgans; 3–5, 11, 15–17 pottery; 6 bone; 7 stone; 8–9, 12–14 copper. From Morgunova (2015).

The genetic turnover identified in horses (Gaunitz et al. 2018) might be associated with the suitability of horses in Repin for long distance travel and warfare, rather than localised pastoral and hunting activity of Botai horses, possibly derived from earlier Khvalynsk domesticates. The likely long-term specialisation based on selective breeding in late Khvalynsk/early Repin is compatible with the existence of horse domestication in the Don–Volga–Ural area since the Khvalynsk–Novodanilovka period.

The adaptation of Yamna horses to other environments, as well as the gene flow over long distances, is suggested by weak geographic patterns among long-range similarities between Europe and East Asia (Zhang, Ni, et al. 2018), and by the rapid expansion and development of certain breeds since the estimated time of domestication (Yoon et al. 2018).

The late stage of the Repin culture, which showed its innovative corded decoration and bosses at the base of the neck (Figure 23), is the one associated with its expansionist trend, which must have begun probably after ca. 3500/3400 BC, and included settlers from the Middle Don migrating to the north into the Upper Don, southwest into the Dnieper region, and south, to the Lower Don and the Lower Volga (Rassamakin 1999).

V.7.2.2. Eastward expansion

In the Volga–Ural region, Repin features are found at transitory camps and burial mounds in the nearby Volga and Ural areas (Figure 24) during the Middle and Late Eneolithic (Morgunova 2015). These findings point to the Repin semi-nomadic culture diffusing into the Cis-Ural region with settlers. Morphometric studies have shown a potential infiltration from the Eneolithic Don–Volga steppes into the Volga Yamna population, while supporting homogeneity of Middle Volga populations during this period (Khokhlov 2016).

Other Middle Eneolithic regional groups like Khvalynsk, Atlantic, Toksk, and Turganik were possibly unified under the new expanding culture, at least in part through cultural diffusion, given the scattered Repin materials and settlements in the area before the synchronous emergence of early Yamna everywhere (Morgunova 2015). This continuity of the material culture—and probably in part of the population—in the eastern steppe could have been facilitated by the sharing of a common steppe habitat and close cultural ties since the Khvalynsk–Novodanilovka expansion, which might have smoothed the transition of local groups to the new steppe economy.

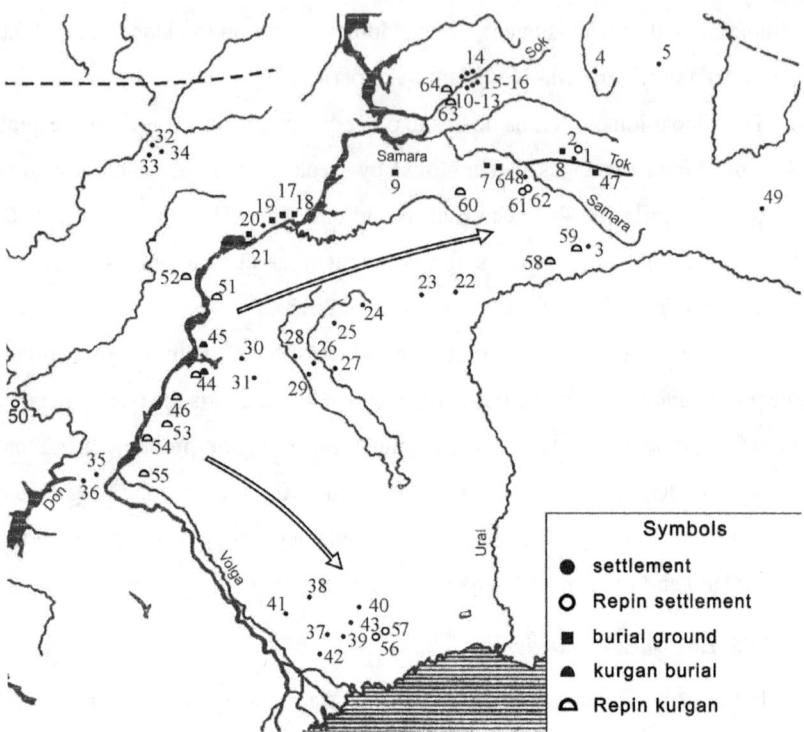

Figure 24. Eneolithic settlements (1–5, 7, 10–16, 20, 22–43, 48, 50), burial grounds (6, 8–9, 17–19, 21, 47, 49) and kurgans (44–46) of the steppe Ural-Volga region: 1 Ivanovka; 2 Turganik; 3 Kuzminki; 4 Mullino; 5 Davlekanovo; 6 Sjezheye (burial ground); 7 Vilovatoe; 8 Ivanovka; 9 Krivoluchye; 10–13 LebyazhinkaI-III-IV-V; 14 Gundorovka; 15–16 Bol. Rakovka I-II; 17–18 Khvalynsk I-II; 19 Lipoviy Ovrag; 20 Alekseevka; 21 Khlopkovskiy; 22 Kuznetsovo I; 23 Ozinki II; 24 Altata; 25 Monakhov I; 26 Oroshaemoe; 27 Rezvoe; 28 Varpholomeevka; 29 Vetelki; 30 Pshenichnoe; 31 Kumuska; 32 Inyasovo; 33 Shapkino VI; 34 Russkoe Truevo I; 35 Tsaritsa I-II; 36 Kamenka I; 37 Kurpezhe-Molla; 38 Istay; 39 Isekiy; 40 Koshalak; 41 Kara-Khuduk; 42 Kair-Shak VI; 43 Kombakte; 44 Berezhnovka I-II; 45 Rovnoe; 46 Politotdelskoe; 47 burial near s. Pushkino; 48 Elshanka; 49 Novoorsk; 50 Khutor Repin; 51 Shumeika; 52 Panitskoe 6b; 53 Skatova; 54 Bykovo I-II; 55 Verkhnegromnoe; 56 settlement Kyzyl-Khak; 57 settlement Kyzyl-Khak II; 58 Boldyrevo; 59 Gerasimova; 60 Orlovka; 61 Petrovka; 62 Skvortsovka; 63 Grachevka; 64 Lopatino I. Image modified from Morgunova (2015).

In the Middle Volga, regions which kept a traditional hunter-fisher economy even after the expansion of Khvalynsk–Novodanilovka witness the emergence of animal husbandry with cattle and sheep–goats and a productive economy, in combination with hunting and fishing, possibly influenced by the 'push' of the initial expansion of Repin. So e.g. in the Lebyazhinka site in the

Sok River (a tributary of the Samara river), with Lebyazhinka III (ca. 5200–4600 BC) compared to Lebyazhinka VI (ca. 4050–3700 BC), whose radiocarbon dates and specific domesticates, broadly related to the Middle and Lower Volga and to the North Caspian region, speak rather in favour of cultural diffusion of domestication to the region before the expansion of late Repin settlers (Korolev et al. 2018).

The presence of regional traits in pottery (a diversity also present in the west), and especially the maintenance of a mainly sheep-herding economy (ca. 65% sheep–goat and only 15% cattle in grave sacrifices) in the Don–Volga–Ural and in the north Caucasian–Caspian early Yamna groups (Anthony 2016), contrasting with the prevalence of cattle herding among Repin settlers and to the west of the Don River (later continued in the Catacomb culture) further supports a limited colonisation wave in the east.

V.7.2.3. Westward colonisation

To the west, demographic pressure and migration seems to have been the main cause of the demise of local cultures. In the north Pontic region, this expansion is considered a true "colonisation" (Suppl. Fig. 8): their demographic impact is seen in the dramatic reduction in territorial extent of Kvityana and Deriïvka cultures, as well as in expanding Repin burial assemblages, with settlements and temporary camps appearing in the Donets basin, in the eastern Azov region, and becoming widely distributed towards the Dnieper (Rassamakin 1999; Anthony 2013, 2007). The late phase of the Konstantinovka variant had continued on the Lower Don during the late Trypillian expansion, as evidenced by the sites of Konstantinovka and Razdorskoe, but with the expansion of Repin settlers the culture ceased to exist.

In the Kuban region, the local Novotitorovka culture emerged, preserving elements of the late Maikop culture. This culture features up to one in four graves with wagons or wagon parts (wooden wheel rims), possibly graves of blacksmiths, a custom common also in late north Pontic steppe cultures. Further connections of the north Pontic area with the Caucasus and of Maikop with the steppe are seen in the imports of arsenical bronzes from Caucasus mines, as well as in the characteristic burials in stone cists beneath grave mounds in the Kemi Oba culture of Crimea featuring Maikop elements (Parzinger 2013).

Contrasting with the characteristic adaptation of the Repin culture to a full pastoral economy, relying heavily on the exploitation of cattle and related secondary products, as well as on horse meat (up to 70% faunal remains in certain sites), north Pontic cultures had specialised throughout the 4th millennium BC in sedentary settlements relying mainly on wild animals, aquatic products, sheep–goat herding, and limited horse-related exploitation. The shift to cattle herding is not detected in the Mikahilovka site, for example, until the emergence of the Yamna culture ca. 3100 BC, a radical adoption of a unique subsistence economy influenced neither by climate nor by environment, which supports cultural belief and economic drivers of new settlers as the main factors (Mileto 2018; Chechushkov and Epimakhov 2018).

Eventually, the Pontic–Caspian steppes became unified under a common culture. The Repin expansion is rightfully considered by many researchers as the early stage of the Yamna culture, since it became culturally and chronologically associated to the synchronous appearance of the early Yamna horizon across the Pontic–Caspian steppes, from the Urals to the southern Buh, and this culture showed little connection to the cultures of the Azov–Black Sea steppes, which it eventually replaced ca. 3300–3000 BC (Figure 25).

All late Repin / early Yamna groups of the north Pontic area absorbed elements from pre-existing local Late Eneolithic formations, although there is a clear remarkably standardised, uniform burial ritual and material culture (see *§VI.1. Early Yamna culture*), opposed to the previous variability in the north Pontic area, which displayed e.g. cromlechs, orthostats, ditches, or sanctuaries in their burials. Similarly, in the Volga–Ural groups regional continuity is also apparent in pottery.

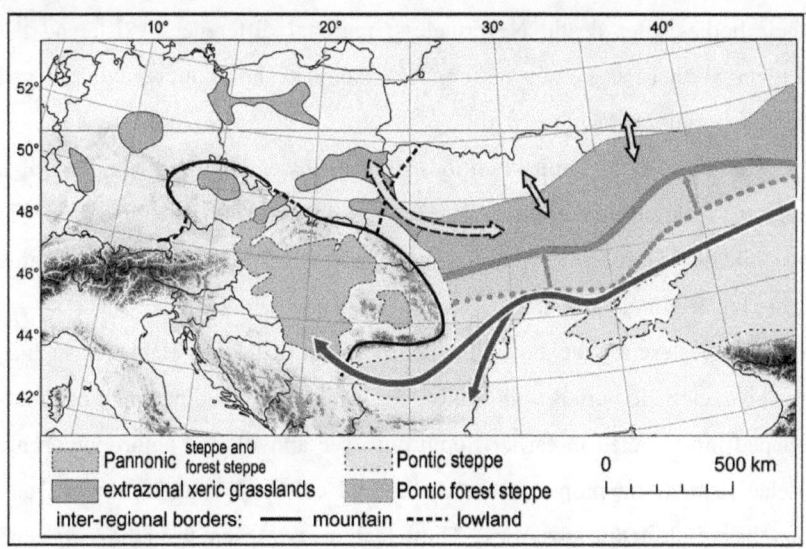

Figure 25. Expansion of the Neolithic north Pontic steppe area (dashed line) to the north during the 4th millennium BC (red arrows), and modern boundary (red line), according to Spiridonovoi and Alöshinskoi (1999). Arrows represent the most common connections and migration routes of steppe and forest-steppe populations. Modified from Kajtoch et al. (2016).

Earliest radiocarbon dates for the start of the early Yamna culture are ca. 3350/3300 BC in both the north Pontic and in the Volga–Ural regions, including the early graves with pottery of the Repin type, with the majority of dates in the north Pontic area lying in the span 3050–2300 BC, although chronologies vary widely in specific regions (Rassamakin and Nikolova 2008).

The earliest radiocarbon dated full-fledged Yamna-like kurgan burial appears precisely among kurgans of the Repin culture ca. 3300–3100 BC, in a region adjoining both the north Caucasian and Volga regions (Shishlina, van der Plicht, and Zazovskaya 2011). The burial ritual continues in part the previous early Khvalynsk tradition of the Volga–Ural area, but with modifications.

Individuals were buried lying on organic mats in grave pits beneath kurgans. Each kurgan contained only one to three individuals, rarely including children, and most graves include adult males, so the majority of the population was excluded from kurgan ceremonies, and we don't know what happened with their bodies after death. Nevertheless, regional differences exist, and the Kuban–northern Caucasus region, for example, shows more children and female burials (Khokhlov 2016).

The survival of Repin traditions on the Lower Don and Middle Volga regions, which gave these early Yamna groups a more archaic appearance (in the so-called Gorodtsov type), further supports that migrations from this area were the origin of early Yamna settlers in the north Pontic area.

The western region shows the incorporation of foreign elements, such as the characteristic burials with wagons, assemblages including small hammer-shaped pins (rooted in earlier Repin pins, see above), and anthropomorphic stelae (a tradition proper of the earlier and contemporaneous Lower Don–Southern Buh steppe territories). These features are absent from the Volga and eastern regions, as are the impressive stratified kurgans, crammed with burials, typical of the Lower Don–Southern Buh territory.

v.7. Common Indo-Europeans

A Common or Classic Indo-European community (Mallory and Adams 2007; West 2007) must have developed during the period of close interaction between late Khvalynsk/early Repin cultures, before the expansion of late Repin/early Yamna caused the divergence evidenced in the Disintegrating Indo-European stage. Based on the ancestry found in Afanasevo individuals (Wang et al. 2019), the population from the late Eneolithic Don–Volga–Ural area probably had a quite homogeneous fully Steppe-like admixture (see above *§iv.2. Indo-Anatolians* and below *§vi.1. Disintegrating Indo-Europeans*) before the colonisation of the north Pontic area.

The split of R1b1a1b1-L23 (TMRCA ca. 4100 BC) into *western* R1b1a1b1a-L51 (TMRCA ca. 3700 BC) and *eastern* R1b1a1b1b-Z2103 lineages (TMRCA ca. 3600 BC, the same for R1b1a1b1b3-Z2106) probably happened early, most likely during the expansion of Khvalynsk clans in the early Eneolithic, which is supported by the disappearance of hg. R1b1a2-V1636 from the region (although it appears in one Yamna individual from the

Caucasus, see below *§vi.1. Disintegrating Indo-Europeans*), reinforcing in turn the concept of a unification of the Don–Volga–Ural region under a single Indo-Anatolian dialect, Common Indo-European, through demic diffusion.

The later successful expansion of R1b1a1b1-L23 subclades, with R1b1a1b1b-Z2103 slightly later than R1b1a1b1a-L51, as well as the prevalent R1b1a1b1b-Z2103 lineages found in eastern Yamna samples and Afanasevo, support the presence of R1b1a1b1a-L51 lineages mainly in the western Don–Volga area, possibly as the majority haplogroup of late Repin. The expansion of the late Repin culture (and later emergence of early Yamna) in the Volga–Ural region was probably then a cultural rather than demic diffusion over populations already genetically and culturally similar.

One Yamna sample from Lopatino II (ca. 3000 BC), possibly of haplogroup R1b1a1b1-L23 (Y410+, L51-)[+], may be thus intermediate between R1b1a1b1-L23 and subclade R1b1a1b1a-L51. The fact that Lopatino II is part of a late Repin kurgan site in the Samara region (see above Figure 24) points to some demic diffusion of Repin clans of R1b1a1b1a-L51 lineages to the east. Further R1b1a1b1a-L51 samples have been reported from central and south Asian sites (Narasimhan et al. 2018), although their actual haplogroups remain dubious. The majority of R1b1a1b1b-Z2103 lineages found to date in the Volga–Ural area, as well as their presence in the Balkans and among western Yamna settlers, points to its original presence probably among Repin settlers of the Middle Don region, too.

This extension of eastern lineages in the Don–Volga–Ural area and genetic homogeneity of the current samples do not let for the moment distinguish the 'Northern' Indo-European community—speaking the dialect ancestral to North-West Indo-European and Tocharian (Oettinger 1997, 2003; Adrados 1998; Mallory and Adams 2007; Mallory 2013; Beekes 2011)—from the 'Southern' or Graeco-Aryan one—speaking the dialect ancestral to Balkan and Indo-Iranian proto-languages (Adrados 1998; Mallory and Adams 2007; West 2007). The cultural division between a western Don–Volga early-to-late Repin culture opposed to an eastern Volga–Ural late Khvalynsk and Samaran cultures before the emergence of early Yamna may be tentatively used to identify a western community of 'Northern' dialect, and an eastern community of 'Southern' one, which is also consistent with the earlier separation of the Northern community and the continued innovations shown by the Southern group.

V.8. Inner Asia

The Kelteminar culture, located in modern Uzbekistan around the Kyzyl Kum desert, represents the colonisation of the Tugias forests and typical steppe close to river deltas and lakes, with a technical tradition derived from the local Mesolithic background. The early stage of the culture (ca. 7th-6th millennium BC), concentrated on the region of the Zeravshan Valley, shows several lithic production systems—including microblades, bladelets, and blades—with at least two techniques, the most common being controlled indirect percussion, the other being the bullet-shaped core method, but there is little evidence for pressure knapping techniques. Therefore, it was probably Mesolithic Ural groups in direct contact with this culture the ones who introduced the technique (Brunet 2012).

During the second stage (5th–4th millennium BC), a significant development of blade and bladelet production is seen, requiring more elaborated skills. Among the tools seen, the Kelteminar arrowhead and the horned trapeze show a wide distribution in parts of Central Asia (Kazakhstan, Russia, Uzbekistan, Turkmenistan, Northern Afghanistan), suggesting a symbolic value, and thus a way to define social identity. During that period, pressure knapping technique becomes prominent in blade production, with new relationships—evidenced by decoration in pottery—probably being developed in Chorasmia with agropastoral communities of southern Turkmenistan (Brunet 2012).

The Atbasar culture (ca. 5th–4th millennium BC) of the forest-steppe zone of Northern Kazakhstan also developed from the local Mesolithic microblade production using bullet-shaped cores as pressure knapping technique. The introduction of few regular blades and new formal tools can be seen during this period, such as points, trapezes, triangular arrowheads with a basal notch, bifacial pieces, and leaf-shaped bifacial points, possibly with different functional and socioeconomic contexts. This and the introduction of pottery with incised or combed decoration, and the domestic horse at the end of the period, suggest migrations in northern Kazakhstan at the same time as those seen among similar post-Mesolithic cultures of eastern Kazakhstan, Altai, and eastern Siberia (Brunet 2012).

The Botai-Tersek culture (ca. 3700–3000 BC) probably emerged from groups of Atbasar foragers in the steppes of northern Kazakhstan who developed a specialised economy as horse riders who hunted essentially horses. Their main diet consisted preferentially of horses, but it included also wild animals like large bovids, elks, deers, bears, etc. The small temporary settlements in the steppes, the evidence for herd-driving hunting techniques, as well as the management and transport of great quantities of horses, together with evidence for bitted and ridden horses prove the appearance of horseback riding ca. 3700–3500 in the northern Kazakh steppes (Brunet 2012).

The Hissar culture (ca. 7th–4th millennium BC) shows a continuation of Mesolithic material culture related to the *Yubetsu* tradition. As in the Atbasar culture, the introduction of new Neolithic components is seen in the late stage, with blade production using indirect percussion, flake production by direct percussion (hard hammer), and the presence of trapezes and polished axes, linked to domestic activities related to leather, skin, and woodworking (Brunet 2012).

v.8. Palaeosiberians

To the east, Neosiberians—from which contemporary Siberians derive—replaced the previous Ancient Palaeosiberian-like populations ca. 9000–2000 BC, restricting their AP-like ancestry to north-east Siberia, represented by an individual from Ol'skaya, Magadan (ca. 1000 BC), who closely resembles present-day Koryaks and Itelmens. In the Cis-Baikal area, thirteen Early Neolithic hunter-gatherers from Shamanka (ca. 5200–4200 BC) are representatives of AEA ancestry, closely related to individuals from the Devil's Gate Cave (ca. 6000–5500 BC). Among reported haplogroups, there are seven samples likely all N1a2-L666, and one C2a1a1a-F3918 (Sikora et al. 2018).

Cis-Baikal populations are replaced likely during the Late Neolithic (Moussa et al. 2016), with samples from Early Bronze Age (ca. 2200–1800 BC) evidencing an almost full population replacement with a resurgence of AP ancestry (up to 50%)—probably from a population migrating from the east and north—and influence from West Eurasian steppe ANE ancestry (ca. 10%) in the Altai region, represented by BA individuals from Afanasevo. Reported haplogroups are all Q1a2a-L712, with one Q1a2a1-L715 and one Q1a2a1c probably suggesting these subclades as those expanding in the EBA (de Barros Damgaard, Martiniano, et al. 2018).

AP ancestry is also found in modern Kets (ca. 40%), speaking a Yeniseian language, with genetic links to Palaeoeskimos, thus connecting Yeniseian genetically with Na-Dene-speaking peoples. The main reported haplogroup of Na-Dene peoples is Q1a2-M25 (ca. 90%), which suggests that its lineages expanded with ancestral Dene-Yeniseian speakers through north Eurasia, most likely during the Late Neolithic (Sikora et al. 2018). Yeniseians, on the other hand, belong to haplogroup Q1b1a-L54 (formed ca. 16100 BC, TMRCA ca. 14000 BC), which may point to the dispersal of certain Palaeosiberian languages with these particular lineages during the Palaeolithic (Huang et al. 2017).

It is unclear which lineage may have spread with AEA ancestry through northern Siberia, along the inner Asian Palaeolithic EHG–ANE–AEA cline, and when, although possibly some N1a1-Tat subclade (formed ca. 13900 BC, TMRCA ca. 9800 BC). In particular, the split into an eastern N1a1a2-Y23747 (TMRCA ca. 4500 BC), found among modern Japanese and Chinese, and a western N1a1a1-F1419 (TMRCA ca. 8800 BC), found among Khakassians and northern Indians, suggests a split around Lake Baikal.

Similarly, its subclade N1a1a1a-L708 (formed ca. 8800 BC, TMRCA ca. 5400 BC) shows a wide Northern Eurasian distribution in the regions east of the Urals. In particular, although basal N1a1a1a-L708 subclades can be found today around the Urals without a particular linguistic connection, its subclade N1a1a1a1a-L1026/L392 (formed ca. 4300 BC, TMRCA ca. 2900 BC) seems to represent expansions through Siberia from around Lake Baikal, often associated with Altaic-speaking peoples. This ancient Altaic connection is reflected in the finding of hg. N1a1a1a1a4-M2019 (formed ca. 4300 BC, TMRCA ca. 1700 BC) through Arctic populations, from Estonians to Tungusic speakers from Yakutia, from Chinese to Hungarians (see *§viii.21.1. Yukaghirs*).

Different expansions of these lineages include, among modern Northern Eurasian populations: N1a1a1a1a2-Z1936 (TMRCA ca. 2300 BC) connects N1a1a1a1a2a1c-Y13850 among peoples from the Trans-Urals region (see *§viii.17. Ugrians and Samoyeds*) with N1a1a1a1a2a-Z1934 in Palaeo-Arctic populations of the Cis-Urals (see *§viii.16.1. Saami and Laplandic peoples*), possibly through a Northern Eurasian forest–taiga route. SNP Y6058 (formed ca. 5300 BC, TMRCA ca. 2900 BC) connects hg. N1a1a1a1a3-Y16323 (TMRCA ca. 2900 BC) of Mongolic speakers and Chukchi of (see *§viii.21.2. Turkic peoples and Mongols*) with Mordvinic and later Balto-Finnic speakers of hg. N1a1a1a1a1-CTS10760 (TMRCA ca. 2100 BC), possibly through more southern forest-steppe–steppe routes of expansion, given the appearance of N1a1a1a1a1c-B479 among Tungusic speakers and Nenets (see *§viii.15. Mordvins and Mari-Permians* and *§viii.16.2. Baltic Finns*).

Late northern Siberian nomadic peoples close to the Arctic region are known to be easily subject to exogamy practices due to their mobility, and are thus associated with plurilingualism and eventual acculturation within few generations of admixture (Karafet et al. 2018). Therefore, it will remain unclear to what extent Palaeo-Laplandic from Lovozero (see *§viii.16. Saami* *and Baltic Finns*) or the language of northern nomadic peoples who adopted Mari-Permic, Ugric or Samoyedic languages remained related to Chukotko-Kamchatkan family thousands of years later, or spoke West Siberian languages like Yeniseian, or other Eurasiatic dialects.

Three samples from the west Siberian forest zone (ca. 6200–4000 BC) are representatives of a mixture of ancestry called "west Siberian hunter-gatherer" (WSHG) ancestry, made up of EHG (ca. 30%), ANE (ca. 50%), and AEA-related ancestry (ca. 20%). This ancestry was also present in the southern steppe and in Turan (BMAC), and formed ca. 80% of the ancestry of an early 3rd millennium BC agropastoralist from Dali, Kazakhstan, contributing to multiple outliers from 2nd millennium sites in Kazakhstan and Turan (Narasimhan et al. 2018).

The widespread presence of this ancestry in west Siberia is compatible with its association with hunter-gatherers of Kelteminar and other central Asian sub-Neolithic cultures (Narasimhan et al. 2018). The presence of an ancestral cline EHG–ANE–AEA ancestry in inner Asia is also supported by the west–to–east gradient formed in the PCA. This ancestral WSHG ancestry, separated from other ancient and present-day populations, is found in Botai (ca. 3600–3100 BC), Okunevo (ca. 2500–1800 BC), central steppe EMBA samples from Sjolpan (ca. 2550 BC), Takhirbai and Gregorievka (ca. 2150 BC), as well as Cis-Baikal EN and EBA populations (de Barros Damgaard, Martiniano, et al. 2018).

Of the three Botai samples published, one (ca. 3600–3100 BC) R1b1a1a-M73, while another (ca. 3300–3100 BC) shows haplogroup N-M231 (Narasimhan et al. 2018). Another individual of hg. R1b1a1-P297 is found in the Bol'shemysskaya culture (ca. 4500–3500 BC). The presence of R1b1a1-P297, and R1b1a1a-M73 in particular, is linked to the previous expansion of the North-Eastern Technocomplex, during the Early and Middle Mesolithic (see *§ii.1. Eurasians*), and thus likely associated with the creation of an ancestral Altaic-speaking population in inner Asia closely related to peoples with WSHG ancestry (see *§viii.21.1. Yukaghirs* and *§viii.21.2. Turkic peoples and Mongols*).

V.9. Afanasevo

Among late Repin settlers migrating to the east, one Trans-Uralian group was especially successful, developing the Afanasevo culture in the Altai region from ca. 3300 BC. The first to propose a common origin of Yamna and Afanasevo based on their shared material culture was I. N. Khlopin, and this hypothesis has been refined to a more archaic cultural phase (the Repin culture), based on archaeological remains, radiocarbon dates, and recently also ancient DNA (Morgunova 2014).

Before the emergence of Afanasevo, the region's most characteristic Eneolithic remains (ca. 4000–3250 BC) were geometrically ornamented pottery found in Botai, Kozhai I, Razboinichij Ostrov, etc. These remains are synchronous with the Repin culture, hence Eneolithisation of the west Siberian region lagged behind that of the Pontic–Caspian steppe, demonstrated also in the absolute lack of copperworking, and scarce, isolated finds of copper goods (such as an adze, and a knife with leaf-shaped blade), which show typological similarities with those of the Pontic–Caspian steppes (Morgunova 2014).

The appearance of Eneolithisation, including domestic animals, and metalworking in the Trans-Urals regions of western Altai and Tian Shan shows a close relationship of the whole process—typology, metal origin, and metalworking—to the early Yamna metallurgy in the Cis-Urals region (see *§VI.1.4. Volga–Ural region*). The appearance of intermediate materials in the territory of eastern and north-eastern Kazakhstan, showing syncretic decoration of Yamna and local Eneolithic cultures of comb-geometric pottery, strengthens the direct connection of the Volga–Cis-Urals area to the Altai precisely during this period (Morgunova 2014).

Burial similarities include the pose of the skeletons, the presence of organic bedding, the sprinkling of ochre, and the forms of the burial pits; differences include the use of stone constructions, and predominance of the south-western and western orientation of the dead (orientation to the east is exceptional). The deposition of pottery increases in relation to early Yamna, and it is marked by its heterogeneity, including its shape (tappered, ovoid, tall body, etc.), ornamental composition showing comb-like stamps, which was no longer typical of the steppe zone, but was common in the late Neolithic in the Urals (Morgunova 2014). Another strong parallel includes the earliest pictorial tradition of petroglyphs, the Yamna–Afanasevo tradition, characterised by the symbolic depiction of sun-headed men and animals (Novozhenov 2012).

Noteworthy features of the material culture are the pottery with swollen body and narrowed neck, which make it possible to assume its relation to late Repin ceramics. Also interesting is the technology used, including moulds, typical of Yamna dishes in the Urals. On the other hand, pottery is marked by considerable syncretism and heterogeneity (Figure 26). Metal is still rarely encountered, but those found show an origin in the Circum-Pontic and Maikop horizon, like those of the Don–Volga–Ural region (Morgunova 2014).

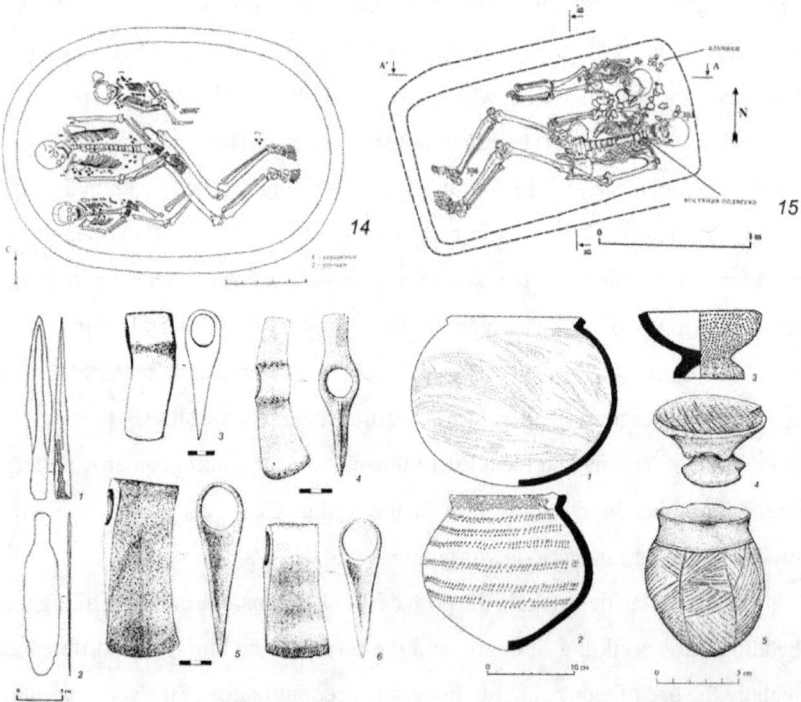

Figure 26. Materials of the Afanasevo type. Top: burial types (after Larin 2005); bottom left: copper products (after Kovaleva et al. 2010, Grushin et al. 2010); bottom right: pottery (after Polyakov 2010, Kovalev, Erdenebaatar 2010). From Morgunova (2014).

The predominant nutrition of buried skeletons is meat, and the main economic occupation of Afanasevo population is sheep and cattle breeding, not known previously in this territory. The presence of sheep, cattle, and horses, now predominant in the region, is typical of the Pontic–Caspian steppes (although the Botai also knew horse-riding before the appearance of Afanasevo). The appearance of Afanasevo must therefore be linked to an early, demographically strong migration wave of late Repin/early Yamna settlers from the Volga–Ural region, following steppe and forest-steppe environments favourable to cattle- and horse-breeding, and possibly in search for new metal deposits (Morgunova 2014).

Paradoxically, though, neighbouring Trans-Urals forest regions including the Botai culture of northern Kazakhstan received practically no influence from Yamna, in contrast to the neighbouring Volosovo–Garino Bor society, adopting metallurgy, productive farming (see *§VIII.15.1. Balanovo*), which suggests that migrants by-passed powerful cultures of combed geometric pottery. Nevertheless, the appearance of censers or 'incense burners' in the Altai ca. 2600–2000 BC, typical of the Catacomb and Poltavka cultures, suggests contacts of late Yamna groups with the region, possibly triggered by later waves of expansion into the Urals (Morgunova 2014).

Afanasevo coexisted with the Okunevo culture for ca. 100 years (ca. 2600–2500 BC), before being fully replaced by it. Okunevo developed for ca. 800 years, until about the 17th–15th centuries BC, when they were replaced by the expanding Fëdorovo culture. The different stages of the Okunevo culture have been radiocarbon dated, the Ujbat stage to the 26th–23rd c. BC, the Chernovaya stage to the 22nd–20th c., and the beginning of the Razliv period is dated ca. 19th–18th c. BC (Poljakov, Svjatko, and Stepanova 2018).

v.9. Pre-Tocharians

Tocharian shows peculiar archaisms and innovations compatible with a development isolated from other Late Indo-European dialects. Its strong differences with neighbouring Indo-Iranian and with other Late Proto-Indo-European dialects in general indicates an extensive period of linguistic separation from the common trunk. The early spread of a group from the late Repin culture into the Altai–Sayan region, emerging as the Afanasevo culture ca. 3300–3100 BC, is compatible with the described early isolation of the Pre-Tocharian group from a Late Indo-European-speaking Yamna community in contact in the Pontic–Caspian steppes (Anthony 2007).

Afanasevo individuals shows full Steppe-like ancestry, coincident with Steppe Eneolithic samples, without sizeable EEF contributions as found later in Yamna (Wang et al. 2019), which is compatible with its origin as an early offshoot from the Don–Volga region [13]. Most published samples from Afanasevo (ca. 3300–2500 BC), the supposed community of Pre-Tocharian speakers, are of haplogroup R1b1a1b1-L23, most probably R1b1a1b1b-Z2103 (Hollard et al. 2018), and many among them possibly of R1b1a1b1b3-Z2106 lineage (Narasimhan et al. 2018). Only three samples are of haplogroup Q-M242, all with a mean radiocarbon date later than 3000 BC, which implies their potential association with emerging Bronze Age cultures from Mongolia, or a resurge of previous populations (see *§viii.21.2. Turkic peoples and Mongols*).

Importantly, no Afanasevo-related ancestry is found in south Asia in the 3rd millennium BC, which discards any important migration through the Inner Asian Mountain Corridor in this period (Narasimhan et al. 2018).

---

[13] The apparent similarity of Afanasevo samples with Yamna individuals from the Kalmykia region between the Don and Volga rivers (among investigated Yamna samples), as well as the similarity in best fits obtained comparable to Yamna from Ukraine or the Caucasus, points to the expansion of Pre-Afanasevo peoples from the Don–Volga area rather than from the Volga–Ural region, suggesting therefore that they were indeed an early offshoot of late Repin.

VI. Early Chalcolithic

VI.1. Early Yamna culture

VI.1.1. North Pontic region

From about 3100 BC and for the next two to three centuries, GAC communities migrated from the Vistula River drainage basin into the area between the Carpathians and the Dnieper, more thoroughly than any of their central European predecessors: they crossed to the eastern bank of the Dnieper, they appeared in the Carpathian basin, and they came into close contact (probably 'face to face') with communities of the Yamna culture (Szmyt 2013).

GAC appeared into the forest-steppe and steppe zone west of the Dnieper ca. 3000–2900 BC, including areas between the Southern Buh and Sinyukha rivers, on the Inhul River, and also on the Dniester–Danube region. At the same time, the Trypillia culture was disintegrating into many regional groups in the forest-steppe and southern forest region between the Prut and the Dnieper (see *§V.6. North Pontic area*). Close interaction in this area is evidenced by mixed grave inventories in at least two parts of the north-western Pontic area, namely the Middle Dnieper and the Siret–Prut–Dniester area, with Yamna settlements showing atypical clay vessels more or less corresponding to GAC style (Szmyt 2013).

Nevertheless, even in the zone of greater migration exchange along the Prut, it is usually possible to draw a line separating the distribution of synchronous settlements, e.g. with GAC settlements occupying territories west of the area, between the Prut and Siret rivers, and Yamna occupying their eastern bank, between the Prut and Dniester. In the steppe zone, contacts in form of adopted pottery ornamentation by Yamna settlers are still less clear, which supports a clear differentiation of both groups (Szmyt 2013).

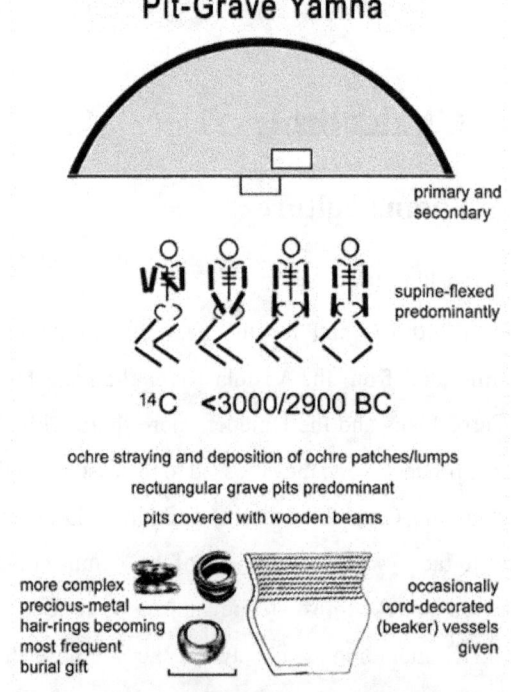

Figure 27. Burial schemes of pit graves found in the Lower Danube region during the Eneolithic according to Frînculeasa, Preda, and Heyd (2015).

After 3000–2900 BC, the majority of pit–graves west of the Black Sea belong to the domination and assimilation of peoples characterised by the Yamna funeral standard (Figure 27), in which the buried—both primary and secondary—were lying supine with the legs bent up in the knees, usually orientated on the west–east direction, in rectangular or sometimes chamber-like grave–pits covered by wooden beams. Poor inventories are the rule (contrasting with previous north Pontic steppe cultures, see IV.2.3. Kurgans), and spiral silver hair rings are the most defining items. Male burials are prevalent, and ochre staining or deposition of lumps is common. Pottery of local origin is rarer than before, and when it appears it is represented by cord-decorated beaker vessels, such as in Coțofeni III pottery ca. 3000–2800 BC (Frînculeasa, Preda, and Heyd 2015). During this late phase, Yamna appears firmly settled in the forest-steppe further north, where they were previously only occasionally found. Larger bands are therefore seen expanding in all directions (Suppl. Fig. 9).

Most Yamna burials in the west Pontic area have radiocarbon dates ca. 2880–2580 BC. Only a small proportion of sites at the Lower Danube shows later dates, with a dilution of the wider Pit-Grave phenomenon. This third stage of pit–graves shows a re-appearance of individuals buried contracted to the side or in extended body position as secondary burials in the mounds. This trend appeared perhaps under the influence of the Catacomb Grave culture or further to the east, or locally at the Lower Danube. This is a period when southeast and central European cultures like Coțofeni, Baden, Ezero A, Globular Amphora and TRB communities were transforming and splitting into successive archaeological cultures, such as Glina, Schneckenberg, Livezile, Makó/Kosihý–Čaka, Ezero B, or Corded Ware proper (Frînculeasa, Preda, and Heyd 2015).

The beginning of the western early Yamna complex is linked to the arrival of a novel set of deeply interlinked social, economic, and ideological innovations. Its five components (Figure 28) play the crucial roles, one due to the next (Frînculeasa, Preda, and Heyd 2015):

1. Subsistence economy based on specialised breeding and herding of cattle only, which leads to the increased use of secondary products in which milk and overall protein-enriched diet supplemented by game and fish (and very low ratio of starch and carbohydrates, as seen in

neglectable caries frequency) have an importance in subsequent changes in peoples' physical appearance and stature (with old anthropometric studies showing that they might have been some inches taller in average than their neighbours).

2. This new economy triggers a higher human mobility: the overall westward migration is a consequence of the ever-lasting search for green pastures for their stock. This mobility may have increased the exchange network, forwarding technical innovations like 'Caucasian metallurgy' of shaft–hole axes, tanged daggers, and previous metal hair rings.

3. Both the new economy and mobility triggered a novel way-of-life, with different land uses and understanding of territory. Peoples become true pastoralists leading a highly mobile way of life, and some segments become true nomads, which alters the social organisation and thus norms, morale, values, symbols and terms, altering the *Weltanschauung* and ideology, as well as religion, which become tradition.

4. A pit–grave under a kurgan becomes a standard in the Yamna custom-set, with its homogenisation reflecting the emerging unifying social norms. Its powerful symbolism is seen as a high landmark, the 'pyramids of the steppe', a monumental and dominant architectural element over ancestor graves (quasi-temples) in an otherwise flat and monotonous 'sea of grass', creating real or virtual ancestry and lineage, and being a sign of possession, and probably claiming territory, as well as delineating the oecumenes of pastoral groups, forming orientation points on the transhumance routes.

5. One key technological innovation made all this possible: the widespread acceptance of the transport complex of wheel and wagon, allowing herders to enter and exploit the deep waterless steppes (the largest part of the steppes) for their stock animals. It also allowed

pastoralists to live in these regions with their families for the longer part of the year, with all their possessions, without the need to keep a base-camp close to a water course.

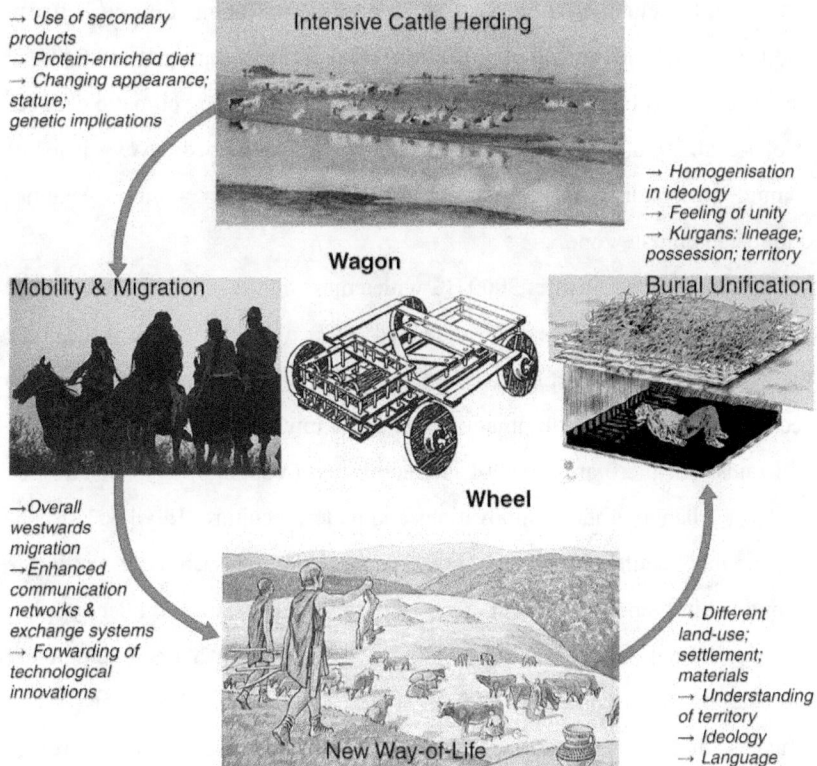

Figure 28. Scheme of interlinked socio-economic-ideological innovations forming the Yamnaya, modified from Frînculeasa, Preda, and Heyd (2015). Images used, from top to bottom and left to right: Cattle on the Puszta of Hrotobagy, by Adrian and Marianne Stokes (Wikimedia); still image from the documentary "Equus: Story of the Horse" (CBC); Bronzezeit keltische Hügelgräber, by Gerhard Beuthner (1930); reconstruction of primary burial (feature 2, Grave 1) of the Hajdúnánás-Tedej-Lyukashalom kurgan by Viktor Szinyei, modified from Pospieszny (2015). Center image: reconstruction of wagon after A.N. Gej (1993), from Novotitorovka wagon burial at Ostannii, kurgan I, burial 150.

VI.1.2. East-central European lowlands

Yamna settlements spread initially west- and southward into the Danube valley. A real current of immigration is noticed from ca. 2950 BC, being the first primary example of large-scale migration in later prehistory, with foreign people flooding over east and east-central European lowlands. More than 500 tumuli and more than 1000 graves have been already studied in this area (Heyd 2012).

A rapid decline in human activities peaked in Central Europe between 4000–3000 BC and recovered only after 3000 BC, accelerating after 2500 BC. This decline has been related to adaptation processes during climatic changes (Kolář et al. 2016; Gardner 2002) – which might have helped the expansion of Yamna settlers into scarcely populated areas closer to a more Atlantic-dominated climate zone.

The area recovered after 3000 BC with a more humid climate that favoured grassland productivity (Harrison and Heyd 2007), at the same time as the horse, the wheel, and pastoralist societies expanded into these areas. Their migration seems not to have been a traumatic event. There might have been local conflicts and raids, but there are signs of interaction with contemporary societies, as well as exchange of ideas, innovations and material culture (Heyd 2012).

The main settlement areas of west Yamna migrants were confined to the steppe habitat, and therefore Yamna settlers (initially) did not occupy, push away, or expel locally settled farming societies. West Yamna tumuli are radiocarbon dated ca. 3000–2500 BC (3100/2900–2500/2475 cal. BC), and contacts with other archaeological cultures confirm that they belong to the first half of the 4th millennium. Distinct from the close contacts between south-eastern European steppes and Pontic–Caspian steppes from earlier periods, it is quite likely that the "infiltration" of small Yamna migrants had begun in the decades, even centuries before the real current of immigration, i.e. at the end of the 4th millennium, as an extension of the north Pontic roaming area (Frînculeasa, Preda, and Heyd 2015).

The massive Yamna migration in south-east Europe is said to have been well organised, either in loose family alliances (the most likely scenario) or in clans, in any case with a clear leadership and structure (Heyd 2012). There was possibly more than one wave of migrations, with cultural differences noted north and south of the Balkans. At least one migration wave seems to have come from the north Pontic steppes, due to the presence of wagons (or parts of wagons) and stelae—characteristic of the Kemi-Oba and neighbouring zones of the Southern Buh–Lower Don steppe—in burial mound cemeteries of Yamna settlements (Kaiser and Winger 2015).

Important settlement areas included (Heyd 2011):

- The first large concentration of Yamna tumuli and burials appeared in the grass and bush-land typical of the steppe-like vegetation and environment, continuous from the north Pontic to the west Pontic area, up to the Dobruja (south of the Danube delta) and north-east Bulgaria.

- The second large concentration, the Tarnava-Rast group, appeared to the west in south-western Romania, on the plains and terraces divided by the lower Danube River. Migrants pushed west, appearing west of the Iron Gates in Jabuke, but the largest number of migrants ended up in the central Carpathian basin.

- Another province was formed by the Upper Thracian Plains south of the Varna Bay, in the Balkan uplands (Kovachevo-Troyanovo), within the region of the Ezero culture (Anthony 2007). It is more influenced by the Mediterranean, and tumuli are widespread, especially in the area between the Maritza and Tundza rivers.

- The Prut–Siret region of well-drained hill and flatlands show tumuli and burials backing onto the eastern Carpathians, in a number smaller than most other western *provinces*. More than distant nomadic settlements, these settlers formed part of a much wider, expanding Yamna group that was originally located further to the east and north-east. This is reflected by the gradually increasing number of tumuli and graves up to their source territories in the Dniester River and the north Pontic steppe

The westernmost group, third in size, lies in the Great Hungarian Plain or Great Alföld, in the central Carpathian Basin, a grassland plain mainly located north and east of the Danube, mainly east of the Tisza River. It covers 50,000 km^2 in Hungary, but reaches also neighbouring modern countries, e.g. Croatia, Serbia and Romania with the regions of Banat and Transylvania forming part of it. The core area of these lowlands is the Hungarian Puszta ('plain'), and Yamna tumuli and burials (Figure 29) are spread all over it, with the largest concentration located in the steppe areas neighbouring the Tisza River. There were originally around 40,000 kurgans in Hungary, but a more recent estimate suggests that there are today less than 2,000 left (Suppl. Fig. 10.A).

Figure 29. A kurgan stands out on a flat 'sea of grass' from the Great Alföld. Hegyes Mound, in the flood-free bank of the Kösely Stream (Nádudvar). Photo by Csaba Tóth, modified from (Tóth, Joó, and Barczi 2015).

Based on the distribution map of kurgans, burials are densest where there were no Boleráz or Baden occupations, although they partially overlapped. Boleráz–Baden groups represented settled, agriculturalist, indigenous groups, while Yamna formed small animal-keeping mobile groups. Where they appear at the same time, the kurgan is always situated on top of a settlement, indicating that they followed Baden and represented a somewhat higher social power and belief system. Apart from burials, no Yamna settlements are known in Hungary, so it is unknown whether they were situated close to the kurgans or somewhere else entirely (Horváth 2016).

Most kurgans are located on the plains, and a smaller portion appear in neighbouring hills and mountains, while in unfavourable areas of sand dunes (Nyírség Region, Danube–Tisza interfluve) kurgans are virtually absent. The highest density of mounds appear in alluvial and loess plains rich in active and abandoned river channels, usually on natural levees, sometimes concentrated along streams forming small or large clusters. Their distribution usually follows a curved line, and vertically mounds usually appear above a certain elevation corresponding to flood-free levees and small aeolian dunes (Tóth, Joó, and Barczi 2015).

In contrast to their compatriots around the Lower Danube and Moldavia, settlers from the Carpathian Basin applied reed mats, textiles, leather, and even furs (possibly even felt and carpets) for the pit walls and floors, and these have been documented outside the grave pit. These colourful decorations – despite the poorly furnished graves and generalized lack of accompanying grave gifts – must have played a distinct role in the Yamna society, as well as the importance of colour combinations and pattern as emplematic and symbolic signs based on associations. Burial chambers prepared in this way were covered with wooden beams, planks, or logs, reminiscent of the few cases in Bulgaria and Romania which show big stone slabs covering the grave pit. All of this is compatible with the importance of the *domus*-idea, as are additional wooden posts, stone frames, and fireplaces or hearths attached (Heyd 2011).

Sizeable concentrations of tumuli are found in steppe areas around the Middle and Upper Danube and its tributaries—such as the 8,000 km^2 wide Little Hungarian Plain (or Little Alföld)—as well as in neighbouring forest-steppe regions close to the Danube, representing a gradual adaptation of Yamna settlers to the forest-steppe region (Horváth et al. 2013; Horváth 2016). Some distant settlements to the west show strong hints of the Yamna culture, such as concentration of tumuli, elements of Yamna burial customs, anthropomorphic statue-stelae, and artefacts with eastern links or origins. These groups include, for example, the north/north-central Middle Elbe–Saale area of east Germany with steppe vegetations, in the shadow of the Harz mountains, or a stripe in the foreland along the east Carpathians and southeast Poland, the border between Romania, the Ukraine, and Poland (Heyd 2011).

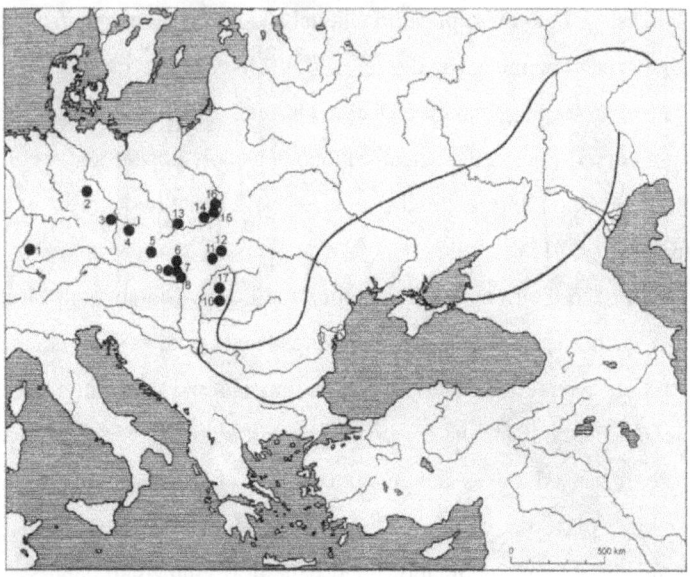

Figure 30. Distribution of artefacts and customs related to the Yamna culture (approximate border delineated up to the Tisza River), after Bátora (2006), fig. 15: 1 – Königshofen, 2 – Nohra, 3 – Kbely u Prahy, 4 – Cerhenice, 5 – Suchohrdly, 6 – Jelsovce, 7 – Nitra-Cermán, 8 – Nitra-Dolné Krskany, 9 – Sala, 10 – Kétegyháza, 11 – Kost'any, 12 – Lesné, 13 – Kietrz, 14 – Miemowo, 15 – Samborzec, 16 – Żuków, 17 – Püspökladány.

Many neighbouring regions with similar environment and landscape are thought to have been likely targets of that westward Yamna migration, although they have not yet yielded archaeological records, such as Black Sea shores of Bulgaria, East Thrace, modern Turkey, and northeast Greece south of the Rhodopes mountains. Interesting are the isolated findings of Yamna material culture in Corded Ware territory to the north (Bátora 2006), probably representing trade contacts or *vanguard* settlements (Figure 30) before the evolution into (and explosive expansion of) Bell Beakers.

An interesting finding is the discovery of a Yamna-like kurgan in Valencina de la Concepción, in southern Iberia (ca. 2875 BC), below which was the body of a man buried with a dagger and Yamna-like sandals, and decorated with red pigment just as Yamna dead were[14]. This suggests that the ideology, lifestyle and death rituals of the Yamna could run far ahead of migrants. The distribution of Yamna findings along the Danube, in central Germany, and up to Chalcolithic Iberia (Suppl. Fig. 10.B) before the emergence of East Bell Beakers—whose emergence happened roughly around the same areas (see below *§VII.7.2. East Bell Beaker group*)—suggests a complex framework of vanguard settlements and intense exchange contacts in central Europe.

Different from all these attested or supposed Yamna territories are some early samples of tumuli of mixed culture (e.g. those including cremation, or foreign material culture, or avoiding certain typical Yamna rites), which may point to Yamna influence on adjacent territories or local 'kurgan' cultures extant from the evolution of the previous Suvorovo–Novodanilovka expansion (see *§IV.2. Khvalynsk–Novodanilovka*).

Remains including a Coţofeni vessel from a Yamna grave in the Dniester (dating to the beginning of the Yamna migrations), and a typical Makó handled pot from Sofievka on the left side of the Dnieper (dating to the mid–3rd millennium) point to Yamna settlements closely connected to the core Yamna territory, thus considered an extension of their normal roaming area, and keeping a close contact among different groups (Heyd 2011).

Close contacts with adjacent cultures can also be seen in the Hungarian group, where for example herders from the Lizevile group in Transylvania seem to have used and economic model of transhumance, with livestock passing the winter and spring in the milder regions of the Great Hungarian Plain, as revealed by certain foreign tumuli in Yamna territory. Such regular visits increased the likelihood of these transhumant herders becoming integrated locally, and during the second quarter of the third millennium the internal coherence of the Yamna ideology had already diminished, which

---

[14] Comments by Volker Heyd (2019).

allowed other Lizevile and other herders to step in and take over locally, initially on a seasonal basis, and then permanently (Gerling et al. 2012).

VI.1.3. The Yamna package

The so-called "Yamna package"[15] (Figure 31) includes eleven components common to the initial western migrants (Harrison and Heyd 2007):

A) The social sphere:

1.   The most important and visible is the round barrow as a personalised monument. Emblematic symbolic signs based on associations, using patterns and colourful decorations, often combining two or more different colours, apart from reed mats, textiles, leather, and even furs for the pit walls and floors, even outside the grave pit, and possibly felt and carpets. Burial chambers are then covered with wooden beams, planks, or logs. These are often combined in Bulgarian and Romanian groups with an anthropomorphic stela covering the grave pit (none are known from Serbia or Hungary). All this reinforces the importance of the *domus*-idea of the tumuli.

2.   The single burial with a typical supine position on its back with flexed legs, usually upright (possibly to the side or in frog-position after a process of decay), often covered in red ochre, in a deep rectangular pit. The most common orientation of pits and skeletons is east–west, with heads in the west, but other directions are also attested.

3.   Social position and gender are systematically marked. Most burials are of adult males, and their percentage is higher in primary graves (so probably very much a *masculine* society). The wooden wagon

---

[15] Package refers to the idea "that a recurring assemblage of well-defined artefacts and customs, visible in the archaeological record, matches an equivalent cluster of social habits, which identify a tightly-knit group. The term 'package' clarifies the idea that it is an arbitrary social choice to create insignia, and that their function is to make and maintain cultural boundaries with others who live in the same area. For this purpose, special clothes and dresses, or drinking and eating habits, are important. A cultural 'package' of this kind is an ideological statement, materialised in a special manner, so its transmission and adoption can be very rapid indeed." (Harrison and Heyd 2007)

marks an elevated social position in the north Pontic area; however, typical of west Yamna migrants are the poorly furnished graves and the general lack of accompanying grave gifts (with complete absence of weaponry and tools). The social expression in west Yamna is thus not manifested in grave goods, but firstly in the labour and communal exertion to erect a tumulus, and secondarily in the efforts to create a burial chamber, the new house of the dead.

4. The creation of a special status for craftsmen (especially metalworkers), especially widespread in the north Pontic region and western migrants. For the first time, metallurgists had a specific social status. The development of the so-called Circum-Pontic Metallurgical Province is also associated with the spread of Yamna—taking over the previous north Pontic industry (see *§IV.3.1. Metalworking*)—including the wide distribution of new methods of copper and arsenical bronze metallurgy, and a set of bronze objects.

5. Hoarding metal objects begins again in steppe cultures, with hoards of shaft–hole axes. Furthermore, the deposition of lumps of ochre in the graves and the fewer secondary burials cut into existing tumuli are typical of the Carpathian basin. This is useful when distinguishing Yamna burials of the Carpathian group from those on the lower Danube and the Prut.

B) The technological sphere:

6. Re-establishment of metallurgy of gold and copper, following a long decline after 3500 BC, but with a different technology of smelting, working and casting in two-piece stone moulds, or 'Caucasian metallurgy' (Sherratt 2004).

7. New weapon designs in copper: the single-edged shaft–hole axe, and the tanged metal dagger.

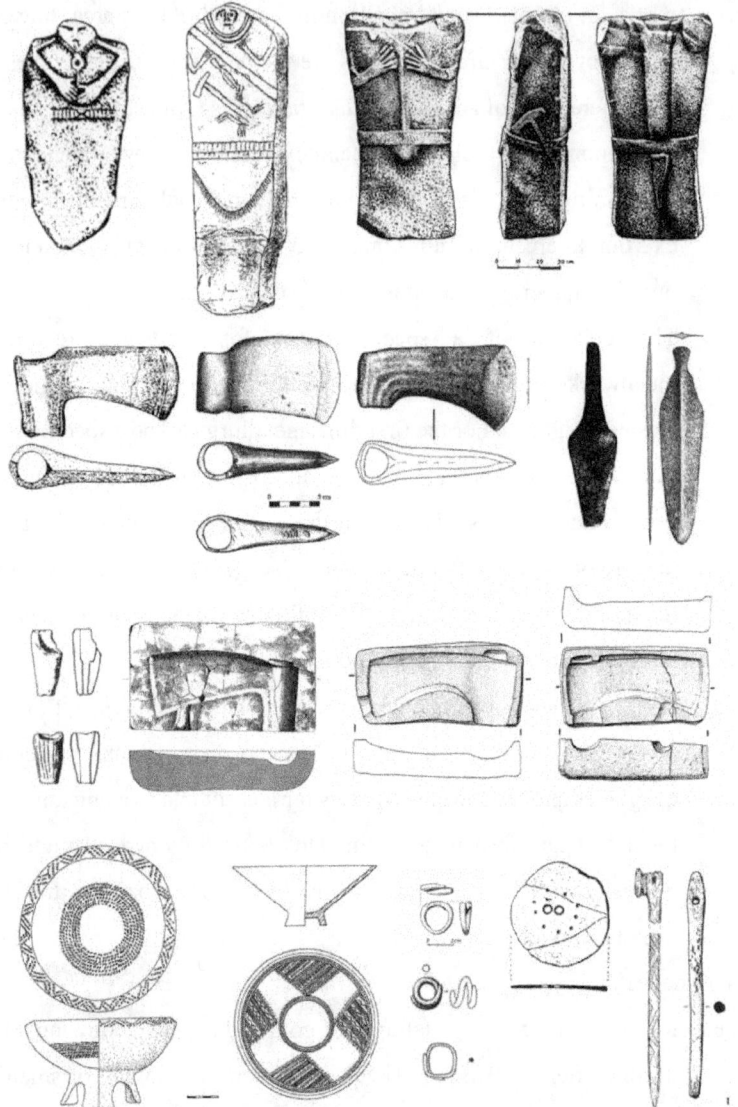

Figure 31. Examples of the western 'Yamna Package' material culture. Image modified from (Harrison and Heyd 2007).

C) The economic sphere:

8. The domesticated horse features importantly in a dedicated pastoral economy which raises herds of cattle, and perhaps flocks of sheep for wool. Domesticated horse is documented since the EBA on both sides

of the Carpathians from bone cheek-pieces, both as transport and as traction animals (Boroffka 2013).

9. Wooden wagons placed in graves as social markers; the westernmost example is the ox-pulled wagon grave of Placidol in northern Bulgaria.

10. The custom of using simple golden, electrum or silver hair rings, a distinctive bone toggle, and decorated bone discs, whose distribution covers all regions of western Yamna. Common adornments are also necklaces and chains of beads or perforated teeth.

11. Widespread use of cord decoration on pottery; the common cross-footed bowls copy models on the eastern Pontic steppes.

The use of pottery in western Yamna points to the import from neighbouring archaeological cultures, such as Coțofeni III vessels in Tarnava; Cernavodă II vessels in the lower Danube; Vučedol-like vessels in Romania and Bulgaria; and Makó and Lizevile-like vessels in Hungary. Most 'original' Yamna vessels include cord-impression techniques, and especially interesting is the classical beaker-like vessel widely distributed in the western regions, as well as the recurring decoration motif of triangles, fringes, or long triangles intermingled with each other, a specific decoration known from the north Pontic area (Heyd 2011).

Animals bones found next to the burial chamber, or part of the meals consumed during the ceremonies found around or on top of the graves, include dogs, sheep–goats, cattle, and horses (which hint at their relevance in the Yamna subsistence economy), and also hunted animals such as deer and birds (Heyd 2011). An important part of the industry related to pastoralism was the production of leather and wool. Examples are found, apart from western settlers, in Kalmykia, where wool and leather are widely used for the production of underlay, pillows, and clothes, with weaving skills showing up in good quality mats; or in Mikhailovka II-III, which shows a high level of leather production, as well as ceramic spindle whorls and weights. Based on scarce findings from Eneolithic steppe cultures, and on the analysis of Khvalynsk posttery ornamented with imprints of textile goods, Yamna findings likely show ultimately a great degree of continuity of an ancestral tradition (Morgunova and Turetskij 2016).

The interaction with previous cultures from south-east Europe may have been resolved in different ways, either violent confrontation, peaceful interaction, or neutral ignorance of each other. Since Yamna settlers occupied the steppe habitat, most economically important territories from neighbouring cultures would have been spared, at least initially, triggering mostly cultural interaction. However, communities derived from the small Suvorovo–Novodanilovka groups that settled the region may have entered into direct contact, which would have been resolved either violently or, perhaps, with rapid assimilation due to the similar economic/social background with comparable lifestyles (Heyd 2011).

Violence and raids must have been present with neighbouring cultures, though, and perhaps the building of a defence-like chain of hill forts along the south shore of the Danube by the Vučedol culture points to such contacts, although this is not the only interpretation possible. On a wider scale, the expansion of the Yamna culture begins a true horizon of transformation and cultural change in many European regions (Harrison and Heyd 2007).

VI.1.4. Volga–Ural region

Unlike western Yamna, where cattle dominates the diet and funerary rites, eastern Yamna in the Volga–Ural and in the northern Caucasian–northern Caspian steppes show a subsistence economy continuing the previous period, based on mainly sheep–goat and (less prominently) cattle, based on remains found in grave sacrifices (Shishlina 2008). The early Volga–Ural Yamna culture is represented by some settlements, small kurgans (ca. 20–25 m in diameter), and Repin-type pottery. The classical or developed phase is represented by the "unification of the funeral ritual, round bottomed pottery, the disappearance of settlements, and the prevalence of wheeled transport" (Morgunova 2002).

Differences between Yamna culture of the Volga–Ural interfluve and west Yamna groups are observed at both the social and economic levels. The traditional development of hereditary social strata in the Volga–Ural region was increasingly based on specific regional developments, such as the common interest in supply of metal objects and wooden products of cattle farming groups, which contributed to greater mobility of the nomadic pastoral population (Morgunova and Fayzullin 2018).

Such specialised production made it possible to raise the prestige of a given activity and individuals producing the necessary vehicles, tools, and weapons. The role of priests and producers (carpenters, blacksmiths) stands out by their unconventional burial rite: isolated skull burials and dismembered sacrifices the former, weapons and tools the latter, as in middle-aged men's burials accompanied by sets of tools for woodworking (axe, adze, big knife, gouge, pin, chisel) in some barrows of the Orenburg oblast. The lack of specific warrior burials points to the likely participation of the whole adult society in battles (Morgunova and Fayzullin 2018).

The construction of monumental kurgans with rich assemblages—such as those of Utevka I, Bodyrevo IV or Krasnosamarskoe IV, or the individual kurgans of Shumaevo II, Kalmytskaya Shishka, Dedurovskiy Mar—support the existence of ruling leaders among the elite, an aristocracy capable of: controlling competition (or supporting alliances) regarding territory, water resources, or raw materials; guiding the tribes; unifying their ritual and promoting the erection of sacred places; and enabling the expansion of homogeneous cultural elements through a huge territory. These elites probably concentrated economic–administrative, military, and religious functions under their charismatic leadership, which would become hereditary, evidenced by the presence of children burials among a majority of adult male elite burials (Morgunova and Fayzullin 2018).

The material culture of the Volga–Ural region (Figure 32) shows a clear connection with that of further eastern groups, including Afanasevo in the Altai region. This connection is evidenced by pieces of copper-containing sandstone from the southern Urals. A Yamna miner was buried in a mining pit ca. 3000 BC in the Kargaly copper ore field, located beyond the headwaters of the Samara River, in south-eastern Kazakhstan. Substantial deforestation near the ore field suggest large-scale copper-ore mining in the Kargaly area, with important mining and smelting operations during the early Yamna period (Parzinger 2013), incrementing in later periods (see *§VII.2.1. Poltavka* and *§VIII.18.1. Sintashta–Potapovka–Filatovka*).

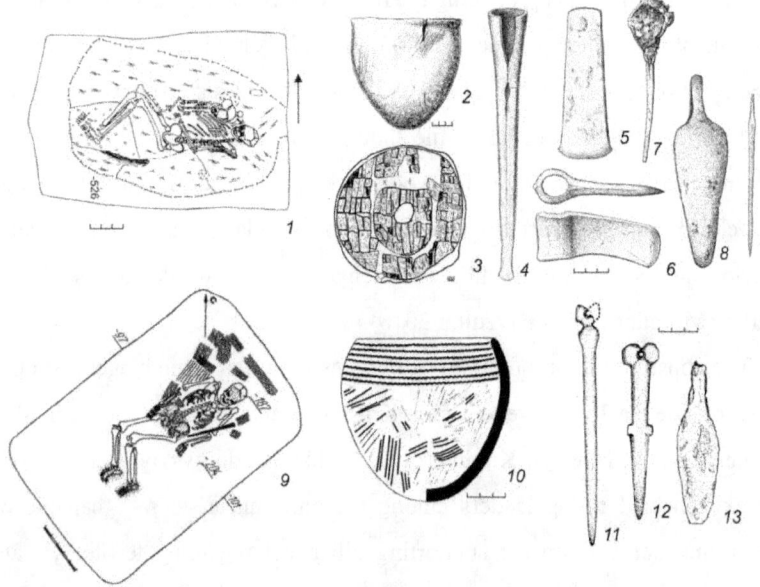

Figure 32. Material culture of Yamna on the Cis-Ural region (after Larin 2005). From Morgunova (2014). Compare with Afanasevo-type materials (see above).

vi.1. Disintegrating Indo-Europeans

Yamna samples had been described as being mainly composed of EHG:CHG ancestry (Jones et al. 2015; Lazaridis et al. 2016). Nevertheless, based on the ancestry of Afanasevo samples without EEF, and on later Yamna samples with EEF contributions, the late Repin expansion to the north Pontic area must have caused a mean EEF increase of ca. 15% to their typical Steppe ancestry inherited from the Khvalynsk–Novodanilovka expansion, varying from a minimum in the east (ca. 13%) to a maximum in the west in Hungary (ca. 17%), with intergroup differences not statistically significant (Suppl. Graph. 7). Further external contributions are found from a source related to Eneolithic Caucasus individuals in Yamna samples from the northern Caucasus (ca. 2900–2400 BC), and especially in a Yamna outlier from Ozera in the north Pontic area (ca. 40%), apart from later Catacomb samples from the same area (Wang et al. 2019), which support admixture with locals through exogamy.

The EEF ancestry found in Yamna points to a mix of NWAN (ca. 80%) and WHG (ca. 20%), and groups as distant from each other as Globular Amphora and Iberian Chalcolithic work as proximate surrogate populations for this kind of contribution (Wang et al. 2019). A similar ancestry may probably be found in Middle and Late Neolithic farmer groups from east-central Europe, such as Funnel Beaker and other post-LBK groups near the Balkans, and perhaps certain late Trypillian groups of the north Pontic forests, based on their close interaction with TRB—although samples from the Verteba cave have shown more EHG contribution (see *§v.6. Late Uralians*). EEF ancestry is also found elevated in Corded Ware samples, possibly up to 50% in certain groups (Wang et al. 2019), with a great part probably due to north Pontic-related Eneolithic interactions.

Compared to the Eneolithic Volga–Ural population, which had probably little or no EEF ancestry, the appearance of this ancestry in Yamna suggests thus an admixture of expanding late Repin groups from the Middle Don with Eneolithic populations of the north Pontic steppe and forest-steppe areas, which are also the likely source (or one of the main sources) of this ancestry found later in Corded Ware migrants. This admixture was probably driven by exogamy during the expansion of the late Repin strict patrilineal society, dominated by male elites, evidenced by the expansion of an overwhelming majority of R1b1a1b1-L23 lineages with Yamna. This is supported by the finding of late Sredni Stog samples as one of the best proxy populations (together with GAC and Iberia Chalcolithic samples) for the extra EEF ancestry found among Yamna peoples [16] . On the other hand, the homogenisation of this new EEF ancestry among all Yamna peoples—with no statistically significant differences between groups with the current number of samples—supports additional intense contacts and "internal exogamy" between Yamna clans of both western and eastern groups, and potentially also the expansion of 'admixed' late Repin settlers to the east.

The eastern Yamna or Volga–Ural–North Caucasian group includes the Volga–Ural variant between the Volga and Ural rivers (with Lower Volga, Middle Volga, and Ural regions), and the North Caucasus variant (right bank of the Volga River region, Kalmykia, and northern Caucasus steppes until the Terek River). Among the dialects spoken in this region was probably the ancestor of Indo-Iranian. Most reported Y-chromosome haplogroups of Yamna samples are from sites in Samara, Kalmykia, and northern Caucasus areas ca. 3100–2500 BC (Haak et al. 2015; Allentoft et al. 2015; Mathieson et al. 2015; Wang et al. 2019): sixteen out of eighteen are of haplogroup R1b1a1b1-L23, with further reported subclades mainly from the R1b1a1b1b-Z2103 trunk, except one sample from Lopatino II, in the R1b1a1b1a-L51 line (see *§v.7. Common Indo-Europeans*). A sample from Karagash in Kazakhstan, of subclade R1b1a1b1b3-Z2106 (de Barros Damgaard, Marchi, et al. 2018),

---

[16] In terms of specific groups, Yamna individuals may be modelled as from samples from the Northern Caucasus Piedmont and the Volga–Ural area, with good fits being Progress Eneolithic (ca. 40–50%)[+] and Khvalynsk–Samara (ca. 35–40%)[+], as well as variable contributions of 'western' populations (ca. 5–15%)[+], with the best proxies for the latter being GAC, Trypillia, Ukraine Neolithic (or Mesolithic), or Varna samples. These 'western' proxies coincide with those of the late Sredni Stog population of the north Pontic area (being themselves also a good fit for Yamna groups)—who do not derive sizeable ancestry from Khvalynsk (see above *§v.6. Late Uralians*)—and are also found elevated in Corded Ware and derived groups (see below *§vi.3. Disintegrating Uralians*).

further supports the connection of eastern Yamna groups of the southern Urals with Afanasevo.

The western Yamna or Southern Buh–Lower Don group included the Don River variant (Lower Don from the Ilovlya River to the mouth of the Don River and valley of the Western Manych River); the Siverskyi Donets variant (right bank of the Siverskyi Donets River between modern Kharkiv and Luhansk cities); the Azov variant (steppe of the Northern Azov Sea coast); the Crimea variant; the Lower Dnieper variant (from the Orel River and the Inhulets River to the Black Sea coast) with the Bilozirka, Nikopol, Kryvyi Rih, Dnieper "Stone stream", Left Bank of the Dnieper, and Black Sea coast regions; the North-Western variant (steppe and forest-steppe borderland on the Middle Dnieper and to the west from it), and the South-Western Variant (between the Buh and Danube rivers).

Local groups of the north Pontic steppe include the Donetsk group, the Middle Dnieper group, the Lower Dnieper and the Azov–Crimea groups, and the Southern Buh group. The kurgans between the Dniester and the Prut Rivers received influences from the main neighbouring regions—such as EBA of central and south-east Europe, Globular Amphora, and Corded Ware, Folteşti 2 and Coţofeni cultures—and two cultural-chronological variants are described: the Early Dniester variant, and the Late Budzhak variant (Rassamakin and Nikolova 2008).

Western Yamna and Danube groups probably spoke the ancestral language to both North-West Indo-European and Palaeo-Balkan languages. While there is scarce data on Y-chromosome haplogroups, based on later European samples it is very likely that these territories hosted R1b1a1b1a-L51 subclades—in particular R1b1a1b1a1a-L151 (TMRCA ca. 2800 BC)—whose lineages are found centuries later spreading with East Bell Beakers (see *§vii.7. North-West Indo-Europeans*), and were probably in the majority among Pre-North-West Indo-European-speakers. It is also likely that West Yamna hosted

R1b1a1b1b-Z2103 lineages, probably associated mainly with Palaeo-Balkan-speaking clans.

Two samples show I2a1b1a2a2a-L699 lineages (formed ca. 6800 BC, TMRCA ca. 4500 BC, one in Kalmykia and one in Bulgaria (an outlier with contributions from NWAN-related ancestry), which supports the presence of this lineage among certain Yamna clans (Mathieson et al. 2018). Similarly, the presence of R1b1a1b-M269 subclade R1b1a1b2-PF7562 among modern populations in the Balkans, Central Europe, Anatolia, and the Caucasus (Myres et al. 2011; Herrera et al. 2012), speaks in favour of clans of this lineage also expanding with late Repin/early Yamna settlers, or potentially of remnant populations of the earlier Suvorovo–Novodanilovka migrations who were part of (or pushed by) expanding Yamna settlers. The finding of hg. R1b1a2-V1636 in Sharakhalsun (ca. 2780 BC) may also suggest that this lineage expanded with Yamna, or alternatively that it belonged to a remnant population of Maikop in the Northern Caucasus Piedmont eventually integrated in Repin or Yamna (see above *§v.2. Early Caucasians*).

The migration of Yamna settlers into Hungary appears to be homogeneous at first, with early samples clustering closely to other west Yamna samples. Two late samples from Hungary show already increased EEF ancestry (ca. 10% more than Yamna, with a close source for this ancestry found probably in neighbouring Baden-like Hungarian samples), at the same time as Catacomb individuals also received further EEF-related contributions (probably from the north Pontic area), whereas a late Yamna sample shows no marked change, and Poltavka shows even less EEF ancestry than preceding eastern groups (Wang et al. 2018). This evolution ca. 3100–2500 BC suggests an initial homogeneous expansion, where Yamna clans either kept close contacts with other steppe clans or displayed little exogamy practices with the local groups they encountered; and a later gradual regional isolation, including interaction and admixture with different local groups.

VI.2. The Transformation of Europe

The so-called "Transformation of Europe" should probably be described as continuing the expansion of the 'steppe package' into western Europe, with expanding European cultures (such as northern European Globular Amphora or Corded Ware cultures, Balkan cultures like Makó/Kosihý–Čaka /Somogyvár) which share common traits (Heyd 2011):

- An essential individualisation in burial ritual, up to individual graves: in wide parts from north-west and west Europe this is associated with the megalithic world, and—still in collective graves—the appearance of individualising marks and personal possessions.
- Mound building as personal memorial over individual graves.
- Gender separation, not only in specific rituals, but also in gender-specific offerings.
- Monumental anthropomorphic stone stelae, appearing first during the Middle Eneolithic in Lower Mikhailovka I, but later widespread in a transect from east to west Europe, from the Kemi-Oba group (succeeding Lower Mikhailovka around Crimea) to Iberia, and being especially relevant in northern Italy and southern France.
- Internationalisation of certain goods, visible e.g. in the Grand Pressigny daggers.
- Symbolic prestige and status objects, also represented in stelae, like copper Remedello daggers and double spirals.
- Assignment of value to flint items according to the raw material they are made of (viz. valorisation of axes made of banded flint in GAC, flints for production of axes in CWC, spread of blade daggers of Grand Pressigny flint in western Europe, etc.).
- Differences in technological advances in the Final Eneolithic become smaller between regions and within them (compared with the production of sophisticated tools by specialised craftsmen in earlier phases).

Reasons for this transformation—as evidenced by the steppe origin of these cultural traits—lie in the reaction of central and western Europe to the expanding economic innovations from the south-east (during the second half of the 4th millennium), associated with the further expansion of the "Secondary Products Revolution", which involves the introduction of the wheel and wagon, plough, sheep wool, and probably also alcoholic beverages, first for the elites, then also to the whole agrarian population. Another reason could have been the influence of Aegean EBA cultures and their intensifying networks of communication, exchange, and even commerce (Heyd 2011).

However, the most important reason for this successful widespread adoption over almost all Europe (especially regarding late influences) must have been the irruption of thousands of Yamna migrants after 3100/3000 BC into the Carpathian Basin, which triggered a 'domino effect' in northern and central Europe, expanding further with CWC, GAC, Baden, and other central and west European cultures in contact with the new immigrants. For example, flint daggers – replacing copper daggers in the mid–4th millennium – reached the whole Balkans, the Aegean and Anatolia in the first quarter of the 3rd millennium (Heyd 2011).

VI.2.1. Tumuli

Western Yamna tumuli were not the first to be erected in their settlement areas (see *§V.4. Steppe package*), but they represent the first real wave of standardised tumulus construction. In neighbouring cultures, sometimes from distant regions, round personalised tumuli emerge. Five areas can be listed around the actual Yamna distribution (Heyd 2011):

- East Banat and west/south-west Transylvania. Tumuli are small and low, stones are used in their construction, and culturally they belong to late Coţofeni and successive cultures such as Lizevile in the west, Şoimuş in the south-west. In other parts of Transylvania, tumuli are constructed as well.

- East-Slovakian burial mounds are ca. 350 tumuli found in the Carpathians of east Slovakia, showing an amalgam of an inner Carpathian culture and Corded Ware tradition originating from the north of the Carpathians, which dates them rather to the mid–3rd millennium, although some poorly equipped ones may be earlier.

- Dalmatia and the east Adriatic coast: a large tumulus province extending all along the Dalmatian coast into its hinterland, reaching into Bosnia and probably Albania, as well as North Italy and Apulia. These are part of the Adriatic province of the Vučedol culture.

- Transdanubia, south-west Slovakia, and the Austrian Burgenland show some coherent cluster of tumuli with other Yamna links, such as a copper dagger, and a hair spiral. Neighbouring the Little Hungarian Plain, culturally it belongs to Makó/Kosihý–Čaka, Vučedol, and early Somogyvár.

- The largest continuous tumulus zone, as in the case of the cord decoration, is the distribution area of the Corded Ware and Single Grave cultures of central and north-central Europe. These are small tumuli, probably receiving the idea through Austria and Moravia, through the northern Carpathians into Poland (the origin of the A-horizon, see *§VI.3.3. East-Central Europe and Globular Amphora*), or more likely south-east Poland through contacts with the rivers San, Prut and Dniester, and the Yamna there.

VI.2.2. Anthropomorphic stelae

Large anthropomorphic stone stelae seem to have first appeared in the Mikhailovka I culture in the second half of the 4th millennium. Mikhailovka I areas were replaced by the Usatovo culture (related to Trypillia), but its culture continued in the Kemi-Oba culture of Crimea. Carved stone stelae appear to have expanded in frequency and elaboration in both territories, and in part of the north Pontic steppes, after about 3300 BC (Anthony 2007).

Strikingly similar stone stelae appeared later in the Caucasus, Troy, and also in central and western Europe, and with special frequency in the Swiss Alps and in the Provence, with examples also in the Iberian Peninsula and northern Germany. A maritime route for some of these cultural expansions has been proposed, which would justify e.g. its early presence in Troy (Anthony 2007).

VI.2.3. Associated cultures

Mainly associated with funerary customs in the Yamna horizon, the use of other carved anthropomorphic stones seems to herald the influence of the Yamna culture in Europe. The first wave of the warrior ideology starts around the mid–4th millennium, probably coinciding with the expansion of late Repin / east Yamna settlers.

Rich single graves, daggers, flint and copper halberds, or anthropomorphic stelae are part of the new Mediterranean trends. Thus, pre-Beaker Italy shows the Gaudo culture (ca 3300 BC), the Remedello and Spilamberto culture (ca. 3400/3300 BC), and the potentially slightly earlier Rinaldone culture. *Statues-menhirs* appear in southern France, and Italian influence is felt in the Alps and south-eastern France in the late 4th millennium, and later in cultures of macro-villages appear in southern Iberia (Jeunesse 2015).

The building of tumuli, the enhancement of gender distinctions, and the internationalisation of special objects made of rare materials as status indicators are seen slightly later. This influence was seen in the Corded Ware/Single Grave culture in central and eastern Europe in the east, Vučedol in the western Balkans, Makó/Kosihý–Čaka/Somogyvár in the Carpathian Basin and even the early Bell Beaker culture in south-western Europe around 2700/2600 BC (Harrison and Heyd 2007). Stone stelae and figurines might have also been used quite differently, or for different purposes, in certain local cultures (Robb 2009; Díaz-Guardamino 2014).

Radiocarbon dates from the north Pontic steppe show the late presence of steppe material cultures in the Carpathian EBA (ca. 2500 BC), in the

Makó/Kosihý–Čaka/Somogyvár–Vinkovci, late Vučedol, and others like Schneckenberg-Glina III, Csepel, or Early Nagyrév. These cultures have been argued to form a cultural unity, and it is proposed that such influence may have come from Yamna settlers on the left bank of the Tisza River (Rassamakin and Nikolova 2008).

The appearance of the Classical Corded Ware culture from the Rhine to the Danube ca. 3000–2750 BC, apart from all these reasons, was facilitated by the previous expansion of the similar Globular Amphora culture, which must have worked as a catalyser, not only because of its similar regional expansion, but also because of its structural similarities with later Bronze Age stages. Globular Amphora was itself rooted in the previous TRB tradition, in the same region (Heyd 2011).

Another common Late Copper Age trait were the sets of weaponry that became associated with individual graves: the battle axe and flint dagger for the Corded Ware; copper shaft–hole axes for Makó, Vučedol, and related groups; and the bow and arrow and copper dagger for the users of Bell Beakers. This weaponry and its symbolism define the idealised image of the Late Copper Age warrior (Heyd 2011).

Signs of this transformation in south-west Europe, from Iberia through Atlantic façade to the Rhine delta, include scattered perforated battle axes of various styles in northern Iberia (3000–2500 BC), and daggers of flint and copper in collective tombs of central Portugal, the Algarve, and Andalusia (3000–2700 BC). The demographic or economic pressure of Yamna migrants must have been responsible for the events in southern and west-central Iberia that led to the creation of macro-villages, i.e. the migration from villages and hamlets into enormous settlements, with their satellites, outlying forts, and cemeteries of megalithic collective tombs (Heyd 2011).

In the end, supra-regional cultures superseding smaller, regional-based cultures of the earlier Copper Age represented a cultural phenomenon that united wide regions of Europe. Influenced by these European trends was born the Proto-Beaker package in west Iberia, expanding quickly into Central Europe, probably triggering cultural adoption, and accompanied only by minor population movements, if at all (Heyd 2011).

vi.2. Late European farmers

Three individuals from the Remedello culture, probably all of haplogroup I2a1a1a1-Y3992[+] (formed ca. 9400 BC, TMRCA ca. 6700 BC), and from Ötzi the Iceman, of haplogroup G2a2a1a2a1a-L166 (ca. 3500–3100 BC), all of northern Italy, show a high affinity with Chalcolithic samples from central Anatolia at Kumtepe. This affinity is higher between them than with earlier Anatolian Neolithic populations, which is against the interpretation of Remedello's ancestry representing a relict population stemming from Neolithic farmers (Hofmanova et al. 2016).

Because of their shared drift with CHG ancestry independent of steppe expansions, and because Kumtepe predates the northern Italian group by some 1,000 years, it has been proposed that they represent a more recent, yet undescribed, gene flow process from Anatolia into Europe. This Anatolian region shows a continued 'eastern' migration found in Anatolian Chalcolithic samples (Kilinc et al. 2016; Lazaridis et al. 2017).

Three Baden samples (ca. 3600–2850 BC) show no contribution of Steppe ancestry (Lipson et al. 2017), with one hg. G2a2b2a1a1c1a-Z1903 (formed ca. 6000 BC, TMRCA ca. 2400 BC), which—together with the genetic picture of Globular Amphora (see *§vi.3. Disintegrating Uralians*)—supports the cultural rather than demic diffusion of concepts related to the Yamna culture during the "Transformation of Europe".

Later cultures emerging in the Balkans near Yamna show contributions from the steppe, though: two of three samples of the Vučedol culture (ca. 2800–2700 BC) show Steppe-related ancestry over a mainly Balkan Chalcolithic population, with one sample from the Vučedol Tell of G2a2a1a2a-Z6488 lineage, and another from Beli Manastir–Popova Zemlja, Croatia (margins of the Vučedol area) of R1b1a2a2-Z2103 subclade (Mathieson et al.

2018). This supports the interpretation of (at least some) Balkan LCA cultures as a mixture of local and steppe populations.

VI.3. Classical Corded Ware culture

VI.3.1. Genesis of the Corded Ware culture

The origin of the Classical Corded Ware culture has been traditionally placed near the Volhynia–Podolia region, related to findings of Lesser Poland, Kuyavia, and adjacent regions of Ukraine and Slovakia, probably ca. 3000/2900 BC, and quite likely directly influenced by the *push* of the Yamna explosive migration to the west, but (at least initially) neither related nor in contact with it (Kristiansen 1989; Anthony 2007; Włodarczak 2008; Kadrow 2008).

The westward expansion of the Yamna culture along the Danube River, south of the Carpathian Mountains and along the upper Tisza River, put this culture in close contact with other "kurgan" cultural systems, south and north of the Carpathians. The Lesser Poland region found itself thus in close contact with communities characterised by new principles of social organisation and a new funeral rite. Around 2800 BC, these changes became evident in different regions of Poland, with the most numerous examples being documented in south-eastern Poland and Kuyavia (Włodarczak 2017).

The new Corded Ware material culture has no straightforward analogies in the world of the Pontic–Caspian steppe communities, though. To the north of the Carpathians, including the first examples of Złota and early Corded Ware, no graves indicating their relationship with communities of the steppe zone have been found. On the contrary, the funerary rites always display a local, central European nature (Włodarczak 2017).

Nevertheless, individual elements typical of the steppe do appear, emphasising the individual, maintaining specific rules of orientation and sharing features like the flexed position of the corpse and the deposition of drinking vessels, weapons, and other specific types of objects as grave goods.

The connection of the Corded Ware culture with Yamna or Bell Beaker and Balkan EBA groups occurs therefore through both the spread of a Yamna package, and the earlier spread of the so-called 'steppe package' (see *§V.4. Steppe package*).

The nature of economic activities of the different communities was variable depending on the environment, with coastal zones of the Baltic Sea, forests and lakeside zones showing an important role of hunting and fishing. However, while natural conditions determined the particular local adaptations, the overall economic structure remained usually dominated husbandry, including herds of bovine and sheep–goat. Even in zones with fertile soils, exploited agriculturally for hundreds or thousands of years prior to their arrival, there was a clear turn towards husbandry, which proved especially attractive to para-Neolithic communities of the not so fertile lands (Włodarczak 2017).

Animals were used for meat consumption, milk, wool, and also as pack animals. The importance of transportation is seen in the well-established roads at the time, and in the ease in travelling long distances. Palynological, zooarchaeological, and geological data, including some features of the material culture (like extensive circulation and short-term encampments along microregions, e.g. river beds) point to a mobile way of life associated with husbandry (Kadrow 2004). The mobility of Corded Ware settlers relied primarily on short-distance shifts, probably repeated multiple times, and was thus a continuation of a model from an earlier period of the Globular Amphora culture (Kośko and Szmyt 2004).

Regarding the impact of animal traction and the wagon, they are present in the archaeological record at least since 3400 BC, but they do not play any visible role in Corded Ware burial rituals, very much in contrast to the previous periods. Finds of horse bones are exceptional discoveries (Pospieszny 2015), unlike in the period before 3000 BC, and no evidence is found for an increased relevance of horse domestication during or in connection with the Corded

Ware culture There is thus no evidence for a widespread use of horseback riding (Włodarczak 2017).

The theories regarding the significant role of the horse in the economic, cultural, or even ethnic changes taking place in this culture are not confirmed Horse bones are not deposited in graves in any form, unlike commonly encountered bones of bovine, goats/sheep, and dog. Burials of horses would only appear later, during the Trzciniec culture (Włodarczak 2017). The economic importance of the domesticated horse was negligible, even lower significance than among Globular Amphora culture communities (Kośko and Szmyt 2004).

All this notwithstanding, the Corded Ware culture brings about a clear change in the structure of networks, a significant widening of scales, connecting formerly distant regions, with common practices and symbols widely exchanged and integrated into the local habits and discourses. This supra-regional network is the result of an increase in mobility, and probably an expansion of patrilineally-related clans (Włodarczak 2017).

VI.3.2. Single Grave

Some of the earliest radiocarbon-dated groups associated with the Corded Ware culture come from new single graves from Jutland in Denmark and Northern Germany, ca. 2900 BC. This Early Single Grave culture is associated with the appearance of individual graves (some time after the decline of megalithic constructions), composed of a small round barrow and a new gender-differentiated burial practice emphasising male individuals orientated west–east (with regional exceptions), combined with the internment with new local battle–axe types (Figure 33): A-Axe, boat-shaped battle–axes with an elongated rib on the upper surface (Furholt 2014).

Figure 33. Neolithic boat axe of the Single Grave culture, from Boberow. Photo by Wolfgang Sauber. Image from Wikimedia Commons.

A 'Corded Ware package', appearing ca. 2900 BC, and available at least partially on many burials, included artefacts such as objects for consumption of drinks (clay beakers or similar vessels) or equipment needed for battle or hunting (stone axe-hammer, flint knife, flint archery accessories such as bow or arrowheads), and less often ornaments made out of bone, copper and amber, as well as tools made of bone or flint (Figure 34). They are therefore an affirmation of battle, hunting and feasting, as well as libations (Włodarczak 2017). However, it is not until ca. 2700 BC when the pure A-Horizon of the Corded Ware group is seen in the region, unifying culturally the 'core Corded Ware province' formed by Jutland and Northern Germany, the Netherlands, Saale, Bohemia, Austria and the Upper Danube regions (Furholt 2014).

In central Europe, mounds had a diameter of 10–20 m and a height of 1.5–2.5 m, with an additional circular groove surrounding the centrally placed grave. The pit was often supplemented by a wooden structure, usually in the form of a box (in some cases by stone structures), and thanks to these additional structures the pit gained the form of a chamber with walls and a roof, "the house of the deceased", where grave goods were also deposited (Włodarczak 2017).

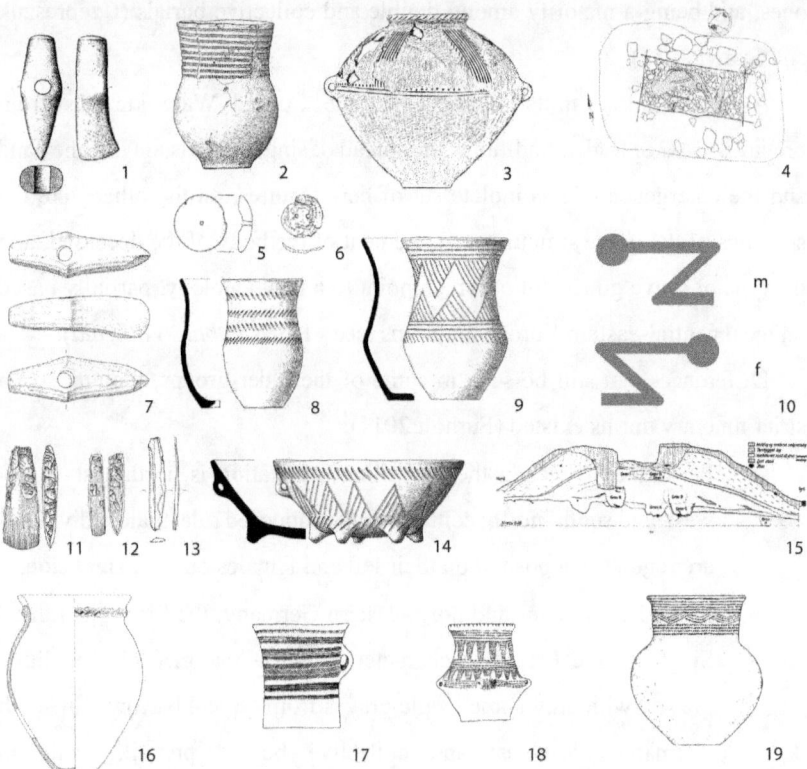

Figure 34. The "Corded Ware package", i.e. elements with a supra-regional distribution: 1. Battle–axe (Type A); 2. Corded beaker (type A); 3. 'Strichbündelamphora'; 4. Single burials below barrows (15), Gender-specific deposition rules (10); 5. Amber ornament disc (all Hübner 2005); 6. Bone ornament disk; 7. Facetted battle–axe; 8. Herringbone-ornamented beaker; 9. Triangle-ornamented beaker (all Dresely 2004); 11. (flint) axe; 12. (flint) chisel; 13. flint blade (all Hü bner 2005); 14. Bowl (Dresely 2004); 16. Short-wave-moulded 'Wellenleisten-'storage vessel (Strahm 1971); 17. Straight-walled vessel; 18. Amphora (both Matthias 1968); 19. Short-necked beaker (Włodarczak 2006). Image modified from Furholt (2014).

Single burials dominated, and the deceased were laid to rest in a foetal position on their side, with men usually on their right, women usually on their left, both looking to the south. The right to the single burial and to the 'grave goods set' (including vessels and weapons) was granted mostly to males (men, and sometimes children, believed to be an inheritance of warrior status), with females showing only some of the objects, and sometimes other more exotic ones, and being a majority among double and collective burials (Czebreszuk and Szmyt 2011).

All these burial traits distinguished the Corded Ware kurgans from previous GAC or Baden traditions: the spread of single burials, on the one hand, and the emergence of a complete set of new features, on the other, like the specifics of the grave structure, arrangement of the body of the deceased, and the type of grave goods, all of which point to a new ideology, partially based on local central-eastern European groups (see *§V.5.2. Lublin–Volhynia*).

Differences can still be seen in some of the older groups, though, so no strict funerary norms existed (Furholt 2014):

- In southern Sweden the prevailing orientation is north-east–south-west, and south–north; contrary to the supposed rule, male individuals are regularly deposited on their left and females on their right side.

- In the Danish Isles and north-eastern Germany, the Final Neolithic / Single Grave Period is characterised by a majority of megalithic graves, with only some single graves from typical barrows. In south Germany, west–east and collective burials prevail, while in Switzerland no graves are found.

- In Kuyavia (south-eastern Poland), Hesse (Germany), or the Baltic, west–east orientation and gender differentiation cannot be proven statistically.

Interesting is the presence of copper-rich burial inventories (some of which are imported from Yamna or Yamna-related cultures), in contrast to neighbouring cultures in the same region (Heyd 2004): for example, central European CWC burials do not show signs of metalworker burials, while west Yamna and later Bell Beaker or Makó/Kosihý–Čaka show a higher specialisation in metallurgy (and still there are proportionally, in comparison with south-eastern Europe, not many metalworking sites).

On the other hand, graves of silex workers and silex commerce is everywhere to be seen associated with the expansion of the culture and its daggers (flint daggers had replaced the earlier, Eneolithic prestige copper daggers from ca. 3600 BC onwards), including west France (with Grand Pressigny type), north Italy (Monte Lessini) and to the north, west of the Rhein (with Bavarian Plattenhornstein type). This exchange network is also seen in the widespread findings of axes from Silesia, or amber from the Baltic region (Heyd 2004).

The presence of semi-products in burials (e.g. flint flakes for arrowhead production), used in activities related to flint acquisition and processing, suggests that this daily routine activity (and not those of metallurgy or blacksmithing) became accented in the Corded Ware culture's funeral ritual (Heyd 2004).

All this points to the continuation in metal-poor central European Corded Ware groups ca. 2900–2500 BC, from an earlier tradition from copper-rich groups, such as those close to the north Pontic steppe (see *§V.6. North Pontic area*). There was, however, a traditional centre of metallurgy around the Podolia and Volhynia regions from ca. 4000 BC (possibly around a large deposit of virgin copper in Volhynia). This centre had potential connections to the Carpathian-Danube circle, based e.g. on the production of "willow–leaf "-shaped jewellery and other types of jewellery, Bytyń-type flat axes with flanges (and later Stublo-type axes from the Vučedol tradition), etc. which eventually served the Corded Ware culture when it occupied the region (Klochko 2013).

The necessity of horizontal social mobility and exchange drove a change in settlement pattern in Central Europe, from a domination of villages to a domination of farmsteads and small hamlets, which took place at the latest around 2800 BC, reflecting a profound change of identity and ideology (Müller et al. 2009): from collective to individual; from the village community to the core family; from regional political organisation to the dispersed identity of far-distant social units (Harrison and Heyd 2007).

As an example, a typical CWC site at Wattendorf-Motzenstein could have at any given time a mean of 4 huts with ca. 20 inhabitants, with each household representing an independent economic unit. Investigation of the activities of the site shows that its inhabitants (Müller et al. 2009):

- undertook intensive cereal and pulse cultivation with a household-orientated processing regime;
- produced essential stone, bone, and ceramic tools in the household;
- used a common area for cultivation, with each household needing ca. 1.5 ha of arable land, and the whole settlement around 9 ha to cover the population needs;
- carried out stockbreeding of cattle, sheep, goat, pig, and horse, and practised hunting;
- collected the necessary raw materials within a distance of less than 20 km;
- occasionally took part in longer-distance expeditions for the acquirement of raw materials;
- planned the settlement layout, with huts arranged along constructed pathways;
- celebrated ritual activities centred on a pinnacle dolomite rock where millstones, broken sherds, and animal bones were deposited under the visual sign of a wooden post, and used miniature wheels and axes for ritual activities amongst everyday life, and within the household.

VI.3.3. East-Central Europe and Globular Amphora

The oldest Corded Ware vessels (the A-Amphorae, which define the "A-Horizon" or "pan-European horizon" of the CWC) come probably from the Złota (or a related) group in Lesser Poland, where a mixed archaeological culture connecting Funnel Beaker, Baden, Globular Amphora and Corded Ware appears ca. 2900–2800 BC. The origins of cord decorations are probably to be found in steppe cultures of the Dniester region, possibly in the Usatovo culture, although it may have appeared under pressure from the Coțofeni culture in the southern parts of the north Carpathian mountains (Furholt 2014).

No cultural (typological) break is seen between earlier Globular Amphorae and the first Corded Ware Amphorae, but rather a *continuum* of traits and characteristics among the recovered vessels (Figure 35). This strengthens the connection of Corded Ware with the Globular Amphora culture. The A-horizon expanded thus probably from Lesser Poland ca. 2800–2700 BC, as seen in the synchronous appearance in local contexts of Poland and neighbouring countries. Compared to the earlier periods, the range of forms used narrowed and was mostly limited to amphorae, beakers, and pots, with large storage vessels no longer used. A pot with a standardised, S-shaped profile, decorated with plastic bands (or various kinds of imprints) became the predominant vessel (Furholt 2014).

Figure 35. Examples of pottery of the Złota culture, showing similarities with Globular Amphorae and Baden cultures of east-central Europe. After Włodarczak (2008).

The first burial mounds associated with Corded Ware, and some of the known settlements of Lesser Poland and the lowland can also be dated to this period. The kurgan-related funeral ritual is associated with Złota cultural communities: single burial graves, along with the habit of interring the deceased in multiple burial graves, but emphasising their individual character by careful deposition of the body and personal nature of the grave goods (Włodarczak 2017).

Additionally, grave goods from Złota groups also display a transitional nature, with materials and stylistics belonging to an older system (e.g. amber products); and others correlated to the 'new world', such as flint products made of the raw materials typical of Lesser Poland's CWC, copper ornaments, stone shaft–hole axes, bone and shell ornaments, and characteristic forms of vessels like beakers and amphoras. Military goods, which would become prevalent in later periods, are present in a moderate number, compatible with their lesser importance (Włodarczak 2017). Interestingly, numerous amber products are found in graves associated with the Globular Amphora culture and Złota groups, but its importance diminished in barrows of the Corded Ware culture, only to increase again during the Bell Beaker period.

Catacomb graves, with an entrance pit and a more extensive niche, and a narrow corridor leading to a vault with interments and grave goods, are also found in this old period (ca. 2900–2800 BC) in three large burial grounds, Grodzisko I, Grodzisko II, and Nad Wawrem, in the vicinity of Sandomierz (Figure 36). Limestone lumbs were used for the construction of a barricade at the entrance of a catacomb, and for making a kind of lining on the floor where the deceased were then laid. There are individual cases of application of ochre and deformation of skulls. These graves were the standard form of burial in south-eastern Poland, and are known in greater number on the left bank of the Vistula River, as well as on the Lublin Upland and western part of Volhynia Upland, and loess uplands within the Subcarpathian zone. Their rich assemblages– including large group of features with metal items – distinguish these communities from others in Central Europe (Włodarczak 2017).

The Złota culture depicts thus the most likely transitional picture between local Late Eneolithic GAC groups and the emerging Final Eneolithic Corded Ware culture, with the development of an original funerary rite, unique material culture, and multi-directional, long-distance contacts: e.g. the import of amber from the north, vessels imported from Baden-related communities, catacomb graves probably connected with areas of eastern Europe (Włodarczak 2017).

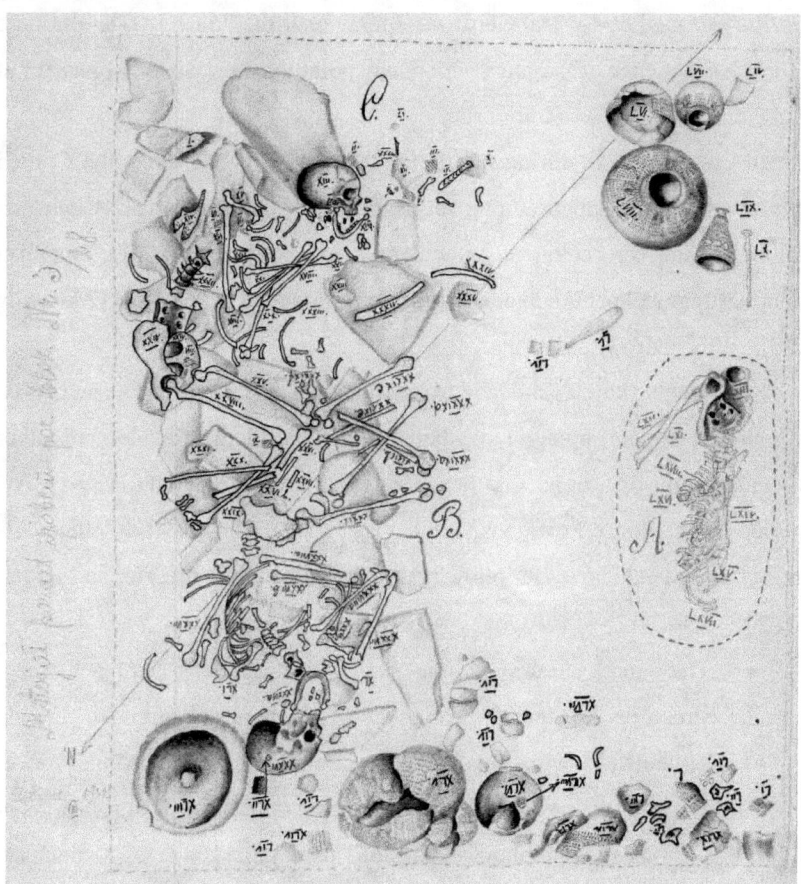

Figure 36. Original drawing of burial no. 325 from cemetery "Grodzisko I" at Złota. 1920s. (archive of State Archaeological Museum in Warsaw). Modified from Włodarczak (2017).

One of the regions with most kurgan findings known is the uplands of south-eastern Poland, probably from an old phase of the culture ca. 2800–2600

BC. This region includes the western Lesser Poland loess uplands, the Carpathian foothills and adjacent Sandomierz Basin, as well as the western edge of the Volhynian Upland, Lublin Upland, and Roztocze. Kurgans were an expression of egalitarianism, but it did not apply to the entire group, only to a specific part; furthermore, kurgan burials of women are rare, and the rich assemblage of one of them suggests that only special women were honoured that way (Włodarczak 2017). In other regions of central Europe, like the territory formed by Central Germany, Moravia, or the Polish Lowlands, there is a small presence of kurgans, which may point to a permanently present but not commonly followed burial rite.

In Lesser Poland, during the first 300 years of its existence, the Corded Ware culture developed among the settlements of the agrarian Baden and Globular Amphora cultures, without mixing (Włodarczak 2001), among a complex regional picture that had formed during the 4th millennium (Zastawny 2015; Wilk 2016).

Settlements show a tendency to smaller sites with short-term occupation, which does not constitute a radical change in the settlement model, but rather a continuation of a trend that began earlier during the late TRB and GAC periods in the Polish Lowlands (Włodarczak 2017). Corded Ware settlements in Lesser Poland show the following characteristics (Czebreszuk and Szmyt 2008):

- They mark a turning point in the history of ancient settlement structures of the Polish Plain. The trend to minimal, mobile settlements (with smaller concentration of ceramics, and fewer elements making use of earthen constructions, such as pits, hearths or postholes) as points of expansion, also attested in GAC settlements (which show more variability in size and finds), acquires its maximum value in the CWC. Later during the BA would the number of settlements grow again.

- It is supposed that in northern Poland more settlements (camps) existed. They were unstable, and only used for a short time, so they left few traces. It is difficult, therefore, to consider such territories as 'scarcely' settled.

- CWC-camps were founded on unstructured (usually sand) soil, in exposed sites like river- and sea-coasts. They preferred settled sites, i.e. those already changed through anthropogenic activity.

- Settlements and burials were located in the same zones, usually in the immediate neighbourhood. In some cases, it is difficult to differentiate settlement from burial fields, since both types of remains are mixed.

The main defining trait of the oldest CWC groups (in the Polish Plain and the Upper Vistula basin) are therefore small kinship groups marked by migrations, with relatively short breaks, building of small encampments of a temporary nature whose relics are extremely difficult to identify. Enduring markers of such migrations were graves, at first isolated at a marked distance from one another, and which did not form cemetery complexes (Czebreszuk and Szmyt 2011). Intergroup links had to be strong, as evidenced by cooperative behaviours determined for different activities: construction of graves, long distance expeditions, exchange, exploitation of natural resources, military conflicts.

In the Kuyavian area, there were no stable spatial barriers separating settlements of the different societies coexisting during the Corded Ware expansion (TRB–Baden, GAC, CWC): the same territory was used by groups of completely different traditions. The long duration of that phenomenon shows the lasting awareness of a separate identity of individual societies grounded in their symbolic behaviour and kinship-based social organisation (Szmyt 2008).

Cohabitation of GAC and CWC is documented in the Polish Lowlands and in the upper Vistula basin (ca. 2800–2600/2500 BC), but in other regions such as Mittleelbe–Saale the emergence of CWC ca. 2800–2700 BC meant the end of the GAC in the area. Both GAC and CWC were supra-regional structures with certain mutual elements of material culture and social behaviour (ritual end economic). Differences included (Szmyt 2008):

- Settlement: semi-settled existence with cemetery complexes playing a stabilising role for GAC; CWC did not possess a stable settlement network, and showed an economy based on mobility (or semi-nomadism) related to animal husbandry.

- Ritual: most CWC burials are single or double, while GAC contains more multiple burials; both show mounds over graves, with GAC mounds being larger, and CWC mounds surrounded by a small ditch with a palisade.

- Corded pottery: technology, morphology, and ornamentation, such as impressions of a 'double' cord in CWC.

- Other objects (flint, stone, bone, tools, weapons) and their means of production.

Flint axes, however, point to mutual traits between the two cultures, although weaponry and tools deposited as grave assemblages include for GAC one flint axe (sometimes more), whereas for CWC it includes a stone axe–hammer, flint knife, and archery accessories (Szmyt 2008).

Social organisation (inferred from burials and materials): CWC shows a status of men as dominant, monopolising rites and social activity as 'heads of family', in relations between neighbouring families, and in decision-making councils at a higher level, i.e. (supra-)regional contacts with other bodies in other parts of the CWC population. The Corded Ware culture has been described as a 'Big Man system', with a warrior elite class (associated with symbolic prestige symbols), with age classes, and composed of family-bound lineages and clans (Strahm 2002). In this sense, the Corded Ware society was a continuation of societies born out of the "Transformation of Europe".

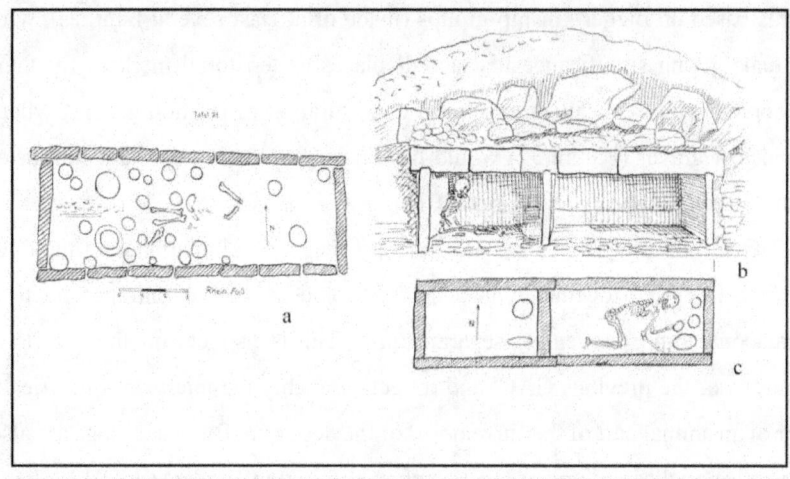

Figure 37. Tumuli of Central Europe during the Corded Ware period. 1a – Farnstädt I; 1b,c – Farnstädt II; 2a – Allstedt, Mallerbacher Feld; 2b – Lißdorf (after Behrends). Modified from Korenevskiy (2012). Bottom image: scene reconstruction from the Jungsteinzeit Bilderreihe by Gerhard Beuthner (1930).

There was a double cultural-social principle of the individual's identity in CWC: kinship (status gained at birth over small migratory kinship groups) and social identity gained through the individual's life. Social organisation in GAC was based on bigger kinship groups (made of at least several families), with rituals giving significance to 'sacred' places as territorial markers for their peoples, with cyclical dispersal and concentration of relation groups, where ancestor graves (Figure 37) would have been extremely important for social stability (Czebreszuk and Szmyt 2011).

One interesting trait of CWC funerary rites is the importance of the dog, which is buried together with a human, sometimes in an entrance pit to a catacomb tomb, or even in a separate burial. This is distinct from the sacrificial burials of the previous GAC, and reflects probably the interment of a friend, or of an animal part of the 'inventory' of the deceased. Burials of dogs are also found sporadically in other CWC groups in Europe (Włodarczak 2017).

In terms of pottery, three areas of regional influences can be distinguished: in south-eastern Poland, beakers display characteristics of the Middle Dnieper culture's forms; in western Pomorze, Great Poland, and Kuyavia, flower-pot-shaped beakers appear with the traits typical of Single Grave culture; in Śląsk, vessels typical of Lusatian, Czech, and Moravian Corded Ware culture are present. These groups may in turn be related to "archaic" manifestations, such as materials from the Złota culture from the Sandomierz Upland, the Rzucewo culture from Pomorze, or the pottery of the forest zone of Mazovia, Masuria, and Podlasia (Włodarczak 2017).

VI.3.4. Circum-Baltic Late Neolithic

The TRB culture is supposed to have reached the Upper Bug ca. 3850/3700 BC, and soon afterwards (ca. 3600/3500 BC) it appeared in the Buh–Dniester frontier, reaching the drainage of the Horyn River, and the Taiga region possibly from the Upper Pripyat. After 3200/3100 BC, TRB colonisers are substituted by GAC societies, with the main process taking place ca. 2950–2350 BC, and reaching as far as Smolensk (ca. 2500 BC). During this time, it seems that some communities returned back to the Vistula–Dnieper region (Klochko and Kośko 1998).

Among the so-called para- or sub-Neolithic traditions surviving into the 3rd millennium in the Forest Zone, the following are important for the development of the Corded Ware culture in the region:

The late stage of the Narva culture, part of the post-Narva phenomenon (dated ca. 4000/3750–3000/2750 BC) shows a highly diversified culture, with syncretic entities such as the Zedmar type, a synthesis of Nava (initially dominant) and Neman (eventually dominant). Other examples include the Šventoji and the Usvyaty cultures, the latter (ca. 3600–2600 BC) with TRB and GAC influences. In the opinion of many scholars, this phase is primarily marked by exogenous influences from the north (from the circle of Pit–Comb Pottery groups) and from central Europe (Funnel Beaker culture, GAC, and CWC). The impact of central European groups is most clearly intelligible in the south-western portion of the culture's range.

The late stage of the Neman culture (ca. 2800–1800 BC) is marked by its evident interaction with GAC and CWC settlers, especially in their area of contact, the drainage basin of the Neman and the Upper Pripyat. Earlier dates of the culture are recorded in Lesser Poland from the mid–4th millennium.

The late stage of the Prick-Comb Pottery culture is represented by the late stage of the Dnieper–Donets culture at the end of the 4th / beginning of the 3rd millennium—expanded from the forest-steppe region on the Dnieper area to the forest zone into part of Volhynia and Polesia (where the last remnants are found)—and the Upper Dnieper culture—possibly a variety of the Dnieper–Donets culture, surviving up to 3500–2500 BC, with its decline paralleled by the expansion of the Middle Dnieper culture.

There is a particularly conspicuous presence of comb ware in the materials of the Late Neolithic Zhizhitska culture (ca. 2450–2200 BC) in the Upper Dvina drainage.

The colonising Neolithic waves are continued by the Circum-Baltic Corded Ware culture, closely related to the traditions of the Single Grave culture and similar traditions of the Northern European Lowlands. After ca. 2900 BC, certain cultural systems with 'corded' traits—genetically related to the catchment area of the south-western Baltic—appear in the drainages of the Neman, Dvina, Upper Dnieper, and even the Volga. These communities are considered the vector of Neolithisation in the Forest Zone.

It is not clear how these 'western' influences affected the Yamna culture to the south, which is identifiable in the forest-steppe zone up to the Dnieper–Inhulets line. For example, in the Yampil Barrow complex between the Buh and Prut rivers, including part of the Podolia region, influences of cultures are seen in the kurgans and their graves, with Kvityana, late Trypillia (Gordineşti or Zhyvotylovka-Volchans'k), GAC, early and middle Yamna, Corded Ware, and Catacomb traits succeeding each other (sometimes with obvious cultural influences between each other, in this period (Włodarczak 2017). Movements in the opposite direction, deep into Corded Ware territory, are also seen by groups of Comb-like decoration, reaching up to the Vistula and Oder (Klochko and Kośko 1998).

Burial customs of Corded Ware settlers in the eastern Baltic (from Lithuania, Latvia, Estonia, and western Belarus) include single graves or small cemeteries for up to 10 individuals, but larger cemeteries are unknown. Interment in a flexed position is common, and grave goods consist of battle–axes, flint axes, and large bladed knives, bone pins and wild boar tusks, with pottery appearing rarely. The absence of barrows, common in Central European CWC and to the south near the Pontic–Caspian area, sets this group apart from others (Piličiauskas et al. 2018).

Organic residue analysis demonstrates that a range of ruminant products, including milk, was preferentially processed in the CWC beakers, representing a radical change with respect to previous sub-Neolithic cultures. Flint blades were probably used for processing meat (Piličiauskas et al. 2018). Corded Ware individuals represent thus a transition from a mainly hunter-gatherer economy to agropastoralists in the Circum-Baltic region, although usually maintaining a mixed economy with a significant role of hunting, fishing, and gathering.

The Pamariu (Rzucewo) culture developed on the south-eastern shorelines of the Baltic, from a basic substratum in the populations of the Narva and Neman cultures, unified after ca. 2800 BC under the TRB, GAC, and CWC cultures, from Gdańsk to the Courland Lagoon (Szmyt 2010). The arrival of the Corded Ware culture in the region has been described as a process of infiltration of small groups in the local culture medium, due to the scarce research areas available to date. Even taking into account the scarcity of findings, there are some similarities in Baltic graves with the grave-set in Lesser Poland, including battle axes, flint artefacts, or crouched position of the body. The most striking finding of this region is probably the presence of permanent settlements within the coastal zone, the relevance of fishing and hunting (especially of marine mammals) for the economy's structure, and the acquisition of amber. There were both large, permanent settlements, situated in the upper zones of surrounding terrain, and short-term campsites located on floodplains (Włodarczak 2017)

One of the best studied sites is Suchacz, in East Prussia, where traces of 16 houses were discovered, consisting of post-frame buildings with sunken floors, typical of the Baltic coastal area (Figure 38). The location of other sites and the nature of dwellings (with occasional traces of structural repairs) point to permanent inhabitation. Permanent and short-term settlements alike were engaged in amber workshops, which continued a Final Eneolithic tradition. The only known large settlements among Corded Ware groups are found in the eastern Baltic, associated with the acquisition of amber. Domestic animals included mainly cattle and pigs, with goat–sheep being less represented than in other CWC groups. There is no confirmed agricultural activity (Włodarczak 2017).

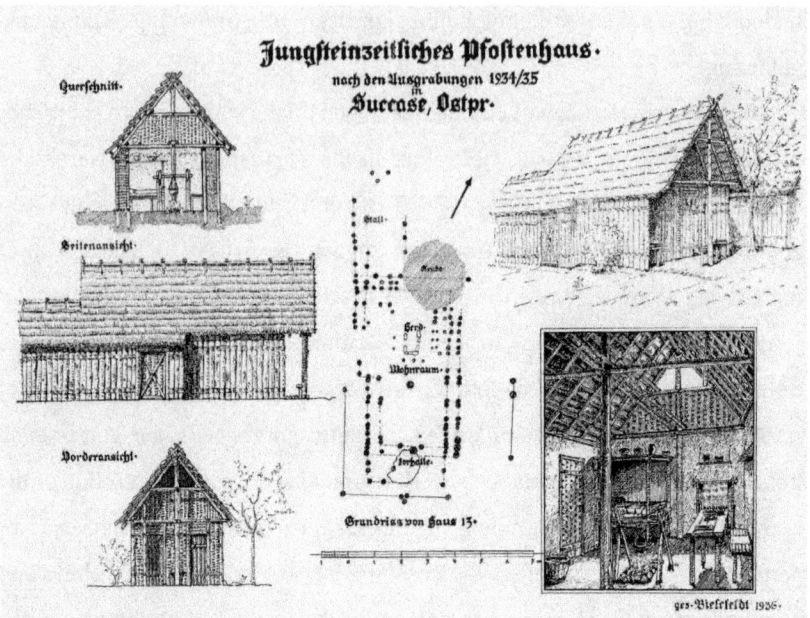

Figure 38. House no. 13 from Suchacz and attempts at its reconstruction, according to Ehrlich (1935). Modified from Włodarczak (2017).

The earliest Corded Ware in Finland is dated to ca. 2900–2800 BC, but the scarce data on the earliest sites preclude a proper estimation of mechanisms through which this culture appeared. There is a clear interaction sphere between the areas surrounding the eastern Gulf of Finland, reaching from Estonia to the areas of present-day Finland and the Karelian Isthmus in Russia, evidenced e.g. by the sharp-butted axes, derived from the Estonian Karlova axe (Nordqvist 2018).

VI.3.5. Contacts with Yamna

Proposed Yamna – Corded Ware cultural contacts are based on certain late 'Yamna–Catacomb' traits in the Baltic drainage basin, such as pit- and catacomb/niche-graves or 'Pontic' traits of funerary structures; production patterns typical of the Donets Basin and Caucasus centres of metallurgy and glass-making; and single finds of insignia-type forms, like Catacomb-type axes (or local imitations), or fluted maces. Other occasional finds include isolated 'Pontic' ritual, settlement, or funerary features (Klochko 2013).

In the Neolithic of the Vistula drainage, there are no prototypes (genetic inspirations) of 'Yamna' designs of grave chambers or single burials underneath barrows that appear in the early stages of CWC development. On the Lowlands, these are deep, straight-walled excavations 0.85–1.0 m deep, dated to 2850–2700 BC. On the Old Uplands of Lesser Poland—where niche chambers dominate—there is a Yamna-like chamber from a feature located underneath a barrow at Koniusza I ca. 2500 BC (Włodarczak 2008). These and similar isolated structures are elements of an exogenous funerary tradition, genetically related to the Yamna culture, likely spread towards the Vistula following left-bank tributaries of the Dniester towards the Bug drainage (Klochko 2013).

Yamna peoples entered the forest-steppe zone to the north of their territory ca. 2700 BC, with their presence being confirmed ca. 2550/2500 BC in the right bank of the Middle Dnieper drainage, and it may be accepted that they did the same in the Dniester–Prut interfluve. Contacts with the GAC population is well documented in terms of GAC cultural remains in Yamna settlements, as well as the use of ochre in certain GAC graves of the area after 2700 BC. The appearance of Corded Ware groups in the region between the Upper Vistula and the Dniester, before 2700 BC, must have culminated with the eastern expansion of Corded Ware peoples that caused the emergence of the Middle Dnieper group, and the decline of the eastern GAC societies (Szmyt 2013).

In the Baltic drainage, 'Yamna' features—leaving aside funerary patterns which are also common to the Złota culture—are thus catacomb cultures from Kraków–Sandomierz (ca. 2700/2600 BC) and Grzęda Sokalska (Roztocze, after ca. 2600 BC), the latter showing closer affinities with forms from Black Sea steppes, suggesting direct contacts of the Inhul group through the Buh drainage (Włodarczak 2008). This assumed route is also supported by the high share of Middle Dnieper culture patterns in the style of Grzęda Sokalska pottery, recorded on both the Bug and lower Vistula rivers.

vi.3. Disintegrating Uralians

The expansion of Proto-Corded Ware peoples was likely not related to the Złota group, given the common ancestry that sampled individuals show with other Globular Amphora samples. Nevertheless, the five individuals sampled from Książnice (ca. 2900–2630) show evidence of an additional gene flow, most likely from an eastern source, related to Steppe (ca. 10%) or EHG ancestry (ca. 8%), which supports contacts through exogamy of this group with Corded Ware, likely during the period of intense GAC–Trypillia–CWC interactions (Schroeder et al. 2019). The only reported paternal lineage from the Złota group is I2a1b1a2b1-L801 in an individual from Wilczyce, which supports the common genetic stock between GAC and Złota, different from Corded Ware peoples, and thus the spread of steppe features proper of the Yamna culture among neighbouring, non-Indo-European populations.

Corded Ware peoples have been described as deriving as much as 75% of their ancestry from a source close to Yamna samples from the Pontic–Caspian region (Allentoft et al. 2015; Haak et al. 2015; Mathieson et al. 2015). However, this shared Steppe ancestry is formed by different layers of admixture, which made both groups eventually converge genetically. The greatest similarity comes from the Steppe ancestry found in Eneolithic peoples from the forest zone (from which Proto-Corded Ware peoples likely derive), acquired through admixture with previous Novodanilovka settlers in the north Pontic forest-steppe region (see *§iv.2. Indo-Anatolians*).

Corded Ware samples published to date have additional EEF ancestry (between 30–50%) with similar values found in late Sredni Stog samples (Wang et al. 2019), as evidenced also in the continuity of both populations in their PCA cluster. In the Volga–Ural area, EEF contribution was probably minimal during the Eneolithic (based on Afanasevo samples), while the ca. 15% found in Yamna was probably due to the admixture of late Repin / early Yamna settlers with late Sredni Stog and other north Pontic groups (see *§vi.1. Disintegrating Indo-Europeans*). A good fit for the ancestry found in early

Corded Ware individuals is the Sredni Stog individual from Oleksandriia (ca. 80–90%)[+] and additional 'local' sources of WHG or EEF ancestry. These differences of CWC individuals with Yamna are more clearly evidenced by the different Y-chromosome bottlenecks undergone in the Khvalynsk/Repin/Yamna communities, predominantly of hg. R1b1a1b-M269, and in the Sredni Stog/Corded Ware groups, with a majority of hg. R1a1a1-M417 lineages.

Among early individuals, one sample from Obłaczkowo (ca. 2700 BC) shows hg. R1b1-L754; samples from Brandúsek (ca. 2900–2200 BC) include two of hg. R1a1a1-M417, and one of hg. I2a1b1b-Y6098 (Olalde et al. 2018); and samples from Jagodno (ca. 2800 BC) show possibly hg. G and J/I (Gworys et al. 2013). Two individuals of the Single Grave culture from Tiefbrunn (ca. 2900–2600 BC) are of hg. R1a1a1-M417, and one from Bergrheinfeld (ca. 2650 BC) of hg. R1a1a1-M417, possibly xR1a1a1b-Z645 (Allentoft et al. 2015).

Individuals from of the Battle Axe group include Baltic Late Neolithic samples: two individuals from Ardu and one from Kursi, Estonia, of hg. R1a1a1b-Z645; one from Kunila (ca. 2450 BC) of hg. R1a1a1b1-Z283, and another from Ardu (ca. 2700 BC); one from Gyvakarai, Lithuania (ca. 2550 BC) of hg. R1a1a1b1a3-Y2395, i.e. pre-R1a1a1b1a3a-Z284 (Saag et al. 2017; Mittnik, Wang, et al. 2018). An individual from Kyndelöse, Denmark (ca. 2500 BC) shows hg. R1a1a1b1a3a-Z284[+], and another from Viby, Sweden (ca. 2550 BC) also shows R1a1a1b-Z645 (Allentoft et al. 2015).

R1a1a1-M417 lineages, in particular R1a1a1b-Z645 (formed ca. 3500 BC, TMRCA ca. 3000 BC), most likely expanded with Corded Ware peoples, with an eastern subclade R1a1a1b2-Z93 (TMRCA ca. 2700 BC) probably expanding with Middle Dnieper and Abashevo, and a western subclade R1a1a1b1-Z283 (TMRCA ca. 2900 BC) expanding explosively with all other European groups (although Single Grave samples show a higher diversity), as evidenced by the similar times of split and TMRCA of R1a1a1b1a-Z282 subclades (TMRCA ca. 2900 BC): R1a1a1b1a-M458 (TMRCA ca. 2700 BC) probably to the west and north; R1a1a1b1a2-Z280 (TMRCA ca. 2600 BC) probably to the north and east, accompanying in part R1a1a1b2-Z93 subclades; and R1a1a1b1a3a-Z289/Z284 (TRMCA ca. 2700 BC) to the north, particularly among Battle Axe peoples.

The presence of a potential R1a1a1-M417 (xR1a1a1b-Z645) sample in Bergrheinfeld and later ones from Esperstedt (see *§vii.1. Western and Eastern Uralians*), all of Single Grave groups, coupled with the initial variability of subclades in early east-central European individuals, may suggest that the early expansion of the Single Grave culture did not undergo the Y-chromosome bottleneck through R1a1a1b-Z645 lineages common in Battle Axe or eastern groups. This is also supported by the prevalent presence of R1a1a1-M417 (xR1a1a1b-Z645) lineages among modern western Europeans, in spite of this subclade spreading from the steppes. Alternatively, these subclades may represent an early wave of Corded Ware groups, before the expansion of the unifying A-horizon, which would not have affected central and central-west Europe as intensely as the east.

The time of potential expansion of R1a1a1-M417 (formed ca. 6600 BC, TMRCA ca. 3500 BC), approximately coinciding with the formation and TMRCA of R1a1a1b-Z645 lineages, in turn close to the expansion of mainly European R1a1a1b1a-Z282 subclades, and mainly eastern R1a1a1b2-Z93 subclades, also supports this division. In fact, the common TMRCA for R1a1a1b1-Z283 and R1a1a1b1a-Z282 suggests an expansion at nearly the same time as peoples of Corded Ware cultures are supposed to have migrated east- and westward, reaching the Middle Elbe–Saale region about 2750 BC. The common TMRCA of 2700 BC for modern Asian lineages gives support to a later successful expansion into Asia centred on the eastern part of the Pontic–Caspian steppes (see *§viii.18.1. Late Indo-Iranians*).

The estimated split of Proto-Uralic into Finno-Ugric and Samoyedic The linguistic estimates for a split of Proto-Uralic into Finno-Ugric and Samoyedic ca. 3000 BC, and of Finno-Ugric into Finno-Permic and Ugric ca. 2500 BC (Janhunen 2009; Kortlandt 2019) fit the known expansion of Proto-Corded Ware first from the north Pontic forest-steppe into east-central Europe (ca. 3000 BC), and then the expansion of Classical Corded Ware into the Baltic (ca. 2800 BC) and to the east into the Volga–Kama region (ca. 2700 BC) with continued contacts of Battle Axe with Abashevo through Fatyanovo reflected in the strong similarity of Finno-Ugric to Proto-Uralic (Kallio 2015).

Among Baltic Late Neolithic individuals, three early samples (ca. 3200–2600 BC) stand out because of their close cluster with the Yamna population: one from Zvejnieki in Latvia, and one from Plinkaigalis; with a slightly later one from Gyvakarai, of hg. R1a1a1b1a3-Y2395, who clusters in an intermediate position between the two outliers and other Corded Ware samples. These outliers are described as forming a clade with Yamna, due to their reduced NWAN ancestry (Mittnik, Wang, et al. 2018), although their EEF-related admixture is ca. 20% or higher (Mathieson et al. 2018). This reduction in NWAN and EEF ancestry—and closer cluster with Yamna—is probably also due to their additional EHG admixture (bringing them closer to Khvalynsk samples, far from the mainly WHG-driven EEF ancestry of Corded Ware), evidenced in the 'northern' shift of this samples on the PCA (Suppl. Graph. 8).

The wide cluster formed by the available West and East Baltic Bronze Age samples, as well as East Baltic and Finland Iron Age samples, which encompass Baltic CWC as intermediate with other Corded Ware samples, confirms the nature of their ancestry as stemming from the admixture of Sredni Stog/Early Corded Ware with WHG:EHG populations from sub-Neolithic populations from the Baltic, rather than through direct exogamy with Yamna groups (see below *§viii.16. Saami and Baltic Finns*).

An additional potential source of similarities of certain CWC groups with Yamna may stem from the shared female population of the north Pontic region, supported by the statistically significant association of mtDNA between (especially west) Yamna and Baltic Corded Ware samples, in contrast to other

Corded Ware groups (Juras et al. 2018). The close traditional connection between the north Pontic area and the eastern Baltic through the Buh–Dnieper–Dniester corridor (Klochko and Kośko 2009), including the Volhynian–Podolian Upland and Polesian Lowland, could have facilitated exogamy with late Sredni Stog or closely related populations, which were also the source of gene flow into Yamna during the colonisation of the north Pontic area by expanding late Repin settlers. In particular, Zvejnieki shows mtDNA hg. U5a1b, associated previously with the north Pontic Neolithic and Maikop, and later with Corded Ware- and Yamna-derived groups (Mathieson et al. 2018; Olalde et al. 2018; Wang et al. 2019).

Exogamy has been argued to be an extended practice among Corded Ware peoples, with many adult women being of non-local origin, based on a recent work on diet and mobility (Sjogren, Price, and Kristiansen 2016), and mtDNA has been documented to be more varied among Corded Ware females than men (Lazaridis et al. 2014). The nature of these Baltic Late Neolithic samples as outliers among Corded Ware peoples is further supported by the close cluster formed between late Sredni Stog individuals and most Corded Ware groups sampled to date, including those of Germany, Poland—where the culture is supposed to have emerged—and the later samples from Sintashta, Potapovka, Andronovo, or Srubna, which suggest a similar genetic picture in the as yet unsampled Middle Dnieper, Fatyanovo–Balanovo and Abashevo cultures. Despite this homogeneity, two Corded Ware outliers from Single Grave and Battle Axe groups cluster closely to EEF and Comb Ware-like populations respectively, showing how admixture easily changes with exogamy.

Analysis of ancient samples has revealed that the plague was a prehistoric disease endemic to the Eurasian steppes, and a European pandemic may have been linked to the expansion of both Yamna and Corded Ware peoples, because they connected vast areas in east-central Europe in a relatively short period. One of the earliest known strains is found outside of the steppe in the Baltic, in the Northern European Plains, and in Croatia in the 3rd millennium

BC (Rasmussen et al. 2015; Andrades Valtueña et al. 2017). Nevertheless, the early finding of a European strain linked to Neolithic populations suggests that an earlier epidemic was probably a source of radical population decline of agricultural groups of central Europe and southern Scandinavia (Müller and Diachenko 2019) before the steppe-associated expansions (see *§v.6. Late Uralians*).

This contemporary population reduction in Europe, coupled with an already smaller population density to the north of the loess belt—contrasting with the greater population size of south-eastern Europe (Müller and Diachenko 2019)—may have provided a disadvantage of central-eastern European lands, and a necessary 'pull' trend for the migration and expansion of Corded Ware (Anthony and Brown 2017). The spread along sparsely populated areas, as well as continuous contacts between clans facilitated by their mobile economy, may have allowed for the genetic homogeneity seen among Corded Ware peoples from west to east, in spite of the proposed generalised practice of exogamy.

VI.4. Middle East

VI.4.1. Maikop–Novosvobodnaya

Shortly after the advent of the Kura–Araxes complex in the mid–4th millennium, the western Caucasus developed its own tradition of dolmens, or megalithic buildings for the dead, as early as 3250 BC. They are often associated with the megalithic traditions of western Europe, and seen thus as a global phenomenon, although they were restricted in this region to a small area in the north-eastern Pontic coast. Dolmens were built of well-squared, heavy stone slabs, placed on their edges and fitted together with precision, positioned to maximise the sunlight on the façade; most facing southwards, some eastwards. Most dolmens were used for multiple interments, and included men and women, young and old, and funerary provisions (Sagona 2017).

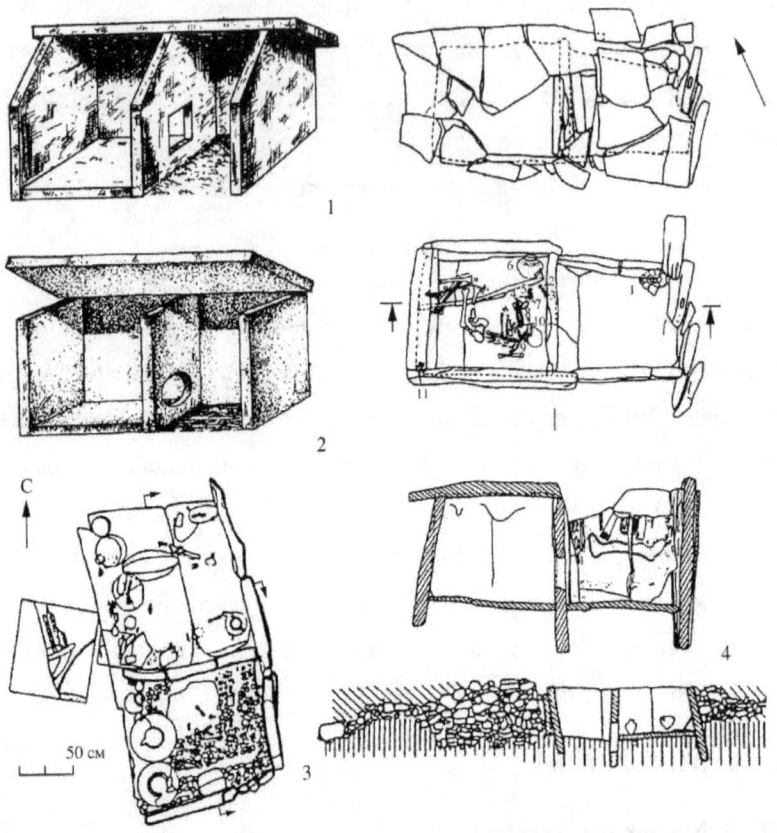

Figure 39. Tombs of the Novosvobodnaya type. 1 – kurgan. 1, 2 – kurgan 2 – excavations N. I. Veselovski; 3 – burial site Klady 31/5; 4 – burial ground klady 28/1 (according to Rezepkin). Modified from Korenevskiy (2012).

The latest stage of the Maikop–Novosvobodnaya historical community is represented by the Novosvobodnaya dolmens, with a span ca. 3300–2800 BC, overlapping with the beginning of west Caucasian dolmens. Unlike the Maikop barrow burials, two rich tombs at Novosvobodnaya were constructed of stone slabs, in a plan that resembles those of the Dolmen tradition, but with distinct characteristics (Figure 39): the interior design and plans differ, with a pair of slabs placed close enough to form a narrow gap, separating two compartments, a paved burial chamber and an antechamber. Masonry is more precise, and large slabs form the roof and overhang the entrance, which faces south-east.

The entire structure is concealed beneath a barrow of stones defined by a kerb (Sagona 2017).

They were not used for collective burials, and the deceased were placed on the right side of the chamber, accompanied by a rich assemblage, although it did not reach the high level of Maikop kurgans. During this period, there is a decline in the number of objects made of precious metals, and an increase in copper objects, with an improvement in copperworking seen in elaborate forms of weapons and tools. Jewellery and metal (golden, but also silver) beads, on the other hand, increase in this late phase (Sagona 2017).

Ceramic evolves from the limited Maikop repertoire to tall-necked jars with incised decoration. New sets of woodworking and leatherworking tools appear, and the lack of impurity in the copper points to an origin of the metal ores in the Balkans. Spearheads from the northern Caucasus are similar to those from the Kura–Araxes culture found in the southern Caucasus and Arslantepe. Their subsistence economy is assumed to be based on stockbreeding of ovicaprids, cattle, and pigs (Sagona 2017).

The other tradition of the area, the Colchian culture, probably developed from indigenous groups that occupied the wetlands and lowlands for over two millennia.

VI.4.2. Kura–Araxes

By 2900 BC, the Kura–Araxes culture was spread throughout much of south-west Asia, which suggests that they competed against northern Mesopotamian societies. The widespread dissemination of this material culture, along with the small size of most sites, the ephemeral nature of their settlements, and their presence in both fertile lowlands and seasonally-inhospitable highlands, suggest that they were formed, at least in part, by mobile pastoralists (Alizadeh et al. 2018).

Kura–Araxes groups primarily inhabited mountains and intermontane valleys of the surrounding highland zone, and they had access to metals, precious and semi-precious stones, stones for tool making, wood, and animal products; resources that were abundant in the mountain area, and essential for Mesopotamian societies. Metallurgical sites like Köhne Shahar suggest that a developing complex of exchange networks and interaction with specialised craft economies facilitated the culture's expansion, by filling a supply vacuum created by the collapse of Uruk colonies (Alizadeh et al. 2018).

In Arslantepe, around 3000 BC, the Palatial system collapsed, apparently by a big fire that destroyed the palace and the whole centralised state forever, including the Mesopotamian-type society, administrative system, and officials. This event seems to be related to the gradual expansion of Kura–Araxes pastoralists, whose seasonal occupation of the site eventually turned into a permanent residence for their chiefs (Frangipane, Manuelli, and Vignola 2017).

New settlements were established, with an outstanding area on top of the mound, consisting of a likely chief hut (ca. 2900 BC) separated from the rest of the settlement by a timber palisade, and an imposing mudbrick building with a reception hall, and store rooms full of vessels and foodstuffs. Pottery had simple shapes, resembling those of north-eastern Anatolian and southern Caucasian origin, no longer central Anatolian, although decoration and technology corresponded to the previous tradition (Frangipane, Manuelli, and Vignola 2017).

The changes from a seasonal occupation to a rural village with mudbrick houses and pottery of the post-Uruk tradition happened ca. 2800 BC, with both communities—the traditional, sedentary one, and the pastoral itinerant community from north-east Anatolia and the southern Caucasus—apparently negotiating, interacting, and clashing, with either side alternatively succeeding, taking control and possession of the mound. The identity of the two groups must have been clearly perceived, and no evidence is found to suggest cultural inclusion or mixing, whereas episodes of conflict are evident (Frangipane 2015).

The social instability and political upheaval are also noticed by the emergence in the region of human sacrifice, usually a resource of hierarchical social structures that accompanied early state-formation processes. In the vacuum of political centralisation that followed the withdrawal of Uruk material and the appearance of Kura–Araxes, instability among smaller polities must have created thus the necessary environment for the introduction of human sacrifice. With the appearance of vast administrative state systems in southern Mesopotamia in the next millennium, this practice disappeared again from the archaeological record (Hassett and Sağlamtimur 2018).

vi.4. Northern Caucasians

Novosvobodnaya (ca. 3600–3300 BC) and late Maikop samples (ca. 3300–3100 BC) in the northern Caucasus foothills also show continuity of ancestry with Maikop samples, falling among the Armenian and Iranian Chalcolithic individuals. Important phylogenetic differences are seen, though: Novosvobodnaya shows hg. J2a1-L26 in two samples from Klady, one of them J2a1a1a2b2a3b1-Y3020, and the other J2a1a1a2b2a3b1a-Y11200 (xY30811-, Z30682-), the same haplogroup found previously in Eneolithic Caucasus (see above *§iv.4. Late Middle Easterners*), and found today mainly in Northeast Caucasian populations; and G2a2a-PF3147 (expanded with Anatolia Neolithic farmers) in one sample from Dlinnaya Polyana; whereas Late Maikop shows hg. L-M20 in one sample from Sinyukha and two from Marinskaya, probably all L2-595 (formed ca. 21000 BC, TMRCA ca. 3200 BC). Haplogroup J1-L255, found previously during the Mesolithic, is also reported for Late Maikop in Marinskaya.

Two late Maikop outliers from the north Caucasus steppe show a higher proportion of Anatolian and Iranian farmer-related ancestry. This may have been driven either by Pontic–Caspian steppe migrations, or by the admixture with local Caucasus populations of AME ancestry. The presence of haplogroup R1a1b-YP1272[+], a typical eastern European lineage, in a sample from Sharakhalsun (ca. 3230 BC), suggests the former as the most likely explanation; the sample from Ipatovo (ca. 3260 BC) shows hg. T1-L206[+], a typically Middle Eastern lineage (Wang et al. 2019).

This *Caucasus Eneolithic*-like ancestry is also continued in Kura–Araxes (Wang et al. 2019), in early samples (ca. 3500–3100 BC) from the south (Kaps, Armenia), one of hg G2b2a2-FGC2964 (formed ca. 13700 BC, TMRCA ca. 1100 BC), found previously in an Iranian Neolithic individual ca. 7300 BC; and in later Kura–Araxes samples (ca. 3100–2800 BC) in the north-east (Velikent, Dagestan), one of hg J1a2b1-Z1842 (formed ca. 5800 BC, TMRCA ca. 4000 BC), a haplogroup probably found later in a west Anatolia Bronze Age sample (ca. 2500 BC), and widely distributed in modern populations of the Middle East, which supports its expansion with Kura–Araxes peoples, and its association with modern speakers of Northeast Caucasian languages. Increased CHG ancestry (ca. 60%) is also seen in other three Early Bronze Age individuals from the Kura–Araxes culture in Armenia dated ca. 3300–2500 BC (Lazaridis et al. 2016), with a late sample from Kalavan (ca. 2550 BC) showing what seems to be the latest finding of hg. R1b1a2-V1636, already part of a non-Indo-European community (see above *§iv.2. Indo-Anatolians*).

An outlier from the Zagros Mountains in Hajji Firuz Tepe also shows elevated Steppe-related ancestry, and clusters between Kura–Araxes and Yamna samples (Narasimhan et al. 2018), consistent with the incorporation of North Caucasus-like populations within the expanding Kura–Araxes groups. The radiocarbon date published (ca. 2465–2286 BC) is compatible with that interpretation, although the collapse of different archaeological layers in the same site has yielded unreliable dates for (at least) one other sample, and it may therefore correspond to a much later date, in particular the Late Bronze Age – Early Iron Age (see *§viii.14. Caucasians and Armenians*).

Based on the territorial expansion of the Kura–Araxes culture, and on the subsequent groups that emerged in its core territories after its demise, the language spread by these southern Caucasian peoples was probably Hurro-Urartian, which may support a connection with North-East Caucasian languages in a hypothetic Alarodian group (Diakonoff and Starostin 1988), at least from a genetic point of view. Territories of the north-western Caucasus, occupied by Maikop, Novosvobodnaya, and Dolmen traditions, would probably then represent evolving North-West Caucasian-speaking peoples.

VI.5. Africa and the Levant

The initial spread of herding into eastern Africa (ca. 3000 BC) coincided with the emergence of a distinctive monumental tradition centred around "pillar sites" built near Lake Turkana, Kenya. These monumental sites more likely served commemorative purposes similar to many of the previous Saharan ceremonial sites before it (see *§V.1. Africa and the Levant*), although they exhibit architecturally distinct elements. These construction changes coincided with the end of the African Humid Period (ca. 3500–3000 BC), which brought about profound changes in environment, economy, and material cultural expression, coupled with a major population collapse due to the decline in favourable climatic conditions (Brierley, Manning, and Maslin 2018).

In the Lake Turkana area, retreating shorelines ca. 3300–2000 BC disrupted fishing practices and exposed new habitats for herbivores, while exchange and/or herder immigration brought cattle and caprines into northwest Kenya, transforming economic strategies to include mobile herding. While previous fishers used local lithic raw materials, the new herders preferred obsidian from varied local, distant, and island sources, which point to extended exchange networks, including boat travel (Hildebrand et al. 2018).

Monumental mortuary expression appearing in the Sahara, Sahel, Nile, and Turkana at the same time as the dramatic environmental shifts—and the accompanying change to herding economy—marks probably a social change rather than the emergence of hierarchical social forms. Distinct forms of commemoration include cattle burials in the central Sahara, megaliths in the eastern Sahara, aggregate cemeteries in the southern Sahara and along the Nile, built mortuary spaces in the Red Sea Hills and around Lake Turkana, and cairn and cremation treatments linked to early pastoralism in central Kenya.

In Mesopotamia, rulers of the Early Dynastic periods (ca. 2900–2350 BC) were leaders of city–states (the so-called 'theocratic temple-states'), where estates of the gods were possibly the property of the king and his ruling family, and the ruler was thus a protector of a city in the name of the city's tutelary deity. Political centralisation dominated in Mesopotamia under the two dynasties of Akkad and Ur, when East Semitic-speaking elites would eventually dominate over the whole society after the rise of Sargon of Akkad. The Mesopotamian king became both a divine figure and a warrior and conqueror (Kristiansen and Larsson 2005).

In the Fertile Crescent, a specialised economy of sheep–goats already evident in the Late Chalcolithic reaches a full development in the mid-3rd millennium BC, perhaps because of commodification of textiles and wool production in Near Eastern polities. The elaborate networks of roads during this period, with a pastoral economic activity growing beyond settled areas and population, are probably the result of interactions between settlements, where individual farmers, labour supply, and flocks of caprines moved across the landscape and were drawn to cities from the surrounding villages or exogenous sources. Urban centres and larger settlements were particularly affected by this demand, with textual evidence from Ebla and Tell Beydar of an institutionalised and centralised pastoral economy, where massive flocks of caprines were directly managed by the palace (Altaweel and Palmisano 2018).

vi.5. Semites and Berbers

East Semitic languages probably entered Mesopotamia from the desert to the west of its core area before 2900 BC, since the first attestations come from Akkadian personal names in Sumerian texts about the 29th century BC. Similarly, the Kish Civilisation—encompassing Semitic states like Ebla and Mari in the north, or Abu Salabikh and Kish in central Mesopotamia—shows probably the first historical record of the language, in the 30th century. Both East Semitic migration events can then be related to the collapse of the Uruk period ca. 3100 BC.

The Bronze Age Levantine population from the site of ʿAyn Ghazal, Jordan (ca. 2490–2300 BC), can be modelled as Levantine Pre-Pottery Neolithic agriculturalists from Motza, Israel, and ʿAyn Ghazal, dated ca. 8300–6700 BC (ca. 58%), with contributions of a population similar to Iran Chalcolithic (ca. 42%), from a more recent period (Lazaridis et al. 2016). It has been suggested that samples from Sidon, Lebanon (ca. 1700 BC) can be modelled as a mixture of Levant Chalcolithic (ca. 48%) and Iran Neolithic or Late Neolithic-related ancestry (Haber et al. 2017; Harney et al. 2018).

The difference between both populations lies then in the Anatolian-related ancestry found in Levant Chalcolithic (ca. 36%), also found in the northern population, which suggests the reintroduction in the south of a population not affected by this Anatolian migration, potentially then from farther south. The lack of relationship between Levant Chalcolithic ancestry and present-day East African Levantine-related ancestry (Harney et al. 2018) further supports a migration of Proto-Semitic from north-east Africa into the southern Levant, and then a back-migration to the south into Arabia and East Africa.

Among Levantine Bronze Age individuals, there is one sample (ca. 2400 BC) of J2b1-M205 lineage (formed ca. 13800 BC, TMRCA ca. 3300 BC), and another (ca. 2100 BC) of hg. J1a2b-Z1828, a haplogroup found previously in an Anatolian sample of the Bronze Age (see *§v.1. Early Semites*). Similarly, there is an individual from a Canaanite burial pit in Tel Shaddud (ca. 1250 BC) reported as of hg. J[17], which clusters—similar to another sample of hg. R1b1a1b-M269 (see below *§viii.12. Greeks and Philistines*)—among modern Levantine populations (van den Brink et al. 2017). The spread of early Semitic peoples was thus probably linked to communities of different local haplogroups, before the known Y-chromosome bottleneck of J1a2a1a2d2b2b-Z2331 lineages, particularly with the early expansion of Central Semitic dialects to the south.

---

[17] Tentative SNP call J obtained with Yleaf, from Wang et al. (2019) supplementary materials.

The predominant East African genetic component of the Horn of Africa points to genetic homogeneity of all populations of the area, whether Afroasiatic (Omotic, Cushitic, Semitic) or Nilotic (van Dorp et al. 2015). They present a great degree of continuity with the ancient Mota individual (ca. 2520 BC), of hg. E1b1a2-M329 (Gallego Llorente et al. 2015). The Maasai from Kenia are also closely related, showing ca. 50% E1b1b-M215 (Pagani et al. 2012), and regional hunter-gatherers from the Horn of Africa—such as the Chabu, the Majang, and the Shekkacho—retain strong genetic affinities with the Mota individual (Gopalan et al. 2019), supporting the original association of this haplogroup and ancestry with indigenous Nilo-Saharan languages.

The Eurasian admixture events inferred from modern populations in Sudan and Ethiopia (Hollfelder et al. 2017) suggest the likely expansion of Egyptian speakers from the north through the Nile, and the likely expansion of Semitic in the Horn of Africa from Arabia is possibly associated with the "eastern"/Iranian farmer-related ancestry found among modern Afar and Somali (Skoglund et al. 2017).

On the basis of genetic data from Pastoralist Neolithic individuals, two phases of admixture can be detected, associated with the spread of pastoralism: the first one (likely ca. 4000–3000 BC) in north-eastern Africa, possibly (based on earlier herding dates from the region) through South Sudan and reaching the Turkana Basin first, although they may have moved via the Horn of Africa; the second one (ca. 2000 BC) between this admixed Early Neolithic Pastoralist group and eastern African foragers, both in the Turkana Basin and during the initial trickle of herding into the south-central Rift Valley (Prendergast et al. 2019).

Descendants of these Early Neolithic Pastoralist and local forager groups gave likely rise to the groups who developed the Pastoralist Neolithic culture traditions of southern Kenya and northern Tanzania. The earliest samples available to date are a man and a woman from Prettejohn's Gully (ca. 2000 BC), whose ancestry profile suggest that they represent an initial (maybe one among many) limited dispersal of herder groups into the south-central Rift Valley that did not leave large numbers of descendants, while the group that gave rise to the Pastoral Neolithic cluster was much more demographically successful than the others. The lineage of this male is E2-M75 (xE2b-M98), while most Neolithic Pastoralists sampled belong to different E1b1b1-M35.1 subclades, apart from E2-M75, E1b1a-V38, and A-M13 lineages, further supporting the gradual and stepped admixture with local forager groups (Prendergast et al. 2019).

Whereas modern Omotic speakers—except those groups among Cushitic and Semitic groups—show strong affinities with the Mota individual, Cushites show continuity of an ancestry already found in a sample from Tanzania (ca. 1050 BC), with higher hunter-gatherer admixture than Bantus and mtDNA from the Rift Valley, suggesting that this Afro-Asiatic branch was widespread to the south before the more recent Bantu expansion, and thus likely to be associated with the Savannah Pastoralist Neolithic (Kießling, Mous, and Nurse 2008).

While ancient Afroasiatic languages were possibly introduced in the area by the expansion of R1b1b-V88 lineages and spread with pastoralism (see *§iv.5. Late Afrasians*), it remains unclear which lineages might have expanded Semitic languages to the region. Given the prevalence of haplogroup T1a1a1b2-CTS2214 near the Gulf of Aden, and the wide ancient distribution of this haplogroup through the Middle East (see *§iv.4. Late Middle Easterners*), the presence of some late subclades in the area probably reflects an acculturation of some East African or Middle Eastern group and subsequent Y-chromosome bottlenecks in the area. The various 'western' (sub-Saharan African admixture) and 'eastern' influences (south Asian and Levantine/European admixture) creating a west–east axis in the Arabian Peninsula support the nature of this region as a sink of different prehistoric migrations rather than a source (Cavadas et al. 2019).

Lexicological studies show that a Common Berber dialect *continuum* must have been still in close contact still around the foundation of Carthage (ca. 800 BC), based on early Phoenician/Punic loans found in most branches (Blažek 2014), which is compatible with the estimated dates for the disintegration of the language (Blažek 2010). This is supported in genetics by the incorporation and further expansion of this Afroasiatic branch under a bottleneck of E1b1b1b1a-M81 lineages, whose late successful expansion is coincident with that date (see above *§iii.4. Northern Africans*).

VII. Late Chalcolithic

VII.1. Eastern Corded Ware expansion

VII.1.1. Central Europe

Pits and traces of houses are more common in Polish sites after 2600–2400 BC (although absent in Lesser Poland), and there is a clear correlation with light sandy soils, on sites usually spread on slightly exposed elevations within low relief, most often of river valleys and lake shores. In certain areas of the Northern European Lowland (such as Mecklenburg, Masuria, the Polish Lowlands, or the Baltic coast area) there is a trend to more spacious sites, a higher number of sunken features, and richer assemblages of movable items, as well as to multi-stage occupation of certain settlement sites on attractive land areas (Włodarczak 2017).

Around the mid–3rd millennium BC, the number of mounds built decreased, and the number of burials dug into existing mounds increased, at the same time as flat cemeteries emerged. Radiocarbon analyses indicate that the expansion of kurgans probably peaked ca. 2600/2500 BC, weakening afterwards, and disappearing completely after ca. 2400/2300 BC, probably representing the diminishing importance of ceremonial–funeral centres as regional landmarks organising the territory of a given group (Włodarczak 2017). Kurgans appearing later, in the Únětice or Strzyżów groups, represent a new, different tradition (see *§VIII.8. Eastern EEBA province*).

The expansion into the forest zone of eastern and north-eastern Poland changed the landscape from hunter-gatherer societies to an economy based primarily on large herds of animals, mainly cattle, while agriculture was hindered by the soils. Settlers occupied primarily the lower zones, often riverbank areas of river valleys, and most remains come from seasonal campsites, with a mix of CWC and Neman traditions. These settlers of the mid–3rd millennium BC were probably related to the Masurian lakeland. The emergence of an allocthonic population is marked by the rare graves of this area, showing infiltrations initially mainly from Lesser Poland, and later (after 2500 BC) from the western Baltic zone (Włodarczak 2017).

There is a strong connection between the rituals of the Single Grave culture and those of the west Baltic region, connected through the Northern European Plains. This relationship becomes clear from the younger phase of the CWC development (ca. 2600–2500 BC) and continues until the demise of the culture. It is marked by specific forms of graves (with implementation of stone structures), and by grave goods typical of the west Baltic region.

VII.1.2. Middle Dnieper

The origins of the Middle Dnieper culture should most likely be traced back to the forest-steppe area of the Dnieper Basin. Concentrated between the Berezina, Dnieper, and Sozh rivers, on the Desna and drainage basin of the Middle Dnieper (from the confluence with the Pripyat in the north to the confluence with the Ros in the south), there are dispersed findings as far as the Middle Pripyat, Upper Neman, and the Seym drainage basin. Most likely, the culture spanned a long period from ca. 2600–1800 BC (Krenke et al. 2013).

However, the earliest Middle Dnieper samples are related to CWC graves between the Upper Vistula and the Bug, containing pottery with Middle Dnieper traits, dated probably ca. 2650/2600 BC or earlier, which establishes the beginning of the culture probably to the west of its core area ca. 2700 BC, with the expansion of the A-horizon (Krenke et al. 2013).

The "Kyiv hoard" and other hoards (like Steblivka) of Corded Ware tribes that populated the Volhynia and western Podolia regions, of willow–leaf metallurgical industry, evidence also the direct connection of the Middle Dnieper peoples to these CWC groups. The battle–axes of the Ingush type, proper of the forest-steppe cultures east of the Vistula river, also mark a completely different direction of inspiration (Klochko and Kośko 2011).

In fact, during the period of ca. 2800–2400 BC, the area of Lesser Poland (with its numerous kurgans and catacomb burials) is considered the western fringe of an area spreading to the east, to the middle Dniester and middle Dnieper river basins, i.e. regions bordering the steppe oecumene. This 'eastern connection' of funeral ritual, raw materials, and stylistic traits of artefacts is also identified in some graves of the Polish Lowlands (Włodarczak 2017).

VII.1.3. Fatyanovo–Balanovo

The Fatyanovo (or Fatyanovo–Balanovo) culture was the easternmost group of the Corded Ware culture, and occupied the centre of the Russian Plain, from Lake Ilmen and the Upper Dnieper drainage to the Wiatka River and the middle course of the Volga. From the few available radiocarbon dates, the oldest ones come from the plains of the Moskva river and from the late Volosovo culture containing also Fatyanovo materials, and in combination they suggest ca. 2700 BC for its appearance in the region, and ca. 2000 BC for their disappearance The Volosovo culture of foragers eventually disappeared (ca. 2300 BC) when the Fatyanovo culture expanded into the Upper and Middle Volga basin (Krenke et al. 2013).

The origin of the Fatyanovo culture is complicated, because it involves at its earliest stage different Corded Ware influences. This is evidenced in neighbouring sites on the Moskva river plains: one, potentially slightly older site, with some materials paralleling the Nida site, of Circum-Baltic and Polish features; and another site, 300 m. downstream, showing a connection with materials from the Khanevo cemetery, in turn a *bridge* to the Middle Dnieper culture. This suggests that groups belonging to different strands of the classic corded ware tradition penetrated the Moscow region. The Balanovo culture to the east seems to have been Fatyanovo's metallurgical heartland, with the Sura–Sviyaga group surviving long after the demise of Fatyanovo (Krenke et al. 2013).

Just before 2500 BC, Corded Ware, Single Grave and Battle Axe, Rzucewo, Middle Dnieper and Fatyanovo–Balanovo cultures arrived at their peak of landscape occupation, domination and coherence. Fatyanovo is exemplified mainly by grave finds, featuring rectangular pits with single burials, with the dead positioned contracted lying on the side, or supine with raised knees, and grave goods of clay pots, animal-tooth pendants, and rarely bronze jewellery. Male graves are identified by stone battle–axes (Parzinger 2013).

VII.1.4. Battle Axe culture

The early, skilfully made Corded Ware culture pots found in Sweden were both imported (in Southern Sweden from north-eastern Estonia and possibly Finland) and made with local clays by skilled potters (in Central Sweden), which supports the relocation of CWC potters from a place where the craft was already well established, namely the eastern part of the Baltic around the Gulf of Finland. The arrival of grog from Sweden at Finnish and Estonian sites may suggest a two-way movement across the Baltic Sea, although it probably represents the tempering of new pots with old ones, either brought to the region through migration or intermarriage, or by commercial contacts (Holmqvist et al. 2018). The CWC seemingly reached east-central Sweden from regions further to the east, where there is evidence of animal husbandry, but only very few signs of plant cultivation (Vanhanen et al. 2019).

The so-called Middle or Intermediate Zone ceramics, attributed to a 2nd wave of Corded Ware migrants from Estonia into Finland, has been recently proposed to be the result of a hybridisation that began soon after the arrival of Corded Ware, at least on the south-eastern coast and the Karelian Isthmus, with influences transmitted towards the inland and the middle-zone. Furthermore, most CWC materials associated with this hybrid pottery come from mixed, multi-period settlement contexts, and also include organic tempers in local groups, which are similar to the so-called Estonian (or late) Corded Ware (Nordqvist 2018). Late Estonian CWC remains show continuity of the preference for terrestrial foodstuffs in the eastern Baltic region, based on domestic animals complemented with agriculture, in contrast with the earlier hunter-gatherer diet (Varul et al. 2019).

There is a close connection of the Corded Ware tradition of the Karelian Isthmus pottery with the central Russian Fatyanovo culture, as well as between the eastern Gulf of Finland and Russian battle axe culture (Nordqvist 2018). These contacts and interactions between eastern CWC groups points to the close cultural connection between them.

There are hundreds of Corded Ware settlements identified to date in eastern Fennoscandia—most of them residential, recurrent activity or camp sites—and thousands of remains, more than in other Scandinavian territory. While the central area of Corded Ware habitation does not seem to include the inland or the northern territories, scattered findings of Corded Ware materials to the east, north, and north-east of these core territories (Nordqvist and Häkälä 2014) may point to isolated vanguard settlers or to imitations of indigenous groups.

In Finland, Corded Ware vessels are associated with beaker-type 'drinking' vessels, often in grave deposits, as well as amphorae and S-shaped pots. Corded Ware settlements, even coastal ones, show reliance on terrestrial ruminants, which could be either domesticated (e.g. cattle) or wild (e.g. elk, forest reindeer), although milk fat residues must have originated from domesticated stock. Therefore, the introduction of animal domestication as a new subsistence strategy can be traced back to at least ca. 2500 BC in Finland (Cramp et al. 2014).

vii.1. Western and Eastern Uralians

Later samples of the Single Grave culture include six individuals from Esperstedt, Saxony-Anhalt (dated ca. 2500–2050 BC), of hg. R1a1a1-M417 (possibly xR1a1a1b-Z645), and one outlier (ca. 2560–2300 BC), with a mean of ca. 71% Steppe ancestry (Mathieson et al. 2015). The outlier has what appears to be a recent contribution from Yamna, clustering closer to Yamna samples than any other Corded Ware sample (except for the Baltic outliers), probably due to exogamy with nearby late Yamna settlers of Hungary or early East Bell Beakers. Five males among them have been inferred to be relatives *via* paternal line (Monroy Kuhn, Jakobsson, and Günther 2017), and one of these, the outlier, is a second-degree relative to the other four.

This interpretation of a recent contribution from Yamna in central Europe is supported by samples of the Corded Ware group from Brandýsek, Bohemia (ca. 2900–2500 BC), which show diminished Steppe ancestry (ca. 40%), and two samples of hg. R1a1a-M198 together with one I2a1b1b-Y6098 (Olalde et al. 2018). The resurgence of this typically Neolithic haplogroup with a marked increase in NWAN ancestry (ca. 45%) seems to suggest a resurgence of local Neolithic groups, which supports the nature of the Esperstedt outlier as an exception among late Corded Ware samples.

Four late Corded Ware samples (ca. 2570–2340 BC) from double burials of related people, among fourteen individuals of a multiple burial in Pikutkowo, Poland, show that they are genetically significantly closer to WHG than to steppe individuals (especially one of the investigated pairs), and can be modelled as an admixture between Corded Ware and local Neolithic populations with hunter-gatherer affinities (such as TRB, ca. 63%). Two samples are of hg. I2a1b1a2b1-L801 (Fernandes et al. 2018), which appeared earlier on GAC samples from Poland (see *§v.6. Late Uralians*), support the resurgence of local lineages among different central European groups at the end of the Corded Ware period.

One sample from Spiginas, Lithuania (ca. 2130–1750 BC) of hg. R1a1a1b1a2b-CTS1211 (of the R1a1a1b1a2-Z280 trunk), of the Battle Axe culture (Mittnik, Wang, et al. 2018), evidences the continuity of typical Corded Ware lineages in the area. Based on later Baltic and Poland Bronze Age samples, this precise subclade probably expanded from this and neighbouring southern areas, or resurged from previous populations of the area (see *§viii.8. Balto-Slavs*).

VII.2. Pontic–Caspian steppes

VII.2.1. Poltavka

The Poltavka culture (ca. 2800–2200 BC), known almost exclusively through its graves, presents a tradition quite similar to Yamna, but with a distinctive style of pottery, and changes in the shape of the grave pit, in details of the mortuary rituals, and in metal tools and weapon styles (Figure 40).

Poltavka cemeteries appear in the same geographic region as the north-eastern group of the early Yamna culture. In the Don–Volga steppes, it overlaps with the Catacomb culture, and its graves there show side chambers—shallow hollows undercut into the side of the grave—equivalent to (although clearly distinct from) Catacomb graves. Five Catacomb-style graves have been found in the Samara Valley, as early as ca. 2750 BC, always as isolated additions to earlier kurgans, suggesting contact and exchange between regions on both sides of the Volga. Most kurgans involved the body placed in a single chamber, with a wide step, beneath a kurgan surrounded by a circular ditch, and graves contained usually an adult male, although adult female central burials have also been found (Kuznetsov and Mochalov 2016; Murphy and Khokhlov 2016).

A coetaneous culture overlapping geographically with Poltavka is the Vol'sko-Lbishche group, with sites on the elevated, forested height west of the Volga, later appearing in Samara. Of the twelve known MBA settlements and seasonal camps in Samara, eight contained Abashevo materials (see *§VIII.17.1.*

Abashevo), while 3 showed Vol'sko-Lbishche pottery. Only two, overlapping with the latter, showed Poltavka pottery, which was 10 times lower in artefact density than succeeding Srubna materials, all of which supports the higher mobility of Poltavka seasonal camps (Kuznetsov and Mochalov 2016).

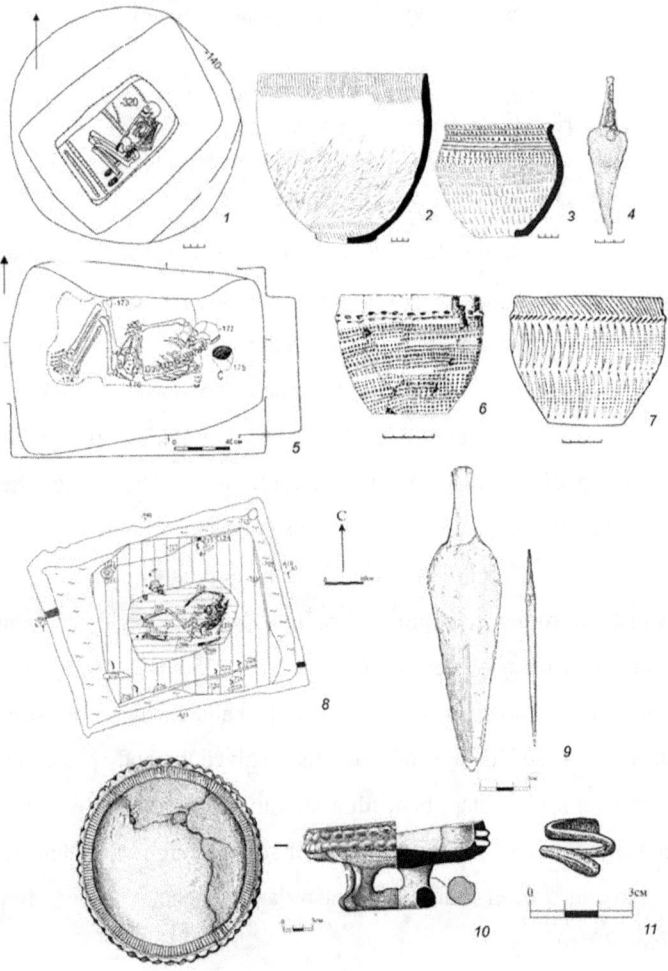

Figure 40. Materials of the late (Poltavka) stage of the Yamna culture in the Volga-Ural region: 1-4 - Orenburg Cis-Urals (Morgunova et al., 2010); 5–7 - Samara Volga region (Vasiliev et al, 2000); 8-11 - The golden kound in the Lower Volga Region (A Passage, 2009). From Morgunova (2014).

Poltavka continues the previous Don–Volga–Ural tradition of predominant sheep-herding economy. Tin from Irtysh sources is found in Troy IIg (ca. 2300–2200 BC), which points to an international east–west trade network in tin across the steppes, and to the connection of Poltavka with Trans-Uralian cultures. In the southern Ural steppes, an intermediate variant called the Tamar-Utkul type has been defined, also called pre-Ural variant of the Late Yamna culture (Bogdanov 2004). Pastoral economy and kurgan burial rituals show continuity with early Yamna.

VI.2.2. Catacomb

Most Yamna burials west of the Black Sea have radiocarbon dates ca. 2880–2580 BC. Only a small proportion of sites at the Lower Danube shows later dates, with a dilution of the wider Pit–Grave phenomenon. This third stage of pit–graves shows a re-appearance of individuals buried contracted to the side or in extended body position as secondary burials in the mounds, perhaps under the influence of the Catacomb Grave culture or further to the east, or locally at the Lower Danube (Frînculeasa, Preda, and Heyd 2015).

By 2500 BC, Yamna is already on the decline, and is gradually transforming everywhere ca. 2600–2400 into the Catacomb Grave culture, while losing grip of settlements in the western Pontic area and retreating thus to its north Pontic core zone. It is hypothesized that the reduction in winter precipitations and an increase in precipitation during the warm season may have caused the increase in productivity of phytocenoses providing more favourable conditions for early cattle breeders, explaining the bloom of the Catacomb culture in the desert steppes (Khomutova et al. 2019).

The Catacomb culture (ca. 2500–1950 BC) is centred on the Dnieper–Azov–Don–Caspian steppes, starting thus during the emergence of the European Early Bronze Age, marking the shift of the centre of gravity from the east-central European lowlands (with west Yamna settlers) to western Europe (with Bell Beakers) and to the Aegean. Methodological problems make it difficult to distinguish late Yamna from Catacomb burials in the Prut,

Carpathian, and Danube areas up to the East Thracian Plain, but it seems established that the culture was centred on the north-west Pontic steppes, with less frequent and intensive infiltrations on the Danube than the previous Yamna culture (Frînculeasa, Preda, and Heyd 2015).

The earliest finds appeared in the area between the Don, Volga, and Caucasus foothills during the late Yamna stage, which is compatible with the burial ritual featuring prominently north Pontic and Kuban tradition of wagon burials, representing members of the social elite. The standard burials are catacomb grave complexes—kurgans with an entrance shaft and burial niches in its side walls—with the dead buried in both crouched and supine positions. The aridity of the previous period continues, as do the seasonal camps with tent-like shelters in the steppe, supporting their cattle-breeding economy. Part of their population probably remained behind in permanent river settlements, engaging in agriculture and pit-breeding, and with scattered complex fortifications and communal grave buildings showing a more complex organisation (Parzinger 2013).

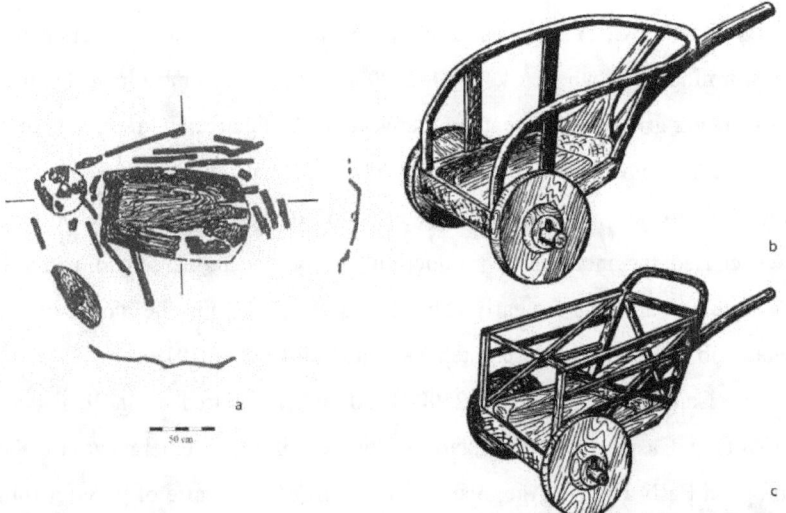

Figure 41. Two-wheeled wagon from burial 27 of the kurgan Tyagunova Mogila (a) and two reconstruction proposals (b–c), by Cherednichenko and Pustovalov (1991), modified from (Pustovalov 2000).

The first two-wheeled vehicles appear in the Black Sea region in Tyagunova Mogila (Figure 41) and Bolshoi Ipatovskyi Kurgan. Both carts are small (diameter up to 60 cm), single-piece disk wheels with an integral nave independently rotating on the axle, and can thus be seen as forerunners of an actual chariot, similar to the vehicles known in the Near East at this time. While the role of the domestic horse in the economy is unclear, there is a clear presence of horses as ritual offerings in burials, suggesting their great social importance (Chechushkov and Epimakhov 2018).

Its subsistence economy, communication, mobility, and exchange patterns are altered for major parts of the population, although a continuous presence of steppe-related settlements is seen in the Lower Danube and Dobruja regions. The rise of metallurgy and the relevance of craftsmen associated with the Early Bronze Age, which began during the Yamna period, acquires probably its full meaning and greatest extent during the Catacomb period, judging by the number of metallurgists' graves in the region during the 3rd millennium, in what has been identified as the Inhul–Donets Early Bronze Age Civilisation (Klochko 2013).

Common bronze artefacts are slender shaft–hole axes, adzes, chisels, daggers with flanged hilts, and different blades, as well as little spirals, beads, and hair rings. Present are also stone clubheads and flint spears and arrowheads. Pottery shows local differences, as in the previous Yamna period, but some cross-regional styles can be distinguished, such as pots with funnel or cylinder necks, deep bowls with short rims, and incense burners (Parzinger 2013).

vii.2. Early Indo-Iranians

Investigated samples from the steppe include six individuals of the Poltavka culture in the Samara region (ca. 2900–2200 BC), seven of the North Caucasus culture (ca. 2800–2500 BC)—in the piedmont steppe of the central northern Caucasus—and six from the Catacomb culture in the Kuban, Caspian, and piedmont steppes (ca. 2600–2200 BC), almost all showing continuation of the typical Steppe ancestry profile of Yamna. In contrast to Poltavka, which shows low EFF contribution (ca. 9%, i.e. between Yamna Kalmykia, of ca. 5%, and Yamna Samara, of ca. 12%), Catacomb shows a slightly higher mean EEF contribution (ca. 17%), with some individuals clearly shifted towards Yamna samples from the North Caucasus and towards Maikop outliers (showing a similar EEF ancestry), which is compatible with genetic continuity in the same territory. All samples with reported Y-chromosome haplogroup (five Poltavka, five North Caucasus, three Catacomb) show R1b1a1b1-L23, and among them two R1b1a1b1a-Z2103 lineages are confirmed in Poltavka; four in North Caucasus, two of them R1b1a1b1b3a-Z2109 positive for Y20993[+] (TMRCA ca. 2800 BC), a subclade of KMS67 (formed ca. 3500 BC, TMRCA ca. 3300 BC), which was found earlier in a Yamna individual from the Samara region; and another R1b1a1b1b3a-Z2109 (formed ca. 3500 BC, TMRCA ca. 3500 BC) among Catacomb samples (Mathieson et al. 2015; Haak et al. 2015; Wang et al. 2019). One sample without a clear cultural adscription from Stalingrad Quarry (ca. 2680 BC) also belongs to hg. R1b1a1b1a-Z2103 (Allentoft et al. 2015).

A sample from Deriïvka (ca. 2900–2700 BC), also of R1b1a1b1a-Z2103 subclade, classified as Late Neolithic (without background information), is a clear outlier, clustering closely with north Pontic Neolithic samples, with contributions of Steppe and NWAN ancestry (Mathieson et al. 2018). This recent admixture of a typical Yamna lineage in the forest-steppe of the north Pontic area—the sample belonging thus probably to the late Yamna or early Catacomb culture—may suggest either the survival of small pockets of local populations among expanding Yamna clans, or an admixture with neighbouring northern peoples from the forest area which expanded to the south. The latter would be justified by the intense contacts of the late Yamna/Catacomb culture with the Middle and Upper Dnieper, evidenced by parallelisms in material culture of Catacomb with the Middle Dnieper culture (Klochko and Kośko 2009).

A Poltavka outlier from the Sok River in Samara (ca. 2900–2500 BC) clusters closely with Central European samples of the Corded Ware culture, and shows hg. R1a1a1b2a-Z94 (formed ca. 2700 BC, TMRCA ca. 2700 BC), a subclade of R1a1a1b2-Z93, widely distributed in the subsequent period in the steppes and Central Asia (Mathieson et al. 2015). This grave was most likely established on top of an older Poltavka cemetery, where a Sintashta cemetery was later found. Nevertheless, assuming the date is correct, it would be the first genetic proof of the intense interaction and admixture in the Volga–Ural region leading to the Sintashta culture, between Abashevo in the forest-steppe area and Poltavka peoples in the steppes. This close interaction with a Uralic-speaking culture makes Poltavka the most likely representative of a Pre-Proto-Indo-Iranian-speaking community.

VII.3. Southern Caucasus

The 3rd millennium BC is represented in the Caucasus by good precipitation and warm temperatures, which promoted a good forest cover in the Bedeni Plateau and the Trialeti region. This probably influenced the dramatic transformation in human behaviour and material culture in the region, represented by the appearance of the so-called Early Kurgan period by the mid–3rd millennium (Karim, Sepideh, and Mohammadi 2018).

This process has been associated with newcomers of a significantly different lifestyle and means of subsistence, possibly associated with a mobile economy, appearing at the same time as the Kura–Araxes traditions disappear. This evidence and the lack of proof of coexistence with the new population suggest a violent end of the culture in the region. In this *push–pull* process, Near Eastern societies form the south had an important role, judging by the cultural changes of Kura–Araxes communities, and the adoption of a sedentary village life by the newcomers (Karim, Sepideh, and Mohammadi 2018).

The appearance of fortifications in certain sites before the end also evidence the increase of intergroup conflicts and militarism during the Early Bronze Age. While the new groups of cattle herding pastoralists with wheeled carts and oxen-pulled wagons appear in the north, the Kura–Araxes communities subsequently moved farther south. It is unclear if the newcomers are part of a southern Caucasus culture – associated with the emerging Trialeti culture –, or if they came from further north, but some settlements seem to have been abandoned without traces of a violent end (Karim, Sepideh, and Mohammadi 2018).

Although the appearance of barrow cultures is abrupt, a transitional period can be observed in certain sites, showing late Kura–Araxes and local elements at the same time. Communities of the Middle Bronze Age I (Martkopi, Early Trialeti) buried their dead beneath large stone mounds, circular in plan, sometimes covered with a layer of earth, and their dimensions likely reflected social status. The most common item in funerary assemblages is pottery. Graves are often multiple (Karim, Sepideh, and Mohammadi 2018).

The Bedeni barrow tradition (in the high country and the lowlands) shows a more diverse range of burials, with the most popular one represented by large timber structures constructed in a deep rectangular or square grave pit, with aboveground mortuary architecture. Barrows show occasionally a gender-based differentiation reminiscent of the Yamna culture, and sometimes wagons are included in the graves, as was typical of some north Pontic groups. Not all burials were monumental or had a rich assemblage (Karim, Sepideh, and Mohammadi 2018).

The few studied Bedeni settlements show villages surrounded by defensive structures—e.g. a stone perimeter wall, or flanking ditches—formed by multiple houses with a plan reminiscent of the Kura–Araxes tradition: square with slightly rounded corners, with an anteroom and a main rectangular room with a central fixed, baked-clay hearth. Innovations in ritual behaviour with respect to the previous period is the appearance of platforms, which apparently replace hearths as focal points; the use of pit digging and filling; and also the burning of abandoned villages (Karim, Sepideh, and Mohammadi 2018).

Bedeni pottery shows coarse wares proper of the previous Kura–Araxes period, but also an innovative and highly developed potting tradition inspired over time by new advances in metalworking, since many vessels have a metallic look about them. There is a high level of woodworking, and sophisticated lithic industry, with a new projectile design different to the tanged and barbed arrowheads of the Kura–Araxes, more effective with grater penetrating power (Karim, Sepideh, and Mohammadi 2018).

vii.3. Southern Caucasians

In the North Caucasus steppes and piedmont, continuity with Steppe ancestry is seen. In the Caucasus region proper and to the south, the latest Kura–Araxes samples available and the subsequent North Caucasus, Dolmen, and Lola samples depict continuity of a typical Caucasus ancestry (see *§v.2. Early Caucasians*), hence the described prehistoric geographical and genetic barrier to steppe invasions (Wang et al. 2019).

Increased CHG ancestry (ca. 60%) is also found among three Early Bronze Age individuals from the Kura–Araxes culture in Armenia dated ca. 3300–2500 BC (Lazaridis et al. 2016), with a late sample (ca. 2550 BC) showing what seems to be a resurge of haplogroup R1b1a-L388 (formed ca. 15100 BC, TMRCA 13600 BC), ancestral to R1b1a1-P297, in the area, with subclade R1b1a2-V1636 (formed ca. 13600 BC, TMRCA ca. 4700 BC).

An outlier from Hajji Firuz Tepe in the north-western Zagros Mountains harbours elevated Steppe-related ancestry and clusters closely to a Yamna outlier from Ozera, Ukraine, and even more closely to Maikop and Armenia Chalcolithic samples (Narasimhan et al. 2018). This is consistent with the incorporation of groups similar to the known Maikop outliers of the northern Caucasus within the expanding Kura–Araxes groups (see *§vi.4. Northern Caucasians*), receiving thus contributions of Maikop and/or Kura–Araxes from the south.

Even though the radiocarbon dates published for the site (in this case ca. 2465–2286 BC) are unreliable, because of the collapse of different archaeological layers, it represents in any case most likely a Caucasus population with Steppe ancestry, rather than a direct migration from the steppe. The invasion of a Maikop-related population to the south is compatible with the described presence of wagons and rich assemblages in Bedeni. Given the uncertain dates, a late steppe-related population cannot be discarded, although Iron Age steppe-related populations, probably incoming from the Balkans, show a more southern cluster in the PCA (see *§viii.14. Caucasians and Armenians*).

VII.4. Aegean Early Bronze Age

In the first centuries of the 3rd millennium BC, new networks of exchange and trade developed, and social complexity increased in Northern Mesopotamia, reaching its peak in the Sumerian Early Dynastic III and Akkadian periods. The new political organisation that consisted at first of more or less independent rival city–states, out of which grew the hegemonial "empire foundation" of the Akkadian period at the end of the 24th century BC. Its centre was in southern Mesopotamia, urbanised since the 4th millennium BC (Ökse 2017).

Humid climatic conditions in the Near East had moved the border of minimal precipitation for dry-farming towards the south, creating a large arable region. No urban centres dating to these centuries are observed in the Upper Tigris, whose settlement structures reflected permanent rural settlements inhabited by egalitarian societies (Ninenvite V) until ca. 2600/2500 BC. These societies changed gradually from a sedentary life to a mobile one with seasonal activities, until the Akkadian supremacy (Ökse 2017).

Flood fills in the valleys of the Euphrates and Tigris indicate an area not suitable for sedentary life from ca. 2650 BC, which created a trend to new settlements on plateaus, or transition to pastoral economies. A settlement decline in the Upper Tigris must have occurred in concert with trade routes gaining importance in the south, and increasing pastoralism in the north. The new system connected regions such as Turkmenistan and Afghanistan in

Central Asia, the Arabian Peninsula, the Indian Ocean as far as the Harappa Culture of the Indus Valley and north-western India, and integrated Levantine-eastern Mediterranean and Anatolian regions, previously considered marginal regions. The last to be absorbed was the area around the Aegean, i.e. western Anatolia and Greece, gradually incorporated since ca. 2750 BC, developing a western nucleus of exchange and trade ca. 2500–2250 BC (Ökse 2017).

The Early Helladic-Cycladic-Minoan II included the following advances: stratified society with many prestige and status objects of the elite, of urbanisation, a three-fold structured settlement system and population growth; quasi-monumental architecture and organised communal works; complex administration and standardised systems of measuring and weighting; economic specialisation and mass production such as wheel-made pottery; and large quantities of copper, gold, and silver, as well as the first tin–bronzes (Heyd 2013).

This period showed a climate favouring agricultural production, based on a mixed small-scale and intensive system, which included Mediterranean polyculture based on grain, wine, and olive oil, apart from figs. There is also an increase in domestic animals and in the use of secondary products like wool, milk, and cheese from the beginning of the 3rd millennium. This contributed to the constant growth of the population and enlargement of urban centres, reaching a population density that would only recur in the late Bronze Age Autarchies were replaced by specialised trading systems, and dependencies were created. Long-distance trade is inferred from exotic objects decorated in the Indus area, and coastal settlements specialised in maritime trading were built in this period. Social hierarchies developed, as well as the notion of territory, political control and 'chiefdoms' in general (Ökse 2017).

The new rural sites established ca. 2400/2300 BC in the Upper Tigris coincide with the period of aridity that moved the minimal border of precipitation again to the north, and probably impacted the socioeconomy of northern Mesopotamia, which forced the Akkadian kings to repopulate the

Upper Tigris region, in order to establish a new agricultural system to provide for food. The new sites showed pottery similar to that produced in the Habur region, indicating a strong relation to the Akkadian territory administered from the Palace at Tell Brak (Ökse 2017).

At the end of this period ca. 2300–2200 BC, a crisis appears in Mesopotamia, then the Levant and Anatolia, and finally the Aegean region, where connections were eventually broken, and trade was cut off. Settlements shrink in size and are abandoned, and demographic levels fall. People of foreign origin take the opportunity of a serious weakening of the whole system to move into these regions in crisis, in the western Aegean during the Early Helladic III period, which lasts ca. 2200–2000 BC, reaching its lowest point ca. 2000 BC in the Middle Helladic (Ökse 2017).

vii.4. Aegeans and Anatolians

Minoans from the Lasithi plateau in the highlands of eastern Crete (ca. 2400–1700 BC) and from the coast of southern Crete (ca. 2200–2700 BC) were a homogeneous population, with an ancestry shared with Bronze Age south-western Anatolians of Harmanören–Göndürle Höyük (2800–2300 BC). All Aegean populations derived most of their ancestry from an Anatolia Neolithic-related population (ca. 62-86%), but with contributions of Iran Neolithic-related ancestry (ca. 9-32%), which was already present during the Neolithic in samples from central Anatolia and in Tepecik-Çiftlik. Two Minoans from Lasithi and one Anatolian individual from the Bronze Age showed haplogroup J-M304, which was rare or non-existent in earlier populations from Greece and western Anatolia, dominated by G2-P287 (Lazaridis et al. 2017):

Minoans show hg. J2a1d-M319 (formed ca. 11000 BC, TMRCA ca. 9700 BC), with haplogroup J2a-M410 found first in the Caucasus, in the Palaeolithic individual from Kotias Klde, and later hg. J2a1-L26 is widely distributed during the Bronze Age from Central Asians in the east to Assyrians and Mycenaeans in the west, which suggests a potential expansion with CHG/Iranian farmer-related ancestry in the Chalcolithic (see *§iv.4. Late*

Middle Easterners). The Minoan sample from southern Crete, of hg. G2a2b2a-P303 (a lineage found in central and eastern European farmers), shows thus continuity of regional male lines.

Based on subsequent migrations, the incoming population of J2a1-L26 subclades may have brought from the east languages ancestral (or related) to Tyrsenian, which would be the common substrate described for Greek and Anatolian. Their integration with other peoples, such as Assyrians in the Bronze Age, proves that the high demographic density and advanced political organisation of Near Eastern cultures may have allowed for different lineages to become integrated into different communities speaking diverse languages, depending on the ruling elites of each period and region. This long-term connection between Anatolia and the Aegean may also support the proposal of a Hatto-Minoan, and its potential relationship with Sumerian (Schrijver 2015).

The south-western Anatolian BA sample shows J1a2b-Z1828 (formed ca. 16000 BC, TMRCA ca. 6100 BC), which was also found in the Levantine Bronze Age (Lazaridis et al. 2017). This hints at ancestral southern Anatolian populations which were probably responsible for the introduction of Levant-related ancestry in the region.

In Anatolia, the first attestation of Anatolian languages is believed to occur in typical masculine personal names found among inscriptions from Armi (ca. 2500–2300 BC) a regional state which enjoyed a privileged relationship with Ebla, a Semitic-speaking Kingdom from south-eastern Anatolia and the northern Levant. The kingdom of Armi was possibly located in the Upper Tigris, in a recently Semiticised area related to silver and copper trade with the north (Bonechi 1990; Archi 2011).

In the late 3rd millennium, the heartland of Hittites probably lay in the upper reaches of the Halys River, in a zone between the Luwian heartland to the southwest, in the Lower Land south of the Tuz Gölü central Anatolia, and the Hattians to the north on the central Anatolian plateau. This tentative location is based on the mixed influence of Hattic and Luwian on early Hittite, and on the presence of Hattic loanwords in prehistoric Hittite *through* an intermediary Luwian language. Poor evidence exists of the Luwian presence in west Anatolia, with few scraps of evidence suggesting that early forms of Carian and Lydian may have been the spoken languages of the area (Melchert 2011).

VII.5. The Balkans

The population of Bronze Age tell settlements from the Carpathian Basin show ritual practices in common with the Mycenaean world, with an official cult practised in specific buildings, like temples destined to serve the entire community, complemented by a family cult, represented by fireplaces and small altar pieces or miniature wagons made of clay (Gogaltan 2012).

There were a potential solar cult (reminiscent of the Zeus/Apollo cult) before its appearance later in the Urnfields culture and in the Nordic Bronze Age; human sacrifices potentially addressed to a deity of war (such as Ares); food offerings potentially for some deity of fertility (like the "Great Mother"); animal idols and drinking vessels; a "hero cult" with weapons and other metal objects, etc. (Gogaltan 2012). All of this strengthens the idea of a common Balkan community, in contact with central European cultures during the Bronze Age.

The peripheries from the Aegean Early Bronze Age developed a dynamic new social and economic system through contacts with the core areas, by way of imitation and innovation. The direct exchange network included the eastern Balkans (northern Aegean), with Bulgaria and western Anatolia; the western Balkans, with the eastern Adriatic coast as well as the inland; and the south-central Mediterranean area, particularly Sicily, Malta, and Apulia, which were eventually under the expansion of Bell Beakers (Heyd 2013).

After west Yamna groups lost their internal coherence and direct contacts with the north Pontic homeland in the mid–3rd millennium, the Aegean became the new cultural model for the Balkan population (Heyd 2013):

From the East Thracian Plain to Troy, after ca. 2500 BC, local cultures become more complex, with graves of local leaders and elites; ritual sites and buried hoards, imports including jewellery of gold and silver, weaponry, and thousands of large and varied golden artefacts; a new dress code that came with widespread dress pins; etc.

In the west the Aegean influence is felt earlier in graves from the East Adriatic coast up to the Danube, perhaps even earlier than 2750 BC, judging by a hierarchically structured settlement of Vučedol and similar sites along the Danube. A trade connection is created between local elites in the Adriatic (including south-eastern Italy) that persists after the demise of the previously dominant Vučedol complex, as witnessed in prestige object imports in regional groups such as Vinkovci in Slavonia and Syrmia, Bubanj Hum III and Armenochóri in eastern Serbia and Macedonia, Belotić-Bela Crkva in Central Serbia, and Cetina along the Adriatic coast.

Overall, a "chiefdom" system based on prestige goods comes into being in both territories, In the east along the Adriatic coast, a maritime trade system develops, as well as local concentration, cultural regionalisation (where rather small areas develop a cultural identity), and a wave of centralisation. Cetina, however, under the influence of the Bell Beaker culture (see *§VIII.9.1. Cetina*), remains apart from these social changes, as evidenced by the absence of prestige goods and its drive to expand to the south in Albania and then in the Peloponnese during the Early Helladic II to III (i.e. ca. 2200 BC).

vii.5. Palaeo-Balkan peoples

One west Yamna individual from Mednikarovo in Bulgaria (ca. 2950 BC), of hg. I2a1b1a2a2a-L699, shows contributions of NWAN-related ancestry, with a clear 'southern' drift in the PCA towards Balkan populations (Mathieson et al. 2018). A similar ancestry (and shift in PCA) is found in a sample of the Vučedol culture from Beli Manastir (ca. 2775 BC), of hg. R1b1a1b1a-Z2103, contrasting with the other available Vučedol sample, which clusters closely to other Balkan Bronze Age populations, in turn clustering closely to Anatolian farmers (see *§viii.11. Thracians and Albanians*).

Other investigated individuals from this period with contributions of Steppe ancestry (general mean ca. 30%) include: two from the EBA barrow necropolis of Beli Breyag (ca. 3400–1600 BC), with hg. I-M170 and I2a1b-M436; two from the same grave-pit, of five tall individuals laid extended on their backs, east–west orientated, head to the East and ochre-stained, from Smyadovo (ca. 3300–3000 BC), of hg. I2a1b1a2-CTS10057 and I2a1b1a2a-L701; one from the Ezero culture in Sabrano (ca. 3100–2900 BC); one from the Kairyaka necropolis under a mound in Merichleri (ca. 3000–2900 BC), buried in a small pit head to the East and ochre-stained, with legs bent at the knees, of hg. I2a1b1a2a2-Y5606; and two from Dzhulyunitsa (ca. 3300–2700 BC), one in a flexed position, of hg. G2a2a1a2-L91 and H2-P96 (Mathieson et al. 2018).

The high NWAN-related ancestry in populations from the Eastern Balkans is explained by the admixture of expanding Yamna settlers with Balkan farmer communities, which had the highest population density of Europe in this period (Müller and Diachenko 2019). While the presence of I2a1b1a2a-L701 in one sample of the LBK from Hungary (ca. 5300–4900 BC) makes this identification unclear, it seems that the simultaneous emergence of I2a1b1-M223 samples in different sites of the Eastern Balkans after the Yamna expansion in the north-west Pontic area must be related to the spread of I2a1b1a2a2a-L699 lineages from Yamna (see *§vi.1. Disintegrating Indo-Europeans*). The lack of this haplogroup in previous samples from the region supports this as the most likely explanation.

The modern distribution involving early R1b1a1b1b-Z2103 lineages includes R1b1a1b1b3-Z2106 subclades R1b1a1b1b3a-Z2108 (formed ca. 3600 BC, TMRCA ca. 3600 BC), and further R1b1a1b1b3a1-Z2110 (formed ca. 3600 BC, TMRCA ca. 3400 BC), found in modern populations from the Balkans and Central Europe; R1b1a1b1b2-L277.1 (formed ca. 2100 BC, TMRCA ca. 2100 BC), also found in the Balkans; and R1b1a1b1b1-L584 (formed ca. 3200 BC, TMRCA ca. 2900 BC), found in Armenian and other

Central European populations. The early split of R1b1a1b1b-Z2103, found widespread also among ancient and modern Indo-Iranians, makes a proper identification of certain lineages with the spread of certain peoples (and specific routes of expansion) difficult without ancient samples.

VII.6. Iberia

The Proto-Beaker package probably emerged in a south-western Iberian region, part of the southern and west-central Iberia that participated in the evolution to complex, huge fortified settlement sites like Los Millares (Figure 42) and Zambujal, and other even larger macro-villages. The regions around the lower and upper Guadiana and the upper Guadalquivir rivers stand out as two of the most densely settled territories in Iberia, probably related to their agricultural potential and their rich copper ore deposits. Macro-villages in this region extend over more than 100 ha (Heyd 2013).

Figure 42. Painting of Copper Age walled settlement of Los Millares, by Miguel Salvatierra Cuenca. Photography by Jose Mª Yuste. From Wikipedia.

This local development must be discussed in light of the "Transformation of Europe", the demographic, cultural, and economic pressure brought about by western Yamna migrants disturbing the equilibrium in central Europe. However, another often discussed influence are the contacts with Early Bronze Age cultures from the Aegean, and probably also the Levant. This east–west Mediterranean connection is evidenced for example by the exchange and use of ivory (Heyd 2013).

The dimensions of the settlement complexes are therefore essentially an expression of the available work force and agricultural wealth of these societies, as well as of the demographic increase observed during the 3rd millennium, all consequence of the new exceedingly productive subsistence strategies. Most Iberian Copper Age communities of any size and geographic location expressed a strong preference for communal values, as shown by collective burial practices and communal organisation of part of the economic production, such as open spaces dedicated to storage pits, specialised metalworking areas, grinding equipment, flint knapping traces, etc. (Risch et al. 2015).

Their connection is also seen in a common universe of decorative motifs, signs, and symbolically meaningful artefacts, apart from the sharing of long distance networks, evidenced by metalworking technology, flint artefacts, and the use of raw materials such as ivory (Risch et al. 2015).

The presence of exotic materials such as ivory, ostrich eggshell, or amber becomes evident, but these foreign elements are most highly concentrated at the largest archaeological sites, such as Valencina de la Concepción or Los Millares, where there was a greater capacity for the mobilisation of work and acquisition of foreign raw materials (Murillo-Barroso and Montero-Ruiz 2017).

Ditched enclosures and fortified settlements coexisted and were thus probably specialised in their economic practices. Both types of settlements were associated with megalithic tombs or subterranean grave structures in the immediate vicinity, *hypogea*, and large pits, with deceased of all ages and sexes buried there over generations. *Tholoi* and natural caves were also used during this time, with differences in funerary rites not representing apparently unsurmountable cultural barriers (Risch et al. 2015).

Nevertheless, the Iberian Copper Age was essentially dominated by a rather mobile residential pattern, with unfortified occupations, often less than 0.5 ha, found in very different topographical positions. Basic features of the subsistence economy include intensive agriculture on the most fertile or humid soils, but also a firmly established husbandry (with milk and wool as by-products), with hunting, gathering and fishing providing important complementary resources. Many of these smaller settlements were probably dependent agricultural communities, compelled to pay tribute to the larger, more complex settlements (Risch et al. 2015).

These Chalcolithic societies thrived in the Final Neolithic–Chalcolithic period, with small and short-lived chiefdoms, transegalitarian and hierarchical polities. The demographic density increased during the pre-Beaker period in the Meseta and in the south-east, peaking during the initial Beaker phase, while the south-western region shows a more discrete growth peaking during the pre-Beaker phase and dropping abruptly ca. 2500 BC, just prior to the Beaker period (Blanco-González et al. 2018).

The introduction of Bell Beaker pottery ca. 2600/2500 BC, quickly expanded through the Atlantic coast, shows a variable importance in the rest of Iberia. In the area of Los Millares, it appears only in certain settlements, while in the south-west Bell Beakers are rare when not completely absent. Furthermore, there is no clear correlation with specific types of habitation, funerary structures, or with metal production (Blanco-González et al. 2018).

The EBA shows changes in south-eastern Spain partially synchronic with those in Italy and the Balkans ca. 23rd/22nd century BC, marking a profound social, political, and ideological evolution. A substantial number of 14C dates confirms that most, if not all, of the Chalcolithic fortified settlements and Late Neolithic–Chalcolithic monumental ditched enclosures had been abandoned by 2200 BC (Blanco-González et al. 2018).

The older networks of symbolical axes made of exotic rocks, flint, ivory and decorated schist plaques, Bell Beaker pottery, etc. collapsed rather abruptly or was reorganised at a much more local scale. In the funerary sphere, this date marks the abandonment of a collective burial rite. However, while the Atlantic coast shows an abrupt de-intensification of human pressure, the central and eastern Iberian regions do not show these fluctuations (Blanco-González et al. 2018).

vii.6. Basque-Iberians

Intensified contacts during the Final Neolithic–Chalcolithic in Iberia–southern France, including common social and economic developments, with the newly created exchange networks and the demographic expansion, are potentially the mark of an expanding common ethnolinguistic community, which may be identified with an Ibero-Basque group (Villar Liébana 2014).

Neolithic individuals from Iberia and France show a large proportion of hunter-gatherer ancestry, continued in a mean of ca. 25% for Middle Neolithic (higher in the north-west, lesser in south-western Iberia) and ca. 30% for Chalcolithic populations, from a source closer to north-western Iberia (Canes1-like) with admixture events that happened most likely between the Early and Middle Neolithic period. While Neolithic samples from Iberia and south-western France include mixed G2-P287, I2-M438, and R1b1b-V88 lineages, the majority of the reported Iberian/France Late Neolithic–Chalcolithic Y-chromosome haplogroups are I2-M438 subclades (Martiniano et al. 2017; Lipson et al. 2017; Gunther et al. 2015; Valdiosera et al. 2018; Olalde et al. 2018; Olalde et al. 2019):

Of the I-M170 lineages, there are at least six I2a1a1-CTS595, of them two I2a1a1a-L158 (formed ca. 16200, TMRCA ca. 9700 BC), concentrated in south-western Iberia, with one sample from the centre. At least one of them is I2a1a1a1-Y3992 (formed ca. 9700 BC, TMRCA ca. 6300 BC), reported from northern Iberia, a lineage shared with north Italian and Balkan Chalcolithic samples, and found in one Early Neolithic sample (ca. 5200 BC), suggesting a spread coinciding with the Neolithic expansion and later migrations from western Europe.

There are at least twenty-seven I2a1b-M436 lineages, also found in one Early Neolithic sample (ca. 5200 BC). Probably most are under I2a1b1b-Y6098 (formed ca. 6000 BC, TMRCA ca. 2900 BC), at least eleven under I2a1b1b1-S23680/S23467 subclades (formed ca. 6000 BC, TMRCA ca. 5700 BC), most of them from the south, but appearing in sites all over Iberia, with one I2a1b1b-Y6098 sample found in a Corded Ware sample from Czechia. Three of them are I2a1b1a2b-Z161 lineages, and also appear scattered all over Iberia in Neolithic and Calcolithic samples.

There are at least eight I2a1a2-M423 lineages in south-western Iberia, northern Iberia, southern France Megalithic, and the Neolithic from the British Isles, being a haplogroup widespread among central European hunter-gatherers and later farmer populations.

There are also seventeen G2a-P15. At least six G2a2b-L30 samples, three within G2a2b2b1a1-F872, two G2a2b2b1a1a-PF3378 (formed ca. 6900 BC, TMRCA ca. 3800 BC), at least one of them G2a2b2a1a1c1a-Z1903 (formed ca. 6100, TMRCA ca. 2500 BC); and seven G2a2a-PF3147 lineages, probably all within the G2a2a1-PF3148 tree, likely continuing an Early Neolithic G2a2a1a3-FGC34625 lineage (formed ca. 8900 BC, TMRCA ca. 7400 BC) that is found in at least two Late Neolithic samples; and there are also three F-M89 (also found during the Early Neolithic and in the Early Bronze Age); and seven H-L901 lineages, at least three of them H2-P96, scattered in different regions.

If there was a Basque-Iberian language, it should be identified with Chalcolithic communities of Iberia and France before the East (or Classic) Bell Beaker expansion. The possibility of a Basque-Iberian language may be supported by the genetic homogeneity and continuity during the whole Neolithic period and during the Chalcolithic. Cultural expansions of western Europe (either with or without population movements), first with the Megalithic culture in the Middle Neolithic, and then with the Proto-Beaker package in the late Chalcolithic, may also be adduced to support this cultural unity. The language of this group should then be associated with the language of early Neolithic farmers, and thus possibly related to their expansion through the Mediterranean (Villar Liébana 2014). The fact that the ancestors of Basques and Iberians were isolated from each other in the Chalcolithic, after the arrival of East Bell Beakers, and were thus surrounded by Indo-European languages but still showed common linguistic traits in the Iron Age seems to support their ancient connection rather than recent areal contacts.

VII.7. Bell Beaker culture

VII.7.1. The Bell Beaker package

The Bell Beaker phenomenon is defined by groups that show a common know-how in technology, especially regarding pottery, copper metallurgy (Amzallag 2009), and flint. No single unified network of know-how transmission can be reconstructed, only local or regional networks (Linden 2015). Despite this, a supra-local homogeneity can be observed in the whole of Europe from ca. 2500 BC "in similar funerary rituals, in the way of interacting with territory, in the way of representing iconography and decorating pottery, and in the way of representing social differences" (Martínez and Salanova 2015). The Bell Beaker phenomenon made thus the previous regional networks of western Europe uniform with identical social codes.

With the advent of radiocarbon dating, the compilation of Bell Beaker pottery dates (Müller and VanVilligen 2001) showed that the most likely origin of the pottery style was Iberia, pointing to high quality, tall beakers of the so-called maritime style. Only later were these dates and the Bell Beaker migrations integrated together in a common paradigm, when it was noted that the expansion of beakers with lower profiles and a more complex decoration, from East Group beakers, were replaced in the Danube area by plain jars, cups and plates. These vessels then dominated in the later developments of the culture (Harrison and Heyd 2007).

The migration of mobile Yamna migrants into the plains of the lower Danube and the central Carpathian basin is noted in small and large groups in the Balkans, establishing pastoral societies as forerunners of Heyd's "Yamna package", with domesticated horses, ox-drawn wagons, and herds of cattle and sheep, and noticed as far as southern and central Germany. The so-called "proto-Bell Beaker package" is composed of essential early elements—such as the Maritime Beaker, copper knives and awls, advanced archery skills and reliance on the bow and arrow, and decorated textiles, perhaps also V perforated buttons—but lacks boars' tusk bow-shaped pendants, stone wrist-guards, and the type of tanged dagger that become identified with the Classical Bell Beaker package later.

This proto-Beaker package arises at the same time ca. 2900–2800 BC in the Tagus river estuary with a high concentration of monotonously decorated early Bell Beakers, associated with a new culture of large fortified settlements, megalithic tombs and collective burials. The proto-package is found for example in the Maritime Beaker, and it expanded ca. 2700–2500 BC, getting enriched through some areas in western Europe (e.g. in the corded Beaker type), but clear internal social boundaries existed in this period. This fashionable Bell Beaker *idea* turned into the classical "Bell Beaker Package" during its expansion to the east, reaching the Rhône and Britany in the 26th century BC, then arriving in central Europe (Suppl. Fig. 10.B), and the Csepel group of the Carpathian basin, around 2500 BC (Harrison and Heyd 2007).

The transformed, classical Beaker package includes (Harrison and Heyd 2007):

1. The Beaker networks on an international scale, a system of power based on knowledge from distant parts (for the relevance of craftsmen in these networks, see below). This is related to the introduction of standardised and rich coloured textiles (dyes and morands), and refined drinking vessels (the bell beaker) used to share alcoholic drinks.

2. The creation of a socially inclusive system of belief, where additional identities (or social positions) are developed for people to acquire, which allows more people to participate (with defined roles) in the ceremonies of social promotion attached to the Beaker package, and thus welcoming more people to an enhanced status, with its new privileges.

3. The warrior self-consciousness, selecting the bow and arrow as status object for men. In its origin, macro-villages were designed for defence by many archers. The choice of archery with the expansion of the package creates a deliberate contrast to earlier styles of combat, which used hafted axes and daggers for close hand–to–hand fighting. Archery allows the warrior to fight at a distance, even from horseback, and can be concealed, which seems antithetical to a code of honour based on individual combat, where rivals face each other within hand's reach.

4. The specific female counterpart to the male warrior, apparent in life on stelae (before the expansion of East Bell Beakers) and in death in the female Beaker graves. This was added probably late due to the Yamna package in western Europe, because it is absent in Iberia but present in Central Europe before its adoption by East Bell Beakers.

5. A specific religious expression linked to sun worship, generalised throughout Europe in the 4th millennium BC. A stela from Sion depicts the rising sun, and East Bell Beaker groups place the dead so that men and women face eastwards towards the sunrise (see below), which continues the western European solar cult.

All these were part of the "Transformation of Europe" ca. 2900–2700, but the Bell Beaker package combined them into a single visual message of power, knowledge, heraldic objects and status, reinforced by a religious element. Local conflict in Sion shows that even Central European Bell Beakers (with an almost full-fledged package) did not show unity, until the expansion of the East Bell Beaker group.

For example, the composite (half-reflexed) bow, which required specialised craftsmanship, expanded to all of Europe with the spread of East Bell Beakers after 2500 BC, but its development (from the original Iberian model) must have happened during the transition of Yamna to Bell Beaker, judging by its depiction in an anthropomorphic stela from a Yamna kurgan in Natalivka ca. 2700–2500 BC (see Figure 43), and its later spread throughout the Eurasian steppe (Klochko and Kośko 1987). The typical proto-Beaker bow, as represented in stelae and bow-shaped pendants, were simple longbows. Double-curved composite bows of small size were more practical and maneuverable, ideal for use from horseback, a warfare technique probably present in eastern Europe at this time (Corboud 2009; Ryan, Desideri, and Besse 2018).

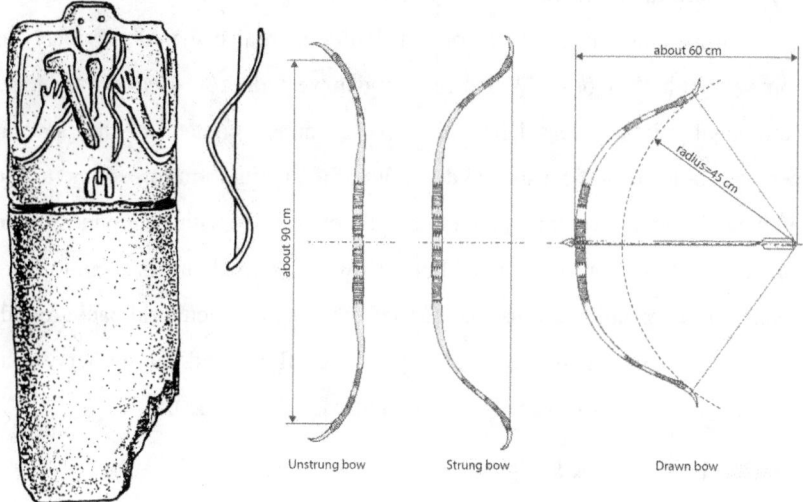

Figure 43. Left: Anthropomorphic stela above a kurgan grave of the Yamna culture (in a territory of previous Lower Mikhailovka culture), ca. 2700–2500 BC, depicting a man fitted out with the attributes of power (battle axe, staff of office, and composite bow), after V.N. Danilenko. Right: Reconstruction of a bow from the Bell Beaker culture based on depictions in stelae by Corboud (2009), modified from Ryan, Desideri, and Besse (2018): Composite bows have a wider middle and a curve at both ends, which compresses the bow allowing for it to be shorter, yet with a more efficient transfer of energy to the arrow, maintaining the power seen in longbows. They were more

complicated to make and required a higher degree of know-how than the simpler,
single-piece wooden yew bows previously prevalent in Europe.

The phenomenon accelerated dramatically when more people being involved seized the opportunity to promote themselves, by adopting the now well-defined package of novelties, so around 2500 BC Bell Beakers expanded explosively (Suppl. Fig. 11). It became thus a pan-European phenomenon, incorporating distant regions following the Atlantic and Mediterranean coasts and the main river systems, such as the Danube, Rhine and Rhône, and their tributaries. Distinct regional traditions were incorporated to different expanding currents, and eventually four large geographical entities could be discerned: the Central European domain or *East Group*, an Atlantic domain, a Mediterranean domain, and a north-west European domain including the Northern European Lowlands and Scandinavia. Within these large entities, different groups could be distinguished, such as the Rhenish/Dutch Beaker or the Northern Italian Beaker groups (Heyd 2013).

With the emblematically decorated Bell Beaker, also shared previously to some extent in the Corded Ware beaker and in western Proto-Beakers, the ideal communal drinking vessel had on average enough content so that several persons could consume a special drink out of it, and its form forced one to use both hands for drinking, and then to hand it over again with both hands, in an almost ritual manner, to the neighbour. However, two elements seem to have been more interesting innovations for newcomers of any cultural background: the dagger idea—no matter if from metal or flint—and the archery idea, materialised in the arrowheads and wristguards (Heyd 2013).

VII.7.2. East Bell Beaker group

The Bell Beaker migrations over all Europe have long been associated with the expansion over central and western Europe of Yamna migrants through the Carpathian basin (Gimbutas 1993). Specific correspondences were found in burial rites, armament, costume, ornaments, technology in general, and also in ranked society, funerary rites, belief in life after death, and in general symbolism.

Through the Upper Danube, west of the main settlement regions in the Hungarian Plains, Yamna migrants spread to southern Germany—following the routes previously created by western Yamna settlers and vanguard groups (see *§VI.1.2. East-central European lowlands*)—where decorated cup styles, domestic pot types, and grave dagger types from the Middle Danube were adopted around 2500 BC (Anthony 2007). Contemporaneous with this migration was the evolution noted in the Classical Bell Beaker culture expanded by the East group (Heyd 2007):

- ranked family-based social structures, rooted on self-sufficient farmsteads;

- a progressive specialisation in stockbreeding and plant cultivation of less demanding species;

- burials following family units, signalled by 'founder' graves;

- and no defensive position, hill forts, or fortifications, unlike later chiefdoms of the Bronze Age, where families and single persons gain power.

This structure allowed for individual and social mobility, increased communication and internal exchange of information, goods, genes, and social values (Heyd 2007).

Findings of regions north of the Alps, from modern east Switzerland to west Hungary, belong to the East (or Middle European) group of the Bell Beaker culture. Special territories of this group include the Bohemian and the Moravian provinces, which are also the source of northern Silesian and Lesser Poland groups, as well as the Csepel group in Hungary. Common traits (Figure 44) of the core East group are (Heyd 2007):

- Specific finds such as the stocky and intricately decorated East Group Beaker, often with metope decoration; the many jars, simple cups, and plates appearing as *Begleitkeramik* (accompanying pottery); and smaller findings like broad 4-hole wristguards, decorated boar tusk bow-shaped pendants, and V-shaped perforated bone buttons.

- Specific spectrum of settlement pottery, more varied than funerary ones, clearly distinguishable from western Bell Beaker groups and Balkan EBA cultures.

- Homogeneous, intensively worked burial custom with many graves organised in necropolis of up to 150 graves (in Moravia and Hungary also with cremation), constrained to certain rules, which had as a result the uncommon and characteristic bipolar gender-differentiated burial where men were buried with their heads to the north, and women with their heads to the south:

  ○ Gender-specific ritual, with men buried on the left side, females on their right side, both sexes in a foetal position.

  ○ Main orientation of the grave pit and corpse is north–south.

  ○ The dead face to the east.

- Settlement type through non-intrusive frame-like, boat-shaped long houses featuring the use of light (which makes them difficult to find).

The Classical Bell Beaker culture associated with this East Group is divided into phases from its initial stage ca. 2500 to the Early Bronze Age innovation waves ca. 2200 BC, which account for some 12-13 generations, assuming that each generation lasted 20–25 years. While eastern groups seem to go steadily through all phases, western ones in southern Germany have shorter initial and later phases, which point to an east–west drift of innovations during this period (Heyd 2007).

Socially, East Bell Beaker graves do not show superior status at regional or supra-regional level. Because the burial goods reflected their use in real life, it is conceivable that views of the afterlife were aligned with the world of real objects and accompanying social categories. Cemeteries show relatively strong kinship ties, signalled by the 'founder' graves. which point to these groups as family units in a wide sense (Heyd 2007).

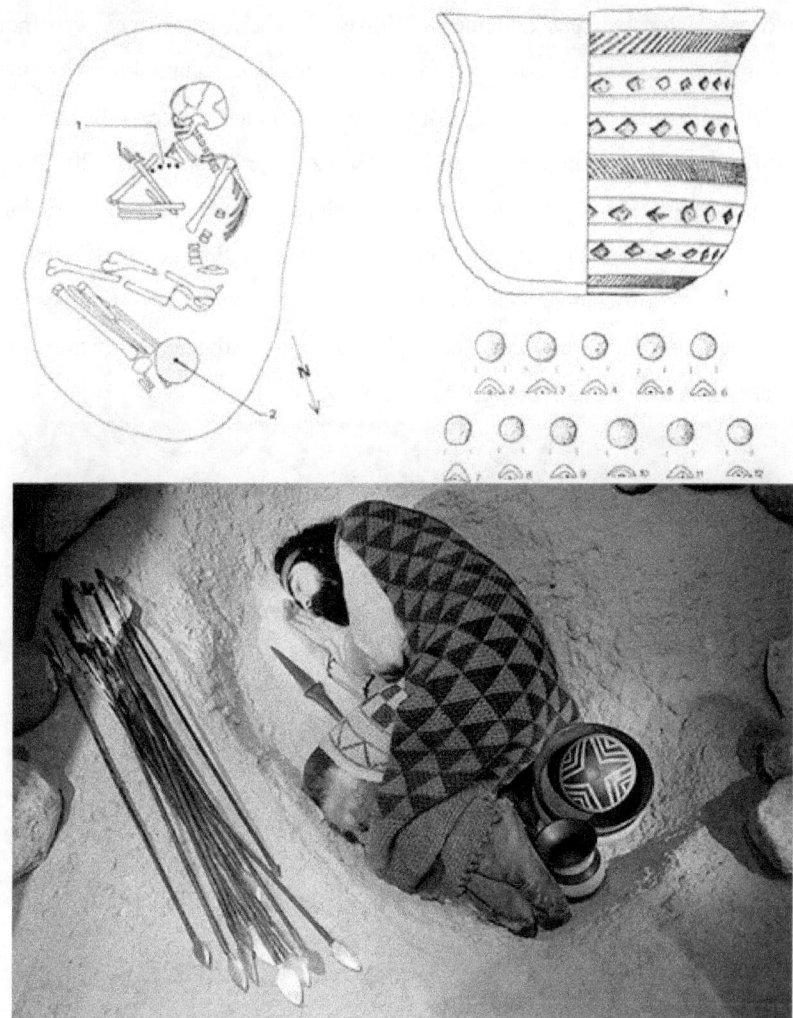

Figure 44. Top: Osterhofen-Altenmark, grave 4: South German sample burial of the East Bell Beaker group, modified from Heyd (2004). Bottom: Reconstruction of a Bell Beaker burial (National Archeological Museum of Spain).

Evidence points to an organisation into isolated communities living in fertile areas in a closely packed farmstead as the smallest unit, consisting of a single dwelling house inhabited by one family unit (with a small number of inhabitants), perhaps with additional farm buildings. These self-sufficient social units were economically specialised to a certain extent. Farmsteads located close to each other, sometimes connected through favourable landscapes, formed real settlements. However, where landscapes were not favourable, no central place is found (which is incompatible with the settlement model of egalitarian farmsteads), and neither are hill forts or fortifications of any kind, proper of later, Bronze Age groups (Heyd 2007).

This egalitarian organisation contrasts with previous hamlets of the region (e.g. in the Corded Ware culture), clusters of a few independent households with their own dwelling houses and adjacent communal farm buildings, or villages, representing middle-sized settlement units. Advantages of this new organisation are (Heyd 2007):

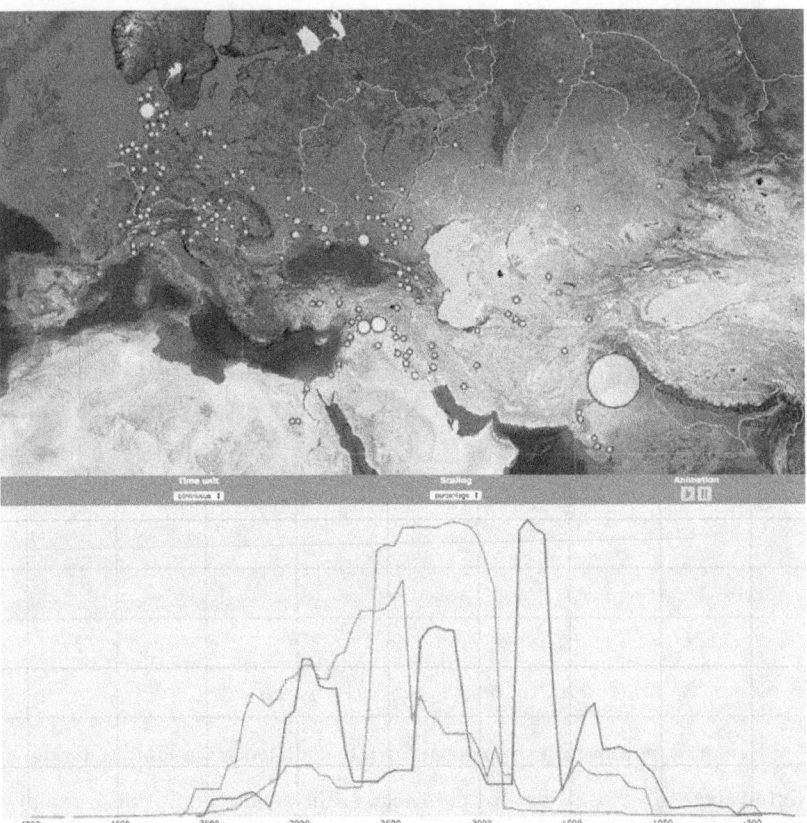

Figure 45. Top image: Map with evidence of animal traction before ca. 2000 BC. Bottom image: frequency of finds of evidence for animal traction (orange), cylinder seals (purple) and potter's wheels (green) in the 4th and 3rd millennium BC (query from the Digital Atlas of Innovations). The data points to an early peak in the expansion of this innovation at the turn of the 4th–3rd millennium BC, while direct evidence supports

a radical increase from around the mid–3ᵗʰ millennium BC until the early 2ⁿᵈ millennium, coinciding with the expansion of East Bell Beakers and related European Early Bronze Age cultures. Data and image modified from Klimscha (2017).

- Economic units are independent and make opportunistic alliances, reacting flexibly to emergencies and dangers from the outside.

- They are more mobile, probably reflecting cattle-breeding and transhumance as the dominant herding strategy, individually and collectively, with less permanent settlements (Figure 45).

- The command of an increased communication system, with internal exchange of information, goods, genes, and social values. They seem to have invested more in mobile possessions, such as copper and cattle, than in settlements.

About prestige goods, which distinguished the buried individuals among these general egalitarian customs, the most common ones are copper tanged daggers, small flint daggers, wristguards, decorated bow pendants, and flint arrowheads in men's graves, together with copper awls and V-perforated bone buttons from women's graves, and gold and amber objects that in both male and female graves (Heyd 2007):

- In comparison with Corded Ware graves, gold objects seem to increase in abundance dramatically, although they represent small quantities even in well-equipped graves, which suggests its origin as rare imported objects. Its presence continues into subsequent Early Bronze Age cultures.

- For the first time, amber goods—until this moment constrained to cultures in contact with the Baltic, like TRB, GAC, and CWC—reach western Europe, and this connection forms the basis for the later 'amber route' of the fully developed Bronze Age. Shells in female burials, associated with neck ornaments made of V-perforated bone buttons, link Bell Beaker to the Mediterranean, probably travelling through the Alps.

• Flint daggers, which spread with the Transformation of Europe, became prevalent in Europe after the expansion of Yamna, and with Bell Beaker they replaced the traditional stone battle axes, which had dominated burials in the region for a thousand years. Copper daggers, in many cases miniatures, are probably to be understood as symbolic artefacts, in a rank similar to the most precious objects, gold and amber artefacts.

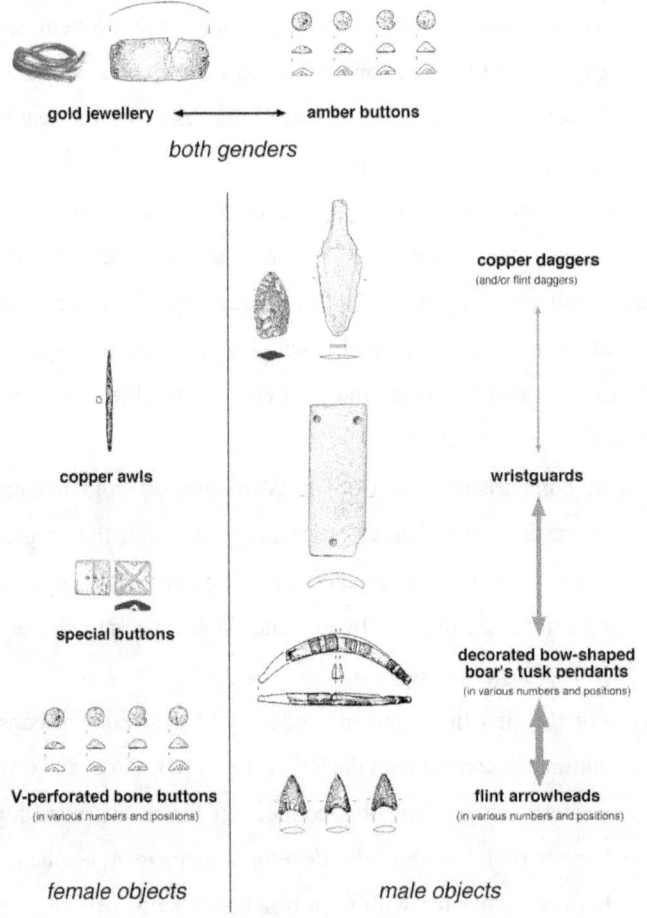

Figure 46. Special grave goods in the Bell Beaker East Group. From top to bottom, rarer to more frequent objects, indicating its relative ranking. From Heyd (2007).

All these grave objects might have had special symbolic and magical values, connected with the social position of their owners, and demonstrating long-distance communication, an international system of values, and a supra-regional trade network. Apparently, the number of artefacts and their association (Figure 46) show a hierarchy of importance of these objects (Heyd 2007).

Similar to the evolution of 'holed' axes of the LBK culture as battle axes in Central Europe (and their cultural successors in the Corded Ware culture), the explosive expansion of arrowheads in East Bell Beaker demonstrate its evolved function and ideology, in contrast to the frequent findings but in small numbers in LBK and in the Danube megalithic traditions. The archery image typical of the "archer's package" and the complete set of archery equipment that appear in certain characteristic "archer burials" (where long wooden bows are supposed to have been interred too) point to a status object of the warrior (Heyd 2007). Specialised archery may be inferred from osteological remains (Ryan, Desideri, and Besse 2018). Horse riding-related osteological marks irrupt in central and western Europe with the expansion of Bell Beakers, as do horse bone remains, which support the suggested expansion of mounted archery.

The closest known specimen to modern horse domesticates comes from Dunaujvarus, an East BBC site in Hungary (end of the 3rd millennium BC), which shows a more archaic branch than those found in two contemporary Sintashta sites. This supports the separation of this lineage with the expansion of Yamna settlers to the west, as horse domesticates kept evolving in the steppes and spread westwards into Europe with later population movements. The early expansion of this branch with Yamna is further supported by the archaic Y-chromosome lineage and mtDNA line of the Dunaujvarus sample (Fages et al.).

The expansion of this archaic branch of horse domesticates with East Bell Beakers is evidenced by its admixture with a source close to Iberian specimens, suggesting long-distance exchange of horses during the Bell Beaker phenomenon; and by the finding of a Pre-Iberian specimen from Els Vilars (ca. 700–550 BC) which shows the same archaic branch (i.e. before the separation of Sintashta horses), hence likely of Iberian/Central European Bell Beaker descent (Fages et al.).

In a few cases, children were also interred with these prestige warrior goods, pointing to its supposed future role, possibly through his father's line, suggesting thus rules of inheritance and dictated social status from birth. Such rich graves for infants existed also in CWC ca. 2700–2500 BC, although there was a clear differentiation according to the age of the deceased (the CWC 'age classes'), which is not found in East Bell Beakers (Heyd 2007).

Like dagger graves (a prestige good) or craftsmen's graves, such warrior graves are rare, and some individuals are even physically larger, showing a better physical constitution. Nevertheless, these special graves occur within family cemeteries, but there are no gross inter-cemetery differences, which point to a loose regional connection between family units coexisting as equals. The internal hierarchy of family units—in the (at least funerary) egalitarian Bell Beaker society—points to an initial stage of stratification on its way to the fully stratified Bronze Age societies (Heyd 2007).

Craftsmen's graves (metalworkers, but also flintworkers) show their role in metal exploitation and prestige good production, especially graves of metalworkers (with copper daggers and wristguards, and special forms of grave construction), showing their particular economic and social importance in the Bell Beaker society. There is a clear direct evolution from the hundreds of craftsmen's graves of the northern and north-western Pontic Yamna and Catacomb graves (ca. 2750–2250 BC), to the other, western European end in the Amesbury Archer in England (Heyd 2013).

VII.7.3. Contacts Bell Beaker – Corded Ware

Settlement areas of both cultures, the Bell Beaker and the Corded Ware culture, especially in the common territories of central Europe, seemed to remain separated. Available data suggest rejection and aversion, but also some form of social discourse between the groups (Heyd 2007).

Neighbouring groups of Bell Beaker, Globular Amphora and Corded Ware cultures of east-central Europe show certain similar artefacts, but made of different materials, and with different interpretations, which might signal imitation among culturally different groups. These cultural differences between Corded Ware and Bell Beaker cultures are maintained over vast distances, from east to west Europe (Czebreszuk and Szmyt 2008), potentially suggesting a strong ethnolinguistic difference (Figure 47). With the interaction of both groups, Corded Ware burials adapted to Bell Beaker customs, and a decline in Corded Ware remains is found in shared areas.

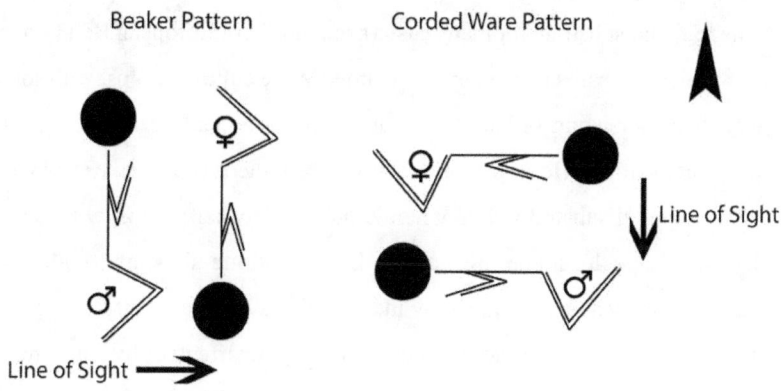

Figure 47. Contrasting continental Bell Beaker and Corded Ware burial patterns. Image from Heise (2014).

The pattern observed is of spatial separation followed by partial integration (dissolution of the spatial-cultural divide), suggesting a land capture by the expanding Bell Beaker culture, and also an ethnic dimension based on cultural expressions and physical anthropology (Heyd 2007). The different strategies of East Bell Beaker groups with local Corded Ware (or derived) groups, as well as other Late Copper Age groups, ranging from full separation to co-operation, shows a different ideological resolution to these interactions in the Late Copper Age, and the creation of new social identities in Central European EBA groups (Bertemes and Heyd 2002):

- the Danubian Early Bronze Age of Southern German groups, with a strong Bell Beaker foundation (based on the exclusion of Corded Ware groups to the south and west of the Middle Danube);

- the Únětice Early Bronze Age, on a strong west Carpathian foundation (based originally on the mix of Moravian Bell Beakers with Transdanubian cultures);

- the Eastern Early Bronze Age (Proto-Mierzanowice), with origins in Epi-Corded Ware cultures (based on a mix of Bell Beaker of Lesser Poland with Carpathian cultures);

- and Bell Beakers from the Polish Plains, mixing with late Corded Ware groups in Eastern Europe to form Trzciniec.

The regional substrate for many eastern and northern European Bell Beaker groups is in many cases formed by late Corded Ware culture groups, with some pottery types persisting in later times, and with individual burials being also used by later settlers. However, in western and southern Bell Beaker territory, previous regional substrates do not herald the Bell Beaker groups, with newer settlements using locations different to Late Neolithic sites, and collective graves being reused or substituted by individual graves (Besse 2014).

In certain small, remote regions, possibly unaffected by the initial explosive expansions along rivers and coasts, Corded Ware groups dominate the records ca. 2500–2200 BC, such as in German Franconia and the Tauber River Valley. Gradually, though, these isolated units decrease in numbers and their ideology seems to have faded, no longer having a supra-regional grip, merging eventually within the succeeding Early Bronze Age cultures (Besse 2014).

Tumulus building was identified by Gimbutas as one of the main cultural manifestations of the Kurgan culture (and "Kurgan people") that spread Indo-European languages during the Neolithic. The practice of mound-building (and single graves) is nevertheless so widespread in time and space that it is hard to associate it with one particular ethnic group (Harding 2011).

vii.7. North-West Indo-Europeans

At the end of the Chalcolithic period, migration waves reached the Carpathian Basin: Yamna from the east, and Proto-Beakers from the north-west (Szécsényi-Nagy et al. 2018), probably following the Danube. Early individuals from the Yamna culture in Hungary show a fully homogeneous admixture with other Yamna groups, with one of four samples showing initial admixture with local EEF populations, hence the slightly elevated contribution (ca. 17%), statistically non-significant. A further contribution is seen in a later individual, with ca. 27% EEF ancestry, and a clear 'central European' shift in the PCA (Wang et al. 2018). Late Chalcolithic samples from Hungary showed mainly NWAN (ca. 86%) and WHG (ca. 14%) ancestry (Olalde et al. 2018).

This admixture of Yamna lineages with local populations represents thus the most likely starting point of emerging East Bell Beakers that expanded from the Middle Danube in all directions (Suppl. Graph. 9). These early stages of admixture are also consistent with the extreme genetic differentiation among early Bell Beaker samples from Hungary (Suppl. Graph. 10), e.g. in Szigetszentmiklós, in the Csepel island (ca. 2450–2150 BC): one of hg. R1b1a2a2-Z2103 shows 75% Steppe ancestry, and clusters closely with the Yamna Hungary EBA sample in the PCA; with him there are one individual of hg. R1b1a1b1a-L51 (ca. 47% Steppe ancestry), one of hg. I2a1a1a-L158 (ca. 59% Steppe ancestry), and another of hg. I2a1b1-M223 (ca. 47% Steppe ancestry), all four males with typical burial position and Bell Beaker assemblages (Olalde et al. 2018).

Slightly later samples from the same site, from the succeeding Proto-Nagyrév culture, show a higher variability of haplogroups in the area (see

§viii.11. Thracians and Albanians), including the presence of a basal lineage R1b1a1b1a1a-L151* (formed ca. 2800 BC, TMRCA ca. 2800 BC), further supporting the nature of the Carpathian Basin and the area around the Tisza river as the *sink* of Yamna settlers. The *source* of expanding Bell Beakers, which show a clear Y-chromosome bottleneck of R1b1a1b1a1a-L151 subclades, was probably closer to the Middle Danube area, between the Devín Gate and the Danube Bend, around the Little Alföld.

Previous to these Early Bronze Age samples from Hungary, where most paternal lineages descend from elite males of Yamna clans, the Carpathian Basin showed typical East European farmer ancestry and lineages, approximately half G2a2-L1259 and half I2-M438 subclades. Subclade I2a1a1a-L158 is found in two samples from Balatonlelle (one ca. 3600–3000 BC, another ca. 3330–2930 BC), with the same haplogroup also reported in later East Bell Beakers from the area (Lipson et al. 2017). Steppe ancestry has been found thus spreading with expanding East Bell Beakers to the west, whereas no substantial contribution is found from Iberian Beaker Complex-associated individuals to central Europe (Olalde et al. 2018).

The expansion of Yamna settlers with a main bottleneck under haplogroup R1b1a1b1a-L51 from western Yamna settlers—evidenced by the expansion of its subclade R1b1a1b1a1a-L151 with Bell Beakers—is also supported by the finding of subclade R1b1a1b1a1-L52* in south-eastern France, and by the distribution of its other subclade R1b1a1b1a2-Z2118 (formed ca. 3700 BC, TMRCA ca. 3100 BC) in south-eastern Europe, hence likely associated with early west Yamna settlers, and the further expansion under SNP Z2116 (formed ca. 3100 BC, TMRCA ca. 2700 BC) into R1b1a1b1a2a-S1161 (TMRA ca. 2700 BC) throughout central and western Europe.

Speakers of the North-West Indo-European language have been recently associated with expanding East Bell Beakers (Mallory 2013), as an offshoot of Yamna, due to the fitting guesstimates for this reconstructed stage. The expansion of the Proto-Beaker archaeological package along the Upper

Danube probably allowed for its cultural diffusion through the river basin among the thousands of Yamna settlers related to the Pannonian Steppe—distributed from the Tisza River and its tributaries and the Middle Danube up to the Austrian Burgenland—just before the renewed impulse of expansion of the classical Bell Beaker folk.

Agricultural substrate loanwords in common to Palaeo-Balkan languages, some of them with potential traits of Afroasiatic nature, are probably due to contacts of expanding Yamna settlers with farming communities of the north Pontic and the Lower Danube regions, which had strong links with hunter-gatherers of haplogroup R1b1b-V88 (Kroonen 2012). The few loanwords of a supposed Vasconic-like origin, apparently concentrated in a later stage, are probably due to contacts with communities derived directly from the early waves of NWAN-related European farmers, probably from the Pannonian Basin, where most non-Indo-European substrate words in North-West Indo-European were likely borrowed. It is therefore likely that the evolution from a Pre-North-West Indo-European stage to a fully-fledged North-West Indo-European language spanned the west Yamna (ca. 3300/3100 BC) to the expanding East Bell Beaker community (ca. 2500-2300 BC).

References

Adamczak, Kamil, Stanisław Kukawka, and Jolanta Małecka-Kukawka. 2016-2017. North-eastern periphery of the Eastern group of the Funnel Beaker culture – 80 years later. In *Papers and materials of the Archaeological and Ethnographic Museum in Łódź*. Łódź.

Adamov, Dmitry, Vladimir M. Guryanov, Sergey Karzhavin, Vladimir Tagankin, and Vadim Urasin. 2015. Defining a New Rate Constant for Y-Chromosome SNPs based on Full Sequencing Data. *The Russian Journal of Genetic Genealogy (Русская версия)* 7 (1):68-89.

Adrados, F.R. 1998. La reconstrucción del indoeuropeo y de su diferenciación dialectal. In *Manual de lingüística indoeuropea*, edited by F. R. Adrados, A. Bernabé and J. Mendoza. Madrid: Ediciones clásicas.

Alizadeh, Karim, Siavash Samei, Kourosh Mohammadkhani, Reza Heidari, and Robert H. Tykot. 2018. Craft production at Köhne Shahar, a Kura-Araxes settlement in Iranian Azerbaijan. *Journal of Anthropological Archaeology* 51:127-143.

Allentoft, Morten E., Martin Sikora, Karl-Goran Sjogren, Simon Rasmussen, Morten Rasmussen, Jesper Stenderup, Peter B. Damgaard, Hannes Schroeder, Torbjorn Ahlstrom, Lasse Vinner, Anna-Sapfo Malaspinas, Ashot Margaryan, Tom Higham, David Chivall, Niels Lynnerup, Lise Harvig, Justyna Baron, Philippe Della Casa, Pawel Dabrowski, Paul R. Duffy, Alexander V. Ebel, Andrey Epimakhov, Karin Frei, Miroslaw Furmanek, Tomasz Gralak, Andrey Gromov, Stanislaw Gronkiewicz, Gisela Grupe, Tamas Hajdu, Radoslaw Jarysz, Valeri Khartanovich, Alexandr Khokhlov, Viktoria Kiss, Jan Kolar, Aivar Kriiska, Irena Lasak, Cristina Longhi, George McGlynn, Algimantas Merkevicius, Inga Merkyte, Mait Metspalu, Ruzan Mkrtchyan, Vyacheslav Moiseyev, Laszlo Paja, Gyorgy Palfi, Dalia Pokutta, Lukasz Pospieszny, T. Douglas Price, Lehti Saag, Mikhail Sablin, Natalia Shishlina, Vaclav Smrcka, Vasilii I. Soenov, Vajk Szeverenyi, Gusztav Toth, Synaru V. Trifanova, Liivi Varul, Magdolna Vicze, Levon Yepiskoposyan, Vladislav Zhitenev, Ludovic Orlando, Thomas Sicheritz-Ponten, Soren Brunak, Rasmus Nielsen, Kristian

Kristiansen, and Eske Willerslev. 2015. Population genomics of Bronze Age Eurasia. *Nature* 522 (7555):167-172.

Altaweel, Mark, and Alessio Palmisano. 2018. Urban and Transport Scaling: Northern Mesopotamia in the Late Chalcolithic and Bronze Age. *Journal of Archaeological Method and Theory.*

Amzallag, Nissim. 2009. From Metallurgy to Bronze Age Civilizations: The Synthetic Theory. *American Journal of Archaeology* 113 (4):497-519.

Andrades Valtueña, Aida, Alissa Mittnik, Felix M. Key, Wolfgang Haak, Raili Allmäe, Andrej Belinskij, Mantas Daubaras, Michal Feldman, Rimantas Jankauskas, Ivor Janković, Ken Massy, Mario Novak, Saskia Pfrengle, Sabine Reinhold, Mario Šlaus, Maria A. Spyrou, Anna Szecsenyi-Nagy, Mari Tõrv, Svend Hansen, Kirsten I. Bos, Philipp W. Stockhammer, Alexander Herbig, and Johannes Krause. 2017. The Stone Age Plague: 1000 years of Persistence in Eurasia. *bioRxiv.*

Anthony, David. 2017. Archaeology and Language: Why Archaeologists Care About the Indo-European Problem--in European Archaeology as Anthropology. In *European archaeology as anthropology: essays in memory of Bernard Wailes* edited by P. J. Crabtree and P. I. Bogucki. Philadelphia: University of Pennsylvania Museum.

Anthony, David W. 2006. Pontic-Caspian Mesolithic and Early Neolithic societies at the time of the Black Sea flood: a small audience and small effects. In *The Black Sea Flood Question: Changes in Coastline, Climate and Human Settlement*, edited by V. Yanko-Hombach, A. S. Gilbert, N. Panin and P. M. Dolukhanov. Dordrecht: Springer.

Repeated Author. 2007. *The Horse, the Wheel, and Language: How Bronze-Age Riders from the Eurasian Steppes Shaped the Modern World*. Princeton and Oxford: Princeton University Press.

Repeated Author. 2013. Two IE phylogenies, three PIE migrations, and four kinds of steppe pastoralism. *Journal of Language Relationship* (9):1-21.

Repeated Author. 2016. The Samara Valley Project and the Evolution of Pastoral Economies in the Western Eurasian Steppes. In *A Bronze Age Landscape in the Russian Steppes. The Samara Valley Project*, edited by D. W. Anthony, D. R. Brown, O. D. Mochalov, A. A. Khokhlov and P. F. Kuznetsov. Los Angeles: The Cotsen Institute of Archaeology Press at UCLA.

Anthony, David W., and Dorcas R. Brown. 2011. The Secondary Products Revolution, Horse-Riding, and Mounted Warfare. *Journal of World Prehistory* 24 (2-3):131-160.

Repeated Author. 2017. Molecular Archaeology and Indo-European linguistics: Impressions from new data. In *Usque ad Radices: Indo-European Studies in Honour of Birgit Anette Olsen*, edited by B. Simmelkjær, S. Hansen, A. Hyllested, A. R. Jørgensen, G. Kroonen, J. H. Larsson, B. N. Whitehead, T. Olander and T. M. Søborg. Copenhagen: Museum Tusculanum Press.

Anthony, David W., and Don Ringe. 2015. The Indo-European Homeland from Linguistic and Archaeological Perspectives. *Annual Review of Linguistics* 1 (1):199-219.

Arbuckle, Benjamin S. . 2009. Chalcolithic Caprines, Dark Age Dairy and Byzantine Beef: A First Look at Animal Exploitation at Middle and Late Holocene Çadir Höyük, North Central Turkey. *Anatolica* 35:179–224.

Arbuckle, Benjamin S., and Emily L. Hammer. 2018. The Rise of Pastoralism in the Ancient Near East. *Journal of Archaeological Research*.

Archi, A. 2011. In Search of Armi. *Journal of Cuneiform Studies* 63:5-34.

Arranz-Otaegui, Amaia, Lara Gonzalez Carretero, Monica N. Ramsey, Dorian Q. Fuller, and Tobias Richter. 2018. Archaeobotanical evidence reveals the origins of bread 14,400 years ago in northeastern Jordan. *Proceedings of the National Academy of Sciences* 115 (31):7925-7930.

Artemova, S. N., D. S. Ikonnikov, and O. Ph. Prikazchikova. 2018. Historical and geo-ecological features of the formation of cultural landscapes of the Upper Possurie and Primokshanie during the Aeneolithic period. *Ekologiya* 6:1-11.

Baker, Jennifer L., Charles N. Rotimi, and Daniel Shriner. 2017. Human ancestry correlates with language and reveals that race is not an objective genomic classifier. *Scientific Reports* 7 (1):1572.

Bátora, Jozef. 2006. *Štúdie ku komunikácii medzi strednou a východnou Európou v dobe bronzovej*. Bratislava: Petrus Publ.

Beekes, Robert S.P. 2011. *Comparative Indo-European Linguistics. An introduction*. 2nd ed. Amsterdam / Philadelphia: John Benjamins.

Bender, M. Lionel. 2007. The Afrasian lexion reconsidered. In *Studies in Semitic and Afroasiatic Linguistics Presented to Gene B. Gragg*, edited by C. L. Miller. Illinois: The University of Chicago.

Beridze, Tengiz. 2019. The 'Wheat Puzzle' and Kartvelians route to the Caucasus. *Genetic Resources and Crop Evolution*.

Bertemes, François, and Volker Heyd. 2002. Der Übergang Kupferzeit / Frühbronzezeit am Nordwestrand des Karpatenbeckens - kulturgeschichtliche und paläometallurgische Betrachtungen. In *Die Anfänge der Metallurgie in der Alten Welt*, edited by M. Bartelheim. Rahden/Westfalen: Leidorf.

Besse, Marie. 2014. Common Ware during the third Millenium BC in Europe. In *Similar but Different: Bell Beakers in Europe*, edited by J. Czebreszuk. Leiden: Sidestone Press.

Biagi, Paolo , and Dmytro Kiosak. 2010. The Mesolithic of the northwestern Pontic region. New AMS dates for the origin and spread of the blade and trapeze industries in southeastern Europe. *Eurasia antiqua: Zeitschrift für Archäologie Eurasiens* 16:21-41.

Bilgi, Ö. 2001. Bİkiztepe Kazılarının 2000 Dönemi Sonuçları. *Kazı Sonuçları Toplantısı* 23:245-254.

Repeated Author. 2005. Distinguished Burials of the Early Bronze Age Graveyard at İkiztepe in Turkey. *İstanbul Üniversitesi Edebiyat Fakültesi Anadolu Araştırmaları Dergisi* XVIII (2):15-113.

Binney, Heather, Mary Edwards, Marc Macias-Fauria, Anatoly Lozhkin, Patricia Anderson, Jed O. Kaplan, Andrei Andreev, Elena Bezrukova, Tatiana Blyakharchuk, Vlasta Jankovska, Irina Khazina, Sergey Krivonogov, Konstantin Kremenetski, Jo Nield, Elena Novenko, Natalya Ryabogina, Nadia Solovieva, Kathy Willis, and Valentina Zernitskaya. 2017. Vegetation of

Eurasia from the last glacial maximum to present: Key biogeographic patterns. *Quaternary Science Reviews* 157:80-97.

Blanco-González, A., K. T. Lillios, J. A. López-Sáez, and B. L. Drake. 2018. Cultural, Demographic and Environmental Dynamics of the Copper and Early Bronze Age in Iberia (3300–1500 BC): Towards an Interregional Multiproxy Comparison at the Time of the 4.2 ky BP Event. *Journal of World Prehistory* 31 (1):1-79.

Blažek, Václav. 2010. On the Classification of Berber. *Folia Orientalia* 47:245-266.

Repeated Author. 2014. Phoenician/Punic loans in Berber languages and their role in chronology of Berber. *Folia Orientalia* 51:275-293.

Bogdanov, S. V. 2004. *Jepoha medi stepnogo priural'ja*. Ekaterinburg: Tipografija UrO RAN.

Bomhard, Alan R. 2017. *The Origins of Proto-Indo-European: The Caucasian Substrate Hypothesis*. Charleston, SC.

Bonechi, M. 1990. Aleppo in età arcaica; a proposito di un'opera recente. *Studi Epigrafici e Linguistici sul Vicino Oriente Antico* 7:15-37.

Boroffka, Nikolaus. 2013. Romania, Moldova, and Bulgaria. In *The Oxford Handbook of the European Bronze Age*, edited by H. Fokkens and A. Harding. Oxford: Oxford University Press.

Brace, S., Y. Diekmann, T. Booth, O. Craig, C. Stringer, D. Reich, M. Thomas, and I. Barnes. 2018. Ancient DNA and the peopling of the British Isles – pattern and process of the Neolithic transition. Paper read at 8th International Symposium on Biomolecular Archaeology ISBA 2018. 18th – 21st September, at Jena, Germany.

Brace, Selina, Yoan Diekmann, Thomas J. Booth, Lucy van Dorp, Zuzana Faltyskova, Nadin Rohland, Swapan Mallick, Iñigo Olalde, Matthew Ferry, Megan Michel, Jonas Oppenheimer, Nasreen Broomandkhoshbacht, Kristin Stewardson, Rui Martiniano, Susan Walsh, Manfred Kayser, Sophy Charlton, Garrett Hellenthal, Ian Armit, Rick Schulting, Oliver E. Craig, Alison Sheridan, Mike Parker Pearson, Chris Stringer, David Reich, Mark G. Thomas, and Ian Barnes. 2019. Ancient genomes indicate population replacement in Early Neolithic Britain. *Nature Ecology & Evolution* 3 (5):765-771.

Bramanti, B., M. G. Thomas, W. Haak, M. Unterlaender, P. Jores, K. Tambets, I. Antanaitis-Jacobs, M. N. Haidle, R. Jankauskas, C.-J. Kind, F. Lueth, T. Terberger, J. Hiller, S. Matsumura, P. Forster, and J. Burger. 2009. Genetic Discontinuity Between Local Hunter-Gatherers and Central Europe's First Farmers. *Science* 326 (5949):137-140.

Brandt, G., W. Haak, C. J. Adler, C. Roth, A. Szecsenyi-Nagy, S. Karimnia, S. Moller-Rieker, H. Meller, R. Ganslmeier, S. Friederich, V. Dresely, N. Nicklisch, J. K. Pickrell, F. Sirocko, D. Reich, A. Cooper, K. W. Alt, and Consortium Genographic. 2013. Ancient DNA reveals key stages in the formation of central European mitochondrial genetic diversity. *Science* 342 (6155):257-61.

Brierley, Chris, Katie Manning, and Mark Maslin. 2018. Pastoralism may have delayed the end of the green Sahara. *Nature Communications* 9 (1):4018.

Brunel, Samantha. 2018. Paléogénomique des dynamiques des populations humaines sur le territoire Français entre 7000 et 2000, Bio Sorbonne Paris Cité (BIOSPC) & Institut Jaques Monod, Sorbonne Paris Cité, Paris.

Brunet, Frédérique. 2012. The Technique of Pressure Knapping in Central Asia: innovation or Diffusion? In *The Emergence of Pressure Blade Making: From Origin to Modern Experimentation*, edited by P. M. Desrosiers. New York: Springer.

Bulatović, Aleksandar. 2014. Corded Ware in the Central and Southern Balkans: A Consequence of Cultural Interaction or an Indication of Ethnic Change? *JIES* 42 (1 & 2).

Campbell, Lyle. 1998. Nostratic: A Personal Assessment. In *Nostratic: Sifting the Evidence*, edited by J. C. Salmons and B. D. Joseph. Amsterdam/Philadelphia: John Benjamins.

Repeated Author. 2015. Do Languages and Genes Correlate? *Language Dynamics and Change* 5 (2):202-226.

Cassidy, Lara. 2018. A Genomic Compendium of an Island: Documenting Continuity and Change across Irish Human Prehistory, School of Genetics & Microbiology, Trinity College Dublin., Dublin.

Cavadas, Bruno, Nicole Pedro, Veronica Fernandes, Joana C Ferreira, Luisa Pereira, François-Xavier Ricaut, Nicolas Brucato, and Farida Alshamali. 2019. Genome-Wide Characterization of Arabian Peninsula Populations: Shedding Light on the History of a Fundamental Bridge between Continents.

Chechushkov, Igor V., and Andrei V. Epimakhov. 2018. Eurasian Steppe Chariots and Social Complexity During the Bronze Age. *Journal of World Prehistory*.

Chekunova, E.M., N.V. Yartseva, M.K. Chekunov, and A.N. Mazurkevich. 2014. The First Results of the Genotyping of the Aboriginals and Human Bone Remains of the Archeological Memorials of the Upper Podvin'e. // Archeology of the lake settlements of IV—II Thousands BC: The chronology of cultures and natural environment and climatic rhythms. Paper read at Proceedings of the International Conference, Devoted to the 50-year Research of the Pile Settlements on the North-West of Russia., 13-15 November, at St. Petersburg.

Clackson, James. 2007. *Indo-European Linguistics. An Introduction*. Cambridge: Cambridge University Press.

Repeated Author. 2013. The Origins of the Indic Languages: the Indo-European model. In *Perspectives on the origin of Indian civilization*, edited by A. Marcantonio and G. N. Jha. New Delhi: D.K. Printworld

Clare, Lee, and Bernhard Weninger. 2016. Early Warfare and its Contribution to Neolithisation and Dispersal. In *Palaeoenviroment and the Development of Early Societies (Şanlıurfa / Turkey, 7 October 2012)*, edited by M. Reindel, K. Bartl, F. Lüth and N. Benecke. Rahden: Marie Leidorf.

Corboud , P. 2009. Les stèles anthropomorphes de la nécropole néolithique du Petit-Chasseur à Sion. *Bulletin d'études préhistoriques et archéologiques alpines* 20:1-89.

Cramp, Lucy J. E., Richard P. Evershed, Mika Lavento, Petri Halinen, Kristiina Mannermaa, Markku Oinonen, Johannes Kettunen, Markus Perola, Päivi Onkamo, and Volker Heyd. 2014. Neolithic dairy farming at the extreme of agriculture in northern Europe. *Proceedings of the Royal Society B: Biological Sciences* 281 (1791).

Cruciani, F., B. Trombetta, D. Sellitto, A. Massaia, G. Destro-Bisol, E. Watson, E. Beraud Colomb, J. M. Dugoujon, P. Moral, and R. Scozzari. 2010. Human Y chromosome haplogroup R-V88: a paternal genetic record of early mid Holocene trans-Saharan connections and the spread of Chadic languages. *Eur J Hum Genet* 18 (7):800-7.

Czebreszuk, J., and M. Szmyt. 2004. Chronology of Central-European Influences within the Western Part of the Forest Zone during the 3rd Millenium BC. In *Проблемы хронологии и этнокультурных взаимодействий в неолите Евразии*, edited by V. I. Timofeev and G. I. Zayceva. Санкт-Петербург: ИИМК РАН.

Czebreszuk, Janusz, and Marzena Szmyt. 2008. Siedlungsformen des 3. Jahrtausends v. Chr. in der polnischen Tiefebene (Kulturen der Trichterbecher, Kugelamphoren und Schnurkeramik). Stand und Perspektiven der Untersuchungen. In *Umwelt – Wirtschaft – Siedlungen im dritten vorchristlichen Jahrtausend Mitteleuropas und Südskandinaviens. Internationale Tagung Kiel 4.–6. November 2005*, edited by W. Dörfler and J. Müller. Neumünster: Wachholtz.

Repeated Author. 2008. What lies behind 'Import' and 'Imitation'? Case Studies from the European Late Neolithic. In *Import and Imitation in Archaeology*, edited by F. Bertemes and A. Furtwängler. Langenweissbach: Beier & Beran.

Repeated Author. 2011. Identities, Differentiation and Interactions on the Central European Plain in the 3rd millennium BC. In *Sozialarchäologische Perspektive: Gesellschaftlicher Wandel 5000-1500 v. Chr. zwichen Atlantik und Kaukasus*, edited by S. Hansen and J. Müller. Darmstadt: philipp von Zabern.

D'Atanasio, Eugenia, Beniamino Trombetta, Maria Bonito, Andrea Finocchio, Genny Di Vito, Mara Seghizzi, Rita Romano, Gianluca Russo, Giacomo Maria Paganotti, Elizabeth Watson, Alfredo Coppa, Paolo Anagnostou, Jean-Michel Dugoujon, Pedro Moral, Daniele Sellitto, Andrea Novelletto, and Fulvio Cruciani. 2018. The peopling of the last Green Sahara revealed by high-coverage resequencing of trans-Saharan patrilineages. *Genome Biology* 19 (1):20.

Damlien, Hege, Inger Marie Berg-Hansen, Ilga Zagorska, Mārcis Kalniņš, Svein V. Nielsen, Lucia U. Koxvold, Valdis Bērziņš, and Almut Schülke. 2018. A technological crossroads: Exploring diversity in the pressure blade technology of Mesolithic Latvia. *Oxford Journal of Archaeology* 37 (3):229-246.

Darmark, Kim. 2012. Surface Pressure Flaking in Eurasia: Mapping the Innovation, Diffusion and Evolution of a Technological Element in the Production of Projectile Points. In *The Emergence of Pressure Blade Making: From Origin to Modern Experimentation*, edited by P. M. Desrosiers. New York: Springer.

de Barros Damgaard, Peter, Nina Marchi, Simon Rasmussen, Michaël Peyrot, Gabriel Renaud, Thorfinn Korneliussen, J. Víctor Moreno-Mayar, Mikkel Winther Pedersen, Amy Goldberg, Emma Usmanova, Nurbol Baimukhanov, Valeriy Loman, Lotte Hedeager, Anders Gorm Pedersen, Kasper Nielsen, Gennady Afanasiev, Kunbolot Akmatov, Almaz Aldashev, Ashyk Alpaslan, Gabit Baimbetov, Vladimir I. Bazaliiskii, Arman Beisenov, Bazartseren

Boldbaatar, Bazartseren Boldgiv, Choduraa Dorzhu, Sturla Ellingvag, Diimaajav Erdenebaatar, Rana Dajani, Evgeniy Dmitriev, Valeriy Evdokimov, Karin M. Frei, Andrey Gromov, Alexander Goryachev, Hakon Hakonarson, Tatyana Hegay, Zaruhi Khachatryan, Ruslan Khaskhanov, Egor Kitov, Alina Kolbina, Tabaldiev Kubatbek, Alexey Kukushkin, Igor Kukushkin, Nina Lau, Ashot Margaryan, Inga Merkyte, Ilya V. Mertz, Viktor K. Mertz, Enkhbayar Mijiddorj, Vyacheslav Moiyesev, Gulmira Mukhtarova, Bekmukhanbet Nurmukhanbetov, Z. Orozbekova, Irina Panyushkina, Karol Pieta, Václav Smrčka, Irina Shevnina, Andrey Logvin, Karl-Göran Sjögren, Tereza Štolcová, Angela M. Taravella, Kadicha Tashbaeva, Alexander Tkachev, Turaly Tulegenov, Dmitriy Voyakin, Levon Yepiskoposyan, Sainbileg Undrakhbold, Victor Varfolomeev, Andrzej Weber, Melissa A. Wilson Sayres, Nikolay Kradin, Morten E. Allentoft, Ludovic Orlando, Rasmus Nielsen, Martin Sikora, Evelyne Heyer, Kristian Kristiansen, and Eske Willerslev. 2018. 137 ancient human genomes from across the Eurasian steppes. *Nature* 557 (7705):369-374.

de Barros Damgaard, Peter, Rui Martiniano, Jack Kamm, J. Víctor Moreno-Mayar, Guus Kroonen, Michaël Peyrot, Gojko Barjamovic, Simon Rasmussen, Claus Zacho, Nurbol Baimukhanov, Victor Zaibert, Victor Merz, Arjun Biddanda, Ilja Merz, Valeriy Loman, Valeriy Evdokimov, Emma Usmanova, Brian Hemphill, Andaine Seguin-Orlando, Fulya Eylem Yediay, Inam Ullah, Karl-Göran Sjögren, Katrine Højholt Iversen, Jeremy Choin, Constanza de la Fuente, Melissa Ilardo, Hannes Schroeder, Vyacheslav Moiseyev, Andrey Gromov, Andrei Polyakov, Sachihiro Omura, Süleyman Yücel Senyurt, Habib Ahmad, Catriona McKenzie, Ashot Margaryan, Abdul Hameed, Abdul Samad, Nazish Gul, Muhammad Hassan Khokhar, O. I. Goriunova, Vladimir I. Bazaliiskii, John Novembre, Andrzej W. Weber, Ludovic Orlando, Morten E. Allentoft, Rasmus Nielsen, Kristian Kristiansen, Martin Sikora, Alan K. Outram, Richard Durbin, and Eske Willerslev. 2018. The first horse herders and the impact of early Bronze Age steppe expansions into Asia. *Science.*

Dergachev, V.A. 2007. *O skipetrakh, o loshadjakh, o voine: etjudy v zashchitu migratsionnoi kontseptsii M. Gimbutas.* Sankt-Peterburg: Nestor-Istorija.

di Lernia, Savino. 2013. Places, monuments, and landscape: evidence from the Holocene central Sahara. *Azania: Archaeological Research in Africa* 48 (2):173-192.

Diakonoff, Igor M., and Sergei A. Starostin. 1988. Hurro-Urartian and East Caucasian Languages. In *Ancient Orient. Ethnocultural Relations.*

Díaz-Guardamino, Marta. 2014. Shaping Social Identities in Bronze Age and Early Iron Age Western Iberia: The Role of Funerary Practices, Stelae, and Statue-Menhirs. *European Journal of Archaeology* 17 (2):329-349.

Drake, Nick A., Roger M. Blench, Simon J. Armitage, Charlie S. Bristow, and Kevin H. White. 2011. Ancient watercourses and biogeography of the Sahara explain the peopling of the desert. *Proceedings of the National Academy of Sciences* 108 (2):458-462.

Dunkel, G.E. 1997. Early, Middle, Late Indo-European: Doing it My Way. *Incontri Linguistici* 20:29-44.

Eisenmann, Stefanie, Eszter Bánffy, Peter van Dommelen, Kerstin P. Hofmann, Joseph Maran, Iosif Lazaridis, Alissa Mittnik, Michael McCormick, Johannes

Krause, David Reich, and Philipp W. Stockhammer. 2018. Reconciling material cultures in archaeology with genetic data: The nomenclature of clusters emerging from archaeogenomic analysis. *Scientific Reports* 8 (1):13003.

Eriksson, Gunilla, Karin Margarita Frei, Rachel Howcroft, Sara Gummesson, Fredrik Molin, Kerstin Lidén, Robert Frei, and Fredrik Hallgren. 2018. Diet and mobility among Mesolithic hunter-gatherers in Motala (Sweden) - The isotope perspective. *Journal of Archaeological Science: Reports* 17:904-918.

Ethier, Jonathan, Eszter Bánffy, Jasna Vuković, Krassimir Leshtakov, Krum Bacvarov, Mélanie Roffet-Salque, Richard P. Evershed, and Maria Ivanova. 2017. Earliest expansion of animal husbandry beyond the Mediterranean zone in the sixth millennium BC. *Scientific Reports* 7:7146.

Fages, Antoine, Kristian Hanghøj, Naveed Khan, Charleen Gaunitz, Andaine Seguin-Orlando, Michela Leonardi, Christian McCrory Constantz, Cristina Gamba, Khaled A. S. Al-Rasheid, Silvia Albizuri, Ahmed H. Alfarhan, Morten Allentoft, Saleh Alquraishi, David Anthony, Nurbol Baimukhanov, James H. Barrett, Jamsranjav Bayarsaikhan, Norbert Benecke, Eloísa Bernáldez-Sánchez, Luis Berrocal-Rangel, Fereidoun Biglari, Sanne Boessenkool, Bazartseren Boldgiv, Gottfried Brem, Dorcas Brown, Joachim Burger, Eric Crubézy, Linas Daugnora, Hossein Davoudi, Peter de Barros Damgaard, María de los Ángeles de Chorro y de Villa-Ceballos, Sabine Deschler-Erb, Cleia Detry, Nadine Dill, Maria do Mar Oom, Anna Dohr, Sturla Ellingvåg, Diimaajav Erdenebaatar, Homa Fathi, Sabine Felkel, Carlos Fernández-Rodríguez, Esteban García-Viñas, Mietje Germonpré, José D. Granado, Jón H. Hallsson, Helmut Hemmer, Michael Hofreiter, Aleksei Kasparov, Mutalib Khasanov, Roya Khazaeli, Pavel Kosintsev, Kristian Kristiansen, Tabaldiev Kubatbek, Lukas Kuderna, Pavel Kuznetsov, Haeedeh Laleh, Jennifer A. Leonard, Johanna Lhuillier, Corina Liesau von Lettow-Vorbeck, Andrey Logvin, Lembi Lõugas, Arne Ludwig, Cristina Luis, Ana Margarida Arruda, Tomas Marques-Bonet, Raquel Matoso Silva, Victor Merz, Enkhbayar Mijiddorj, Bryan K. Miller, Oleg Monchalov, Fatemeh A. Mohaseb, Arturo Morales, Ariadna Nieto-Espinet, Heidi Nistelberger, Vedat Onar, Albína H. Pálsdóttir, Vladimir Pitulko, Konstantin Pitskhelauri, Mélanie Pruvost, Petra Rajic Sikanjic, Anita Rapan Papeša, Natalia Roslyakova, Alireza Sardari, Eberhard Sauer, Renate Schafberg, Amelie Scheu, Jörg Schibler, Angela Schlumbaum, Nathalie Serrand, Aitor Serres-Armero, Beth Shapiro, Shiva Sheikhi Seno, Irina Shevnina, Sonia Shidrang, John Southon, Bastiaan Star, Naomi Sykes, Kamal Taheri, William Taylor, Wolf-Rüdiger Teegen, Tajana Trbojević Vukičević, Simon Trixl, Dashzeveg Tumen, Sainbileg Undrakhbold, Emma Usmanova, Ali Vahdati, Silvia Valenzuela-Lamas, Catarina Viegas, Barbara Wallner, Jaco Weinstock, Victor Zaibert, Benoit Clavel, Sébastien Lepetz, Marjan Mashkour, Agnar Helgason, Kári Stefánsson, Eric Barrey, Eske Willerslev, Alan K. Outram, Pablo Librado, and Ludovic Orlando. 2019. Tracking Five Millennia of Horse Management with Extensive Ancient Genome Time Series. *Cell*.

Fedyunin, I. V. 2015. The Mesolithic of the Forest-Steppe Don Area: Retrospective and Prospective Reviews1. *Archaeology, Ethnology and Anthropology of Eurasia* 43 (1):16-27.

Feldman, Michal, Eva Fernández-Domínguez, Luke Reynolds, Douglas Baird, Jessica Pearson, Israel Hershkovitz, Hila May, Nigel Goring-Morris, Marion Benz, Julia Gresky, Raffaela A. Bianco, Andrew Fairbairn, Gökhan Mustafaoğlu, Philipp W. Stockhammer, Cosimo Posth, Wolfgang Haak, Choongwon Jeong, and Johannes Krause. 2019. Late Pleistocene human genome suggests a local origin for the first farmers of central Anatolia. *Nature Communications* 10 (1):1218.

Fernandes, D. M., D. Strapagiel, P. Borówka, B. Marciniak, E. Żądzińska, K. Sirak, V. Siska, R. Grygiel, J. Carlsson, A. Manica, W. Lorkiewicz, and R. Pinhasi. 2018. A genomic Neolithic time transect of hunter-farmer admixture in central Poland. *Scientific Reports* 8 (1):14879.

Flegontov, P., P. Changmai, A. Zidkova, M. D. Logacheva, N. E. Altinisik, O. Flegontova, M. S. Gelfand, E. S. Gerasimov, E. E. Khrameeva, O. P. Konovalova, T. Neretina, Y. V. Nikolsky, G. Starostin, V. V. Stepanova, I. V. Travinsky, M. Triska, P. Triska, and T. V. Tatarinova. 2016. Genomic study of the Ket: a Paleo-Eskimo-related ethnic group with significant ancient North Eurasian ancestry. *Sci Rep* 6:20768.

Frangipane, Marcella. 2015. Different types of multiethnic societies and different patterns of development and change in the prehistoric Near East. *Proceedings of the National Academy of Sciences* 112 (30):9182-9189.

Frangipane, Marcella, Federico Manuelli, and Cristiano Vignola. 2017. Arslantepe, Malatya: Recent Discoveries in the 2015 and 2016 Seasons. In *The Archaeology of Anatolia Volume II: Recent Discoveries (2015-2016)*, edited by S. R. Steadman and G. McMahon. Newcastle: Cambridge Scholars Publishing.

Fregel, Rosa, Fernado L. Mendez, Youssef Bokbot, Dimas Martin-Socas, Maria D. Camalich-Massieu, Maria C. Avila-Arcos, Peter A. Underhill, Beth Shapiro, Genevieve L Wojcik, Morten Rasmussen, Andre E. R. Soares, Joshua Kapp, Alexandra Sockell, Francisco J. Rodriguez-Santos, Abdeslam Mikdad, Jonathan Santana, Aioze Trujillo-Mederos, and Carlos D. Bustamante. 2017. Neolithization of North Africa involved the migration of people from both the Levant and Europe. *bioRxiv*.

Fregel, Rosa, Fernando L. Méndez, Youssef Bokbot, Dimas Martín-Socas, María D. Camalich-Massieu, Jonathan Santana, Jacob Morales, María C. Ávila-Arcos, Peter A. Underhill, Beth Shapiro, Genevieve Wojcik, Morten Rasmussen, André E. R. Soares, Joshua Kapp, Alexandra Sockell, Francisco J. Rodríguez-Santos, Abdeslam Mikdad, Aioze Trujillo-Mederos, and Carlos D. Bustamante. 2018. Ancient genomes from North Africa evidence prehistoric migrations to the Maghreb from both the Levant and Europe. *Proceedings of the National Academy of Sciences* 115 (26):6774-6779.

Frînculeasa, Alin, Bianca Preda, and Volker Heyd. 2015. Pit-Graves, Yamnaya and Kurgans at the Lower Danube: Disentangling late 4th and early 3rd Millennium BC Burial Customs, Equipment and Chronology. *Praehistorische Zeitschrift* 90/1-2, 2015, 45-113. *Praehistorische Zeitschrift* 90 (1-2):45-113.

Fu, Q., H. Li, P. Moorjani, F. Jay, S. M. Slepchenko, A. A. Bondarev, P. L. Johnson, A. Aximu-Petri, K. Prufer, C. de Filippo, M. Meyer, N. Zwyns, D. C. Salazar-Garcia, Y. V. Kuzmin, S. G. Keates, P. A. Kosintsev, D. I. Razhev, M. P. Richards, N. V. Peristov, M. Lachmann, K. Douka, T. F. Higham, M. Slatkin,

J. J. Hublin, D. Reich, J. Kelso, T. B. Viola, and S. Paabo. 2014. Genome sequence of a 45,000-year-old modern human from western Siberia. *Nature* 514 (7523):445-9.

Fu, Q., C. Posth, M. Hajdinjak, M. Petr, S. Mallick, D. Fernandes, A. Furtwangler, W. Haak, M. Meyer, A. Mittnik, B. Nickel, A. Peltzer, N. Rohland, V. Slon, S. Talamo, I. Lazaridis, M. Lipson, I. Mathieson, S. Schiffels, P. Skoglund, A. P. Derevianko, N. Drozdov, V. Slavinsky, A. Tsybankov, R. G. Cremonesi, F. Mallegni, B. Gely, E. Vacca, M. R. Morales, L. G. Straus, C. Neugebauer-Maresch, M. Teschler-Nicola, S. Constantin, O. T. Moldovan, S. Benazzi, M. Peresani, D. Coppola, M. Lari, S. Ricci, A. Ronchitelli, F. Valentin, C. Thevenet, K. Wehrberger, D. Grigorescu, H. Rougier, I. Crevecoeur, D. Flas, P. Semal, M. A. Mannino, C. Cupillard, H. Bocherens, N. J. Conard, K. Harvati, V. Moiseyev, D. G. Drucker, J. Svoboda, M. P. Richards, D. Caramelli, R. Pinhasi, J. Kelso, N. Patterson, J. Krause, S. Paabo, and D. Reich. 2016. The genetic history of Ice Age Europe. *Nature* 534 (7606):200-5.

Furholt, Martin. 2014. Upending a 'Totality': Re-evaluating Corded Ware Variability in Late

Neolithic Europe. *Proceedings of the Prehistoric Society* 80:67-86.

Repeated Author. 2017. Massive Migrations? The Impact of Recent aDNA Studies on our View of Third Millennium Europe. *European Journal of Archaeology* 21 (2):159-191.

Gakuhari, Takashi, Shigeki Nakagome, Simon Rasmussen, Morten Allentoft, Takehiro Sato, Thorfinn Korneliussen, Blánaid Ní Chuinneagáin, Hiromi Matsumae, Kae Koganebuchi, Ryan Schmidt, Souichiro Mizushima, Osamu Kondo, Nobuo Shigehara, Minoru Yoneda, Ryosuke Kimura, Hajime Ishida, Yoshiyuki Masuyama, Yasuhiro Yamada, Atsushi Tajima, Hiroki Shibata, Atsushi Toyoda, Toshiyuki Tsurumoto, Tetsuaki Wakebe, Hiromi Shitara, Tsunehiko Hanihara, Eske Willerslev, Martin Sikora, and Hiroki Oota. 2019. Jomon genome sheds light on East Asian population history. *bioRxiv*:579177.

Gallego Llorente, M., E. R. Jones, A. Eriksson, V. Siska, K. W. Arthur, J. W. Arthur, M. C. Curtis, J. T. Stock, M. Coltorti, P. Pieruccini, S. Stretton, F. Brock, T. Higham, Y. Park, M. Hofreiter, D. G. Bradley, J. Bhak, R. Pinhasi, and A. Manica. 2015. Ancient Ethiopian genome reveals extensive Eurasian admixture in Eastern Africa. *Science* 350 (6262):820.

Gamba, C., E. R. Jones, M. D. Teasdale, R. L. McLaughlin, G. Gonzalez-Fortes, V. Mattiangeli, L. Domboroczki, I. Kovari, I. Pap, A. Anders, A. Whittle, J. Dani, P. Raczky, T. F. Higham, M. Hofreiter, D. G. Bradley, and R. Pinhasi. 2014. Genome flux and stasis in a five millennium transect of European prehistory. *Nat Commun* 5:5257.

Gardner, A.R:. 2002. Neolithic to Copper Age woodland impacts in northeast Hungary? Evidence from the pollen and sediment chemistry records. *The Holocene* 12 (5):541-553.

Gaunitz, Charleen, Antoine Fages, Kristian Hanghøj, Anders Albrechtsen, Naveed Khan, Mikkel Schubert, Andaine Seguin-Orlando, Ivy J. Owens, Sabine Felkel, Olivier Bignon-Lau, Peter de Barros Damgaard, Alissa Mittnik, Azadeh F. Mohaseb, Hossein Davoudi, Saleh Alquraishi, Ahmed H. Alfarhan, Khaled A. S. Al-Rasheid, Eric Crubézy, Norbert Benecke, Sandra Olsen, Dorcas Brown,

David Anthony, Ken Massy, Vladimir Pitulko, Aleksei Kasparov, Gottfried Brem, Michael Hofreiter, Gulmira Mukhtarova, Nurbol Baimukhanov, Lembi Lõugas, Vedat Onar, Philipp W. Stockhammer, Johannes Krause, Bazartseren Boldgiv, Sainbileg Undrakhbold, Diimaajav Erdenebaatar, Sébastien Lepetz, Marjan Mashkour, Arne Ludwig, Barbara Wallner, Victor Merz, Ilja Merz, Viktor Zaibert, Eske Willerslev, Pablo Librado, Alan K. Outram, and Ludovic Orlando. 2018. Ancient genomes revisit the ancestry of domestic and Przewalski's horses. *Science* 360 (6384):111-114.

Gerling, C., E. Bánffy, J. Dani, K. Köhler, G. Kulcsár, A.W.G. Pike, V. Szeverényi, and V. Heyd. 2012. Immigration and transhumance in the Early Bronze Age Carpathian Basin: the occupants of a kurgan. *Antiquity* 86 (334):1097-1111.

Gibbs, Kevin, and Peter Jordan. 2013. Bridging the Boreal Forest: Siberian Archaeology and the Emergence of Pottery among Prehistoric Hunter-Gatherers of Northern Eurasia. *Sibirica* 12 (1):1-38.

Gimbutas, Marija. 1963. The Indo-Europeans: Archeological Problems. *American Anthropologist* 65 (4):815-836.

Repeated Author. 1977. The first wave of eurasian pastoralists into copper age europe. *JIES* 5 (4):277-338.

Repeated Author. 1993. The Indo-Europeanization of Europe: the intrusion of steppe pastoralists from south Russia and the transformation of Old Europe. *Word* 44 (2):205-222.

Gogaltan, Florin. 2012. Ritual Aspects of the Bronze Age Tell-Settlements in the Carpathian Basin. A Methodological Approach. *Ephemeris Napocensis* 22:7-56.

González-Fortes, G., F. Tassi, E. Trucchi, K. Henneberger, J. L. A. Paijmans, D. Díez-del-Molino, H. Schroeder, R. R. Susca, C. Barroso-Ruíz, F. J. Bermudez, C. Barroso-Medina, A. M. S. Bettencourt, H. A. Sampaio, A. Grandal-d'Anglade, A. Salas, A. de Lombera-Hermida, R. Fabregas Valcarce, M. Vaquero, S. Alonso, M. Lozano, X. P. Rodríguez-Alvarez, C. Fernández-Rodríguez, A. Manica, M. Hofreiter, and G. Barbujani. 2019. A western route of prehistoric human migration from Africa into the Iberian Peninsula. *Proceedings of the Royal Society B: Biological Sciences* 286 (1895):20182288.

González-Fortes, Gloria, Eppie R. Jones, Emma Lightfoot, Clive Bonsall, Catalin Lazar, Aurora Grandal-d'Anglade, María Dolores Garralda, Labib Drak, Veronika Siska, Angela Simalcsik, Adina Boroneanţ, Juan Ramón Vidal Romaní, Marcos Vaqueiro Rodríguez, Pablo Arias, Ron Pinhasi, Andrea Manica, and Michael Hofreiter. 2017. Paleogenomic Evidence for Multi-generational Mixing between Neolithic Farmers and Mesolithic Hunter-Gatherers in the Lower Danube Basin. *Current Biology* 27 (12):1801-1810.

Gopalan, Shyamalika, Richard E. W. Berl, Gillian Belbin, Christopher Gignoux, Marcus W. Feldman, Barry S. Hewlett, and Brenna M. Henn. 2019. Hunter-gatherer genomes reveal diverse demographic trajectories following the rise of farming in East Africa. *bioRxiv*:517730.

Grugni, V., V. Battaglia, B. Hooshiar Kashani, S. Parolo, N. Al-Zahery, A. Achilli, A. Olivieri, F. Gandini, M. Houshmand, M. H. Sanati, A. Torroni, and O. Semino. 2012. Ancient migratory events in the Middle East: new clues from the Y-chromosome variation of modern Iranians. *PLoS One* 7 (7):e41252.

Grugni, Viola, Alessandro Raveane, Linda Ongaro, Vincenza Battaglia, Beniamino Trombetta, Giulia Colombo, Marco Rosario Capodiferro, Anna Olivieri, Alessandro Achilli, Ugo A. Perego, Jorge Motta, Maribel Tribaldos, Scott R. Woodward, Luca Ferretti, Fulvio Cruciani, Antonio Torroni, and Ornella Semino. 2019. Analysis of the human Y-chromosome haplogroup Q characterizes ancient population movements in Eurasia and the Americas. *BMC Biology* 17 (1):3.

Guilaine, Jean. 2017. The Neolithic Transition: From the Eastern to the Western Mediterranean. In *Times of Neolithic Transition along the Western Mediterranean*, edited by O. García-Puchol and D. C. Salazar-García: Springer.

Gunther, T., C. Valdiosera, H. Malmstrom, I. Urena, R. Rodriguez-Varela, O. O. Sverrisdottir, E. A. Daskalaki, P. Skoglund, T. Naidoo, E. M. Svensson, J. M. Bermudez de Castro, E. Carbonell, M. Dunn, J. Stora, E. Iriarte, J. L. Arsuaga, J. M. Carretero, A. Gotherstrom, and M. Jakobsson. 2015. Ancient genomes link early farmers from Atapuerca in Spain to modern-day Basques. *Proc Natl Acad Sci U S A* 112 (38):11917-22.

Günther, Torsten, Helena Malmström, Emma Svensson, Ayça Omrak, Federico Sánchez-Quinto, Gülşah M. Kılınç, Maja Krzewińska, Gunilla Eriksson, Magdalena Fraser, Hanna Edlund, Arielle R. Munters, Alexandra Coutinho, Luciana G. Simões, Mário Vicente, Anders Sjölander, Berit Jansen Sellevold, Roger Jørgensen, Peter Claes, Mark D. Shriver, Cristina Valdiosera, Mihai G. Netea, Jan Apel, Kerstin Lidén, Birgitte Skar, Jan Storå, Anders Götherström, and Mattias Jakobsson. 2017. Genomics of Mesolithic Scandinavia reveal colonization routes and high-latitude adaptation. *bioRxiv*.

Gworys, Bohdan, Joanna Rosińczuk-Tonderys, Aleksander Chrószcz, Maciej Janeczek, Andrzej Dwojak, Justyna Bazan, Mirosław Furmanek, Tadeusz Dobosz, Małgorzata Bonar, Anna Jonkisz, and Ireneusz Całkosiński. 2013. Assessment of late Neolithic pastoralist's life conditions from the Wroclaw–Jagodno site (SW Poland) on the basis of physiological stress markers. *Journal of Archaeological Science* 40 (6):2621-2630.

Haak, W., O. Balanovsky, J. J. Sanchez, S. Koshel, V. Zaporozhchenko, C. J. Adler, C. S. Der Sarkissian, G. Brandt, C. Schwarz, N. Nicklisch, V. Dresely, B. Fritsch, E. Balanovska, R. Villems, H. Meller, K. W. Alt, A. Cooper, and Consortium Members of the Genographic. 2010. Ancient DNA from European early neolithic farmers reveals their near eastern affinities. *PLoS Biol* 8 (11):e1000536.

Haak, W., I. Lazaridis, N. Patterson, N. Rohland, S. Mallick, B. Llamas, G. Brandt, S. Nordenfelt, E. Harney, K. Stewardson, Q. Fu, A. Mittnik, E. Banffy, C. Economou, M. Francken, S. Friederich, R. G. Pena, F. Hallgren, V. Khartanovich, A. Khokhlov, M. Kunst, P. Kuznetsov, H. Meller, O. Mochalov, V. Moiseyev, N. Nicklisch, S. L. Pichler, R. Risch, M. A. Rojo Guerra, C. Roth, A. Szecsenyi-Nagy, J. Wahl, M. Meyer, J. Krause, D. Brown, D. Anthony, A. Cooper, K. W. Alt, and D. Reich. 2015. Massive migration from the steppe was a source for Indo-European languages in Europe. *Nature* 522 (7555):207-11.

Haber, Marc, Claude Doumet-Serhal, Christiana Scheib, Yali Xue, Petr Danecek, Massimo Mezzavilla, Sonia Youhanna, Rui Martiniano, Javier Prado-Martinez, Michał Szpak, Elizabeth Matisoo-Smith, Holger Schutkowski, Richard

Mikulski, Pierre Zalloua, Toomas Kivisild, and Chris Tyler-Smith. 2017. Continuity and admixture in the last five millennia of Levantine history from ancient Canaanite and present-day Lebanese genome sequences. *bioRxiv*.

Haber, Marc, Massimo Mezzavilla, Anders Bergström, Javier Prado-Martinez, Pille Hallast, Riyadh Saif-Ali, Molham Al-Habori, George Dedoussis, Eleftheria Zeggini, Jason Blue-Smith, R. Spencer Wells, Yali Xue, Pierre A Zalloua, and Chris Tyler-Smith. 2016. Chad Genetic Diversity Reveals an African History Marked by Multiple Holocene Eurasian Migrations. *The American Journal of Human Genetics* 99 (6):1316-1324.

Hänsel, B., and S. Zimmer. 1994. *Die Indogermanen und das Pferd: Festschrift für Bernfried Schlerath*. Edited by S. Bökönyi and W. Meid, *Archaeolingua*. Budapest: Archaeolingua Alapítvány.

Harding, Anthony F. 2011. The tumulus in European prehistory: covering the body, housing the soul. In *Burial mounds in the copper and Bronze ages (Central and Eastern Europe – Balkans – Adriatic – Aegean, 4th-2nd millennium B.C.)*, edited by E. Borgna and S. Müller-Celka. Lyon: Maison de l'Orient.

Harney, Éadaoin, Hila May, Dina Shalem, Nadin Rohland, Swapan Mallick, Iosif Lazaridis, Rachel Sarig, Kristin Stewardson, Susanne Nordenfelt, Nick Patterson, Israel Hershkovitz, and David Reich. 2018. Ancient DNA from Chalcolithic Israel reveals the role of population mixture in cultural transformation. *Nature Communications* 9 (1):3336.

Harper, Thomas K., Aleksandr Diachenko, Yuri Ya Rassamakin, and Douglas J. Kennett. 2019. Ecological dimensions of population dynamics and subsistence in Neo-Eneolithic Eastern Europe. *Journal of Anthropological Archaeology* 53:92-101.

Harrison, Richard, and Volker Heyd. 2007. The Transformation of Europe in the Third Millennium BC: the example of 'Le Petit-Chasseur I + III' (Sion, Valais, Switzerland). *Praehistorische Zeitschrift* 82 (2).

Hassett, Brenna, and Haluk Sağlamtimur. 2018. Radical 'royals'? Burial practices at Başur Höyük and the emergence of early states in Mesopotamia. *Antiquity* 92 (363):640-654.

Heggarty, Paul. 2015. Ancient DNA and the Indo-European Question. In *Diversity Linguistics Comment. Language structures throughout the world*. Leipzig: Max Planck Institute for Evolutionary Anthropology.

Heise, Marc E. 2014. Heads North or East? A re-examination of Beaker Burials in Britain, School of History, Classics and Archaeology, University of Edinburgh, Edinburgh.

Hellenthal, G., G. B. Busby, G. Band, J. F. Wilson, C. Capelli, D. Falush, and S. Myers. 2014. A genetic atlas of human admixture history. *Science* 343 (6172):747-51.

Herrera, K. J., R. K. Lowery, L. Hadden, S. Calderon, C. Chiou, L. Yepiskoposyan, M. Regueiro, P. A. Underhill, and R. J. Herrera. 2012. Neolithic patrilineal signals indicate that the Armenian plateau was repopulated by agriculturalists. *Eur J Hum Genet* 20 (3):313-20.

Heyd, Volker. 2004. Soziale Organisation im 3. Jahrtausend v. Chr. entlang der oberen Donau: Der Fall Schnurkeramik und Glockenbecher. *Das Altertum* 49 (3):183-214.

Repeated Author. 2007. Families, Prestige Goods, Warriors & Complex Societies: Beaker Groups of the 3rd Millennium cal BC Along the Upper & Middle Danube. *Proceedings of the Prehistoric Society* 73:327-379.

Repeated Author. 2011. Yamnaya Groups and Tumuli west of the Black Sea. In: Ancestral Landscapes. In *Ancestral Landscape. Burial mounds in the Copper and Bronze Ages. Proceedings of the International Conference held in Udine, May 15th-18th 2008*, edited by E. Borgna and S. Mülller Celka. Lyon: TMO.

Repeated Author. 2012. Yamnaya gropus and tumuli west of the Black Sea. *Travaux de la Maison de l'Orient et de la Méditerranée. Série recherches archéologiques* 58 (1):535-555.

Repeated Author. 2013. Europe 2500 to 2200 BC: Between Expiring Ideologies and Emerging Complexity. In *The Oxford Handbook of the European Bronze Age*, edited by H. Fokkens and A. Harding. Oxford: Oxford University Press.

Repeated Author. 2013. Europe at the Dawn of the Bronze Age. In *Transition to the Bronze Age*, edited by V. Heyd, G. Kulcsár and V. Szeverényi. Budapest: Archaeolingua.

Repeated Author. 2016. Das Zeitalter der Ideologien: Migration, Interaktion und Expansion im prähistorischen Europa des 4. und 3. Jahrtausends v. Chr. In *Transitional Landscapes? The 3rd Millennium BC in Europe. Proceedings of the International Workshop "Socio-Environmental Dynamics over the Last 12,000 Years: The Creation of Landscapes III (15th – 18th April 2013)" in Kiel*, edited by M. Furholt, R. Großmann and M. Szmyt. Bonn: Dr. Rudolf Habelt.

Repeated Author. 2017. Kossinna's smile. *Antiquity* 91 (356):348-359.

Heyd, Volker, and Katharine Walker. 2004. The First Metalwork and Expressions of Social Power. In *The Oxford Handbook of Neolithic Europe*, edited by C. Fowler, J. Harding and D. Hofmann. Oxford: Oxford University Press.

Heyd, Vollker. 2007. When the West meets the East: The Eastern Periphery of the Bell Beaker Phenomenon and its Relation with the Aegean Early Bronze Age. In *Between the Aegean and Baltic Seas*, edited by I. Galanaki. Liège: Aegaeum.

Hildebrand, Elisabeth A., Katherine M. Grillo, Elizabeth A. Sawchuk, Susan K. Pfeiffer, Lawrence B. Conyers, Steven T. Goldstein, Austin Chad Hill, Anneke Janzen, Carla E. Klehm, Mark Helper, Purity Kiura, Emmanuel Ndiema, Cecilia Ngugi, John J. Shea, and Hong Wang. 2018. A monumental cemetery built by eastern Africa's first herders near Lake Turkana, Kenya. *Proceedings of the National Academy of Sciences* 115 (36):8942-8947.

Hofmanova, Z., S. Kreutzer, G. Hellenthal, C. Sell, Y. Diekmann, D. Diez-Del-Molino, L. van Dorp, S. Lopez, A. Kousathanas, V. Link, K. Kirsanow, L. M. Cassidy, R. Martiniano, M. Strobel, A. Scheu, K. Kotsakis, P. Halstead, S. Triantaphyllou, N. Kyparissi-Apostolika, D. Urem-Kotsou, C. Ziota, F. Adaktylou, S. Gopalan, D. M. Bobo, L. Winkelbach, J. Blocher, M. Unterlander, C. Leuenberger, C. Cilingiroglu, B. Horejs, F. Gerritsen, S. J. Shennan, D. G. Bradley, M. Currat, K. R. Veeramah, D. Wegmann, M. G. Thomas, C. Papageorgopoulou, and J. Burger. 2016. Early farmers from across Europe directly descended from Neolithic Aegeans. *Proc Natl Acad Sci U S A* 113 (25):6886-91.

Hollard, Clémence, Vincent Zvénigorosky, Alexey Kovalev, Yurii Kiryushin, Alexey Tishkin, Igor Lazaretov, Eric Crubézy, Bertrand Ludes, and Christine

Keyser. 2018. New genetic evidence of affinities and discontinuities between bronze age Siberian populations. *American Journal of Physical Anthropology* 167 (1):97-107.

Hollfelder, Nina, Carina M. Schlebusch, Torsten Günther, Hiba Babiker, Hisham Y. Hassan, and Mattias Jakobsson. 2017. Northeast African genomic variation shaped by the continuity of indigenous groups and Eurasian migrations. *PLOS Genetics* 13 (8):e1006976.

Holmqvist, Elisabeth, Åsa M. Larsson, Aivar Kriiska, Vesa Palonen, Petro Pesonen, Kenichiro Mizohata, Paula Kouki, and Jyrki Räisänen. 2018. Tracing grog and pots to reveal Neolithic Corded Ware Culture contacts in the Baltic Sea region (SEM-EDS, PIXE). *Journal of Archaeological Science* 91:77-91.

Horváth, Csaba Barnabás. 2014. The story of two northward migrations - origins of Finno-Permic and Balto-Slavic languages in northeast Europe, based on Y-chromosome haplogroups. *European Scientific Journal* 2:531-538.

Horváth, Tünde. 2016. 4000-2000 BC in Hungary: The Age of Transformation. In *The Carpathian Basin and the Northern Balkans between 3500 and 2500 BC: Common Aspects and Regional Differences*, edited by C. I. Popa. Alba Iulia: Mega.

Horváth, Tünde, János Dani, Ákos Pető, Łukasz Pospieszny, and Éva Svingor. 2013. Multidisciplinary Contributions to the Study of Pit Grave Culture Kurgans of the Great Hungarian Plain. In *Transition to the Bronze Age: Interregional Interaction and Socio-Cultural Change at the Beginning of the Third Millennium BC in the Carpathian Basin and Surrounding Regions*, edited by V. M. Heyd, G. Kulcsár and V. Szeverényi. Budapest: Archaeolingua.

Huang, Yun-Zhi, Horolma Pamjav, Pavel Flegontov, Vlastimil Stenzl, Shao-Qing Wen, Xin-Zhu Tong, Chuan-Chao Wang, Ling-Xiang Wang, Lan-Hai Wei, Jing-Yi Gao, Li Jin, and Hui Li. 2017. Dispersals of the Siberian Y-chromosome haplogroup Q in Eurasia. *Molecular Genetics and Genomics*.

ISOGG. 2018. Y-DNA Haplogroup Tree. Version: 13.270 . Date: 13 November 2018. International Society of Genetic Genealogy.

Ivanova, M., B. De Cupere, J. Ethier, and E. Marinova. 2018. Pioneer farming in southeast Europe during the early sixth millennium BC: Climate-related adaptations in the exploitation of plants and animals. *PLoS ONE* 13 (5):e0197225.

Ivanova, Svetlana V., and Gennadiy N. Toschev. 2015. The Middle-Dniester Cultural Contact Area of Early Metal Age Societies. The Frontier of Pontic and Baltic Drainage Basins in the 4Th/3Rd-2Nd Millennium Bc. In *Baltic-Pontic Studies*.

Janhunen, Juha. 2009. Proto-Uralic: What, where, and when? – The quasquicentennial of the Finno-Ugrian society. *SUST* 258:57-78.

Jaruf, Pablo. 2017. Santuarios subterráneos en el Calcolítico palestiniense (ca. 4500-3800/3600 a.C.). In *Si un Hombre desde el Sur... / Šumma Awīlum ina Šūtim...: Escritos de Alumnos, Colegas y Amigos en Homenaje a Bernardo Gandulla*, edited by I. Milevski, L. Monti and P. Jaruf. Buenos Aires: Universidad de Buenos Aires.

Jensen, Theis Z. T., Jonas Niemann, Katrine Hoejholt Iversen, Anna K. Fotakis, Shyam Gopalakrishnan, Mikkel H. S. Sinding, Martin R. Ellegaard, Morten E.

Allentoft, Liam T. Lanigan, Alberto J. Taurozzi, Sofie Holtsmark Nielsen, Michael W. Dee, Martin N. Mortensen, Mads C. Christensen, Soeren A. Soerensen, Matthew J. Collins, Tom Gilbert, Martin Sikora, Simon Rasmussen, and Hannes Schroeder. 2018. Stone Age "chewing gum" yields 5,700 year-old human genome and oral microbiome. *bioRxiv*:493882.

Jeunesse, Christian. 2015. Das Aufkommen der Ideologie des Kriegers im westlichen Mittelmeerraum in der zweiten Hälfte des vierten Jahrtausends v. Chr. *Das Altertum* 60:263-282.

Repeated Author. 2017. From Neolithic kings to the Staffordshire hoard. Hoards and aristocratic graves in the European Neolithic: the birth of a 'Barbarian' Europe? In *The Neolithic in Europe. Papers in Honour of Alasdair Whittle*, edited by P. Bickle, V. Cummings, D. Hofmann and J. Pollard. Oxford & Philadelphia: Oxbow.

Jones, E. R., G. Gonzalez-Fortes, S. Connell, V. Siska, A. Eriksson, R. Martiniano, R. L. McLaughlin, M. Gallego Llorente, L. M. Cassidy, C. Gamba, T. Meshveliani, O. Bar-Yosef, W. Muller, A. Belfer-Cohen, Z. Matskevich, N. Jakeli, T. F. Higham, M. Currat, D. Lordkipanidze, M. Hofreiter, A. Manica, R. Pinhasi, and D. G. Bradley. 2015. Upper Palaeolithic genomes reveal deep roots of modern Eurasians. *Nat Commun* 6:8912.

Jones, Eppie R., Gunita Zarina, Vyacheslav Moiseyev, Emma Lightfoot, Philip R. Nigst, Andrea Manica, Ron Pinhasi, and Daniel G. Bradley. 2017. The Neolithic Transition in the Baltic Was Not Driven by Admixture with Early European Farmers. *Current Biology*.

Juras, Anna, Maciej Chyleński, Edvard Ehler, Helena Malmström, Danuta Żurkiewicz, Piotr Włodarczak, Stanisław Wilk, Jaroslav Peška, Pavel Fojtík, Miroslav Králík, Jerzy Libera, Jolanta Bagińska, Krzysztof Tunia, Viktor I. Klochko, Miroslawa Dabert, Mattias Jakobsson, and Aleksander Kośko. 2018. Mitochondrial genomes reveal an east to west cline of steppe ancestry in Corded Ware populations. *Scientific Reports* 8 (1):11603.

Kador, Thomas, Lara M. Cassidy, Jonny Geber, Robert Hensey, Pádraig Meehan, and Sam Moore. 2018. Rites of Passage: Mortuary Practice, Population Dynamics, and Chronology at the Carrowkeel Passage Tomb Complex, Co. Sligo, Ireland. *Proceedings of the Prehistoric Society*:1-31.

Kadrow, S. 2004. Problemy archeologicznej identyfikacji zjawisk nomadyzmu i pastoralizmu na przełomie epok kamienia i brązu w Europie. In *Nomadyzm a pastoralizm w międzyrzeczu Wisły i Dniepru (neolit, eneolit, epoka brązu)*, edited by A. Kośko and M. Szmyt. Poznań: Archaeologia Bimaris.

Kadrow, Sławomir. 2008. Settlements and subsistence strategies of the Corded Ware Culture at the beginning of the 3rd millenium BC in Southeastern Poland and in Western Ukraine. In *Umwelt - Wirtschaft - Siedlungen im dritten vorchristlichen Jahrtausend Mitteleuropas und Südskandinaviens. Internationale Tagung Kiel 4.-6. November 2005*, edited by W. Dörfler and J. Müller. Neumünster: Wachholtz.

Repeated Author. 2016. Exchange of People, Ideas and Things between Cucuteni-Trypillian Complex and Areas of South-Eastern Poland. In *Cucuteni culture within the European Neo-Eneolithic context*, edited by C. P. Ciprian and D. Nicola. Romania: Complexul Muzeal Județean Neamț.

Repeated Author. 2018. South-Eastern Group of Funnel Beaker culture. *Prace i Materiały Muzeum Archeologicznego w Łodzi. Seria Archeologiczna* 47:255-266.

Kaiser, Elke, and Katja Winger. 2015. Pit graves in Bulgaria and the Yamnaya Culture. *Praehistorische Zeitschrift* 90 (1-2).

Kajtoch, Łukasz, Elżbieta Cieślak, Zoltán Varga, Wojciech Paul, Miłosz A. Mazur, Gábor Sramkó, and Daniel Kubisz. 2016. Phylogeographic patterns of steppe species in Eastern Central Europe: a review and the implications for conservation. *Biodiversity and Conservation* 25 (12):2309-2339.

Kallio, Petri. 2002. Prehistoric Contacts between Indo-European and Uralic. In *Proceedings of the Thirteenth Annual UCLA Indo-European Conference*, edited by K. Jones-Bley, M. E. Huld, A. D. Volpe and M. R. Dexter. Washington, DC: Institute for the Study of Man.

Repeated Author. 2014. The Diversification of Proto-Finnic. In *Fibula, Fabula, Fact: The Viking Age in Finland*, edited by J. A. Frog and C. Tolley Helsinki.

Repeated Author. 2015. Nugae Indo-Uralicae. *JIES* 43 (3 & 4):368-375.

Karafet, Tatiana M., Ludmila P. Osipova, Olga V. Savina, Brian Hallmark, and Michael F. Hammer. 2018. Siberian genetic diversity reveals complex origins of the Samoyedic-speaking populations. *American Journal of Human Biology* 0 (0):e23194.

Karim, Alizadeh, Maziar Sepideh, and M. Rouhollah Mohammadi. 2018. The End of the Kura-Araxes Culture as Seen from Nadir Tepesi in Iranian Azerbaijan. *American Journal of Archaeology* 122 (3):463-477.

Kashuba, Natalija, Emrah Kırdök, Hege Damlien, Mikael A. Manninen, Bengt Nordqvist, Per Persson, and Anders Götherström. 2019. Ancient DNA from mastics solidifies connection between material culture and genetics of mesolithic hunter–gatherers in Scandinavia. *Communications Biology* 2 (1):185.

Kefi, Rym, Meriem Hechmi, Chokri Naouali, Haifa Jmel, Sana Hsouna, Eric Bouzaid, Sonia Abdelhak, Eliane Beraud-Colomb, and Alain Stevanovitch. 2018. On the origin of Iberomaurusians: new data based on ancient mitochondrial DNA and phylogenetic analysis of Afalou and Taforalt populations. *Mitochondrial DNA Part A* 29 (1):147-157.

Khokhlov, Aleksander. 2018. Предварительные результаты антрополого-генетических исследований материалов Волго-Уралья периода неолита-ранней бронзы международной группой ученых. Paper read at XIV Conference on Samaran Archaeology, 27-28th January, at Samara.

Khokhlov, Aleksandr A. 2016. Demographic and Cranial Characteristics of the Volga-Ural Population in the Eneolithic and Bronze Age. In *A Bronze Age Landscape in the Russian Steppes. The Samara Valley Project*, edited by D. W. Anthony, D. R. Brown, O. D. Mochalov, A. A. Khokhlov and P. F. Kuznetsov. Los Angeles: Cotsen Institute of Archaeology Press.

Khokhlova, Olga, Nina Morgunova, Alexander Khokhlov, and Alexandra Golyeva. 2018. Dynamics of paleoenvironments in the Cis-Ural steppes during the mid-to late Holocene. *Quaternary Research*:1-15.

Khomutova, Tatiana E., Natalia N. Kashirskaya, Tatiana S. Demkina, Tatiana V. Kuznetsova, Flavio Fornasier, Natalia I. Shishlina, and Alexander V. Borisov.

2019. Precipitation pattern during warm and cold periods in the Bronze Age (around 4.5-3.8 ka BP) in the desert steppes of Russia: Soil-microbiological approach for palaeoenvironmental reconstruction. *Quaternary International*.

Kießling, Roland, Maarten Mous, and Derek Nurse. 2008. The Tanzanian Rift Valley area. In *A Linguistic Geography of Africa* edited by B. Heine and D. Nurse. Cambridge: Cambridge University Press.

Kilinc, G. M., A. Omrak, F. Ozer, T. Gunther, A. M. Buyukkarakaya, E. Bicakci, D. Baird, H. M. Donertas, A. Ghalichi, R. Yaka, D. Koptekin, S. C. Acan, P. Parvizi, M. Krzewinska, E. A. Daskalaki, E. Yuncu, N. D. Dagtas, A. Fairbairn, J. Pearson, G. Mustafaoglu, Y. S. Erdal, Y. G. Cakan, I. Togan, M. Somel, J. Stora, M. Jakobsson, and A. Gotherstrom. 2016. The Demographic Development of the First Farmers in Anatolia. *Curr Biol* 26 (19):2659-2666.

Kılınç, Gülşah Merve, Dilek Koptekin, Çiğdem Atakuman, Arev Pelin Sümer, Handan Melike Dönertaş, Reyhan Yaka, Cemal Can Bilgin, Ali Metin Büyükkarakaya, Douglas Baird, Ezgi Altınışık, Pavel Flegontov, Anders Götherström, İnci Togan, and Mehmet Somel. 2017. Archaeogenomic analysis of the first steps of Neolithization in Anatolia and the Aegean. *Proceedings of the Royal Society B: Biological Sciences* 284 (1867).

Klejn, Leo S., Wolfgang Haak, Iosif Lazaridis, Nick Patterson, David Reich, Kristian Kristiansen, Karl-Göran Sjögren, Morten Allentoft, Martin Sikora, and Eske Willerslev. 2017. Discussion: Are the Origins of Indo-European Languages Explained by the Migration of the Yamnaya Culture to the West? *European Journal of Archaeology*:1-15.

Klimscha, Florian. 2017. Transforming Technical Know-how in Time and Space. Using the Digital Atlas of Innovations to Understand the Innovation Process of Animal Traction and the Wheel. *eTopoi* 6:16-63.

Klochko, V. I., and A Kośko. 2011. Społeczności kultur ceramiki sznurowej i stepu nadczarnomorskiego (jamowej oraz katakumbowej) w systemie organizacji szlaków bałtycko-pontyjskiego międzymorza. In *Między Bałtykiem a Morzem Czarnym. Szlaki międzymorza IV–I tys. przed Chr*. edited by M. Ignaczak, A. Kośko and M. Szmyt. Poznań.

Klochko, V. I., and A. Kośko. 2009. The societies of Corded Ware cultures and those of Black Sea steppes (Yamnaya and Catacomb Grave cultures) in the route network between the Baltic and Black Seas. *Baltic-Pontic-Studies* 14:269-301.

Klochko, Viktor. 2013. The Baltic drainage basin in the reconstruction of the Mental Map of Central Europe Held in common by Northern-Pontic Early-Bronze Civilization Communities: 3200-1600 BC. An outline of the research programme. In *The Ingul-Donets Early Bronze Civilization as Springboard for Transmission of Pontic Cultural Patterns to the Baltic Drainage Basin 3200-1750 BC*. Poznań.

Repeated Author. 2013. Complex of Metal Goods between the Vistula and Dnieper rivers at the turn of the 4th/3rd to the 3rd millennium BC. Concept of the Carpathian - Volhynia "Willow Leaf" metallurgy centre. In *The Ingul-Donets Early Bronze Civilization as Springboard for Transmission of Pontic Cultural Patterns to the Baltic Drainage Basin 3200-1750 BC*. Poznań.

Klochko, Viktor, and Aleksander Kośko. 1987. A Late Neolithic Composite Bow. *Journal of the Society of Archer Antiquaries. London* 30:15-23.

Repeated Author. 1998. "Trzciniec" – Borderland of Early Bronze Civilization of Eastern and Western Europe? In *The Trzciniec Area of the Early Bronze Age Civilization: 1950-1200 BC*. Poznan.

Repeated Author. 2009. Transit routes between the Baltic and Black seas: early development stages – from the 3rd to the middle of the 1st millennium BC. An outline of research project./Routes between the seas: Baltik-Boh-Bug-Pont from the 3rd to the middle of the 1st millennium BC. Poznan, 2009, (BPS 14), P. 7 – 16. In *Routes between the seas: Baltik-Boh-Bug-Pont from the 3rd to the middle of the 1st millennium BC*. Poznan.

Kobusiewicz, Michael. 2002. The problem of the Palaeolithic-Mesolithic transition on the Polish Plain: the state of research. In *Hunters in a changing world. Environment and Archaeology of the Pleistocene - Holocene Transition (ca. 11000 - 9000 B.C.) in Northern Central Europe*, edited by T. Terberger and B. V. Eriksen. Greifswald: Workshop of the U.I.S.P.P.-Commission XXXII.

Koivulehto, J. . 1991. *Uralische Evidenz für die Laryngaltheorie*. Vol. 566. Viena: Österreichische Akademie der Wissenschaften.

Koivulehto, Jorma. 2003. Frühe Kontakte zwischen Uralisch und Indogermanisch im nordwestindogermanischen Raum. In *Languages in Prehistoric Europe*, edited by A. Bammesberger and T. Vennemann. Heidelberg: Universitätsverlag Winter.

Kolář, Jan, Petr Kuneš, Péter Szabó, Mária Hajnalová, Helena Svitavská Svobodová, Martin Macek, and Peter Tkáč. 2016. Population and forest dynamics during the Central European Eneolithic (4500–2000 BC). *Archaeological and Anthropological Sciences*.

Korenevskiy, S.N. 2012. *Rozhdenie kurgana: pogrebalnye pamjatniki eneoliticheskogo vremeni Predkavkaz'ja i Volgo-Donskogo mezhdurech'ja*. Moskva: TAUS.

Korolev, A., M. Kulkova M, V. Platonov, N. Roslyakova, A. Shalapinin, and Y. E. Yanish. 2018. Archaeological Materials of Eneolithic Settlements in Forest-Steppe Zone of the Volga Region: A Source for Diet and Chronology. *Radiocarbon* 60 (5):1587-1596.

Korolev, Arkadiy, Anna Kochkina, and Dmitry Stashenkov. 2019. КЕРАМИКА ГРУНТОВОГО МОГИЛЬНИКА ЕКАТЕРИНОВСКИЙ МЫС (ПО МАТЕРИАЛАМ РАСКОПОК 2013-2016 ГГ.). *ПОВОЛЖСКАЯ АРХЕОЛОГИЯ* 1 (27):18-32.

Kortlandt, Frederik. 1990. The spread of the Indo-Europeans. *Journal of Indo-European Studies* 18 (2):131-140.

Repeated Author. 2002. The Indo-Uralic verb. In *Finno-Ugrians and Indo-Europeans: Linguistic and literary contacts*. Maastricht: Shaker.

Repeated Author. 2019. On the reconstruction of Proto-Uralic. In *Petri Kallio Rocks. Liber Semisaecularis 7.2.2019*, edited by T. S. Junttila and J. Kuokkala. Helsinki: Printall.

Kośko, Aleksander, and Marzena Szmyt. 2004. Hodowla w systemach gospodarki Niżu: IV-III tys. BC (kultury: pucharów lejkowatych i amfor kulistych). In *Nomadyzm a pastoralizm w międzyrzeczu Wisły i Dniepru (neolit, eneolit, epoka brązu)* . edited by A. Kośko and M. Szmyt. Poznań: Archaeologia Bimaris.

Kotova, Nadezhda. 2016. The contacts of the Eastern European steppe people with the Balkan population during the transition period from Neolithic to Eneolithic. In *Der Schwarzmeerraum vom Neolithikum bis in die Früheisenzeit (6000-600 v. Chr.). Kulturelle Interferenzen in der Zirkumpontischen Zone und Kontakte mit ihren Nachbargebieten,* edited by V. Nikolov and W. Schier. Rahden/Westfalia: Marie Leidorf.

Kotova, Nadezhda S. 2008. *Early Eneolithic in the Pontic Steppes, British Archaeological Reports International Series 1735.* Oxford: John and Erica Hedges.

Kozłowski, Stefan Karol. 2009. *Thinking Mesolithic.* Oxford: Oxbow Books.

Krause-Kyora, Ben, Julian Susat, Felix M. Key, Denise Kühnert, Esther Bosse, Alexander Immel, Christoph Rinne, Sabin-Christin Kornell, Diego Yepes, Sören Franzenburg, Henrike O. Heyne, Thomas Meier, Sandra Lösch, Harald Meller, Susanne Friederich, Nicole Nicklisch, Kurt W. Alt, Stefan Schreiber, Andreas Tholey, Alexander Herbig, Almut Nebel, and Johannes Krause. 2018. Neolithic and medieval virus genomes reveal complex evolution of hepatitis B. *eLife* 7:e36666.

Krauß, Raiko, Clemens Schmid, David Kirschenheuter, Jonas Abele, Vladimir Slavchev, and Bernhard Weninger. 2017. Chronology and development of the Chalcolithic necropolis of Varna I. *Documenta Praehistorica* XLIV:282-300.

Krenke, N., I. Erschov, E. Erschova, and A. Lazukin. 2013. Corded ware, Fatjanovo and Abashevo culture sites on the flood-plain of the Moskva River. *Sprawozdania Archeologiczne* 65:413-424.

Kristiansen, Kristian. 1989. Prehistoric Migrations - the Case of the Single Grave and Corded Ware Cultures. *Journal of Danish Archaeology* 8 (1):211-225.

Kristiansen, Kristian, Morten E. Allentoft, Karin M. Frei, Rune Iversen, Niels N. Johannsen, Guus Kroonen, Łukasz Pospieszny, T. Douglas Price, Simon Rasmussen, Karl-Göran Sjögren, Martin Sikora, and Eske Willerslev. 2017. Re-theorising mobility and the formation of culture and language among the Corded Ware Culture in Europe. *Antiquity* 91 (356):334-347.

Kristiansen, Kristian, and Thomas B. Larsson. 2005. Rulership in the Near East and the eastern Mediterranean during the Bronze Age. In *The Rise of Bronze Age Society. Travels, Transmissions and Transformations.* Cambridge: Cambridge University Press.

Kroonen, Guus. 2012. Non-Indo-European root nouns in Germanic: Evidence in support of the Agricultural Substrate Hypothesis. In *A Linguistic Map of Prehistoric Northern Europe,* edited by R. Grünthal and P. Kallio. Helsinki: Suomalais-Ugrilaisen Seura.

Kuznetsov, Pavel F., and Oleg D. Mochalov. 2016. The Samara Valley in the Bronze Age: A Review of Archaeological Discoveries. In *A Bronze Age Landscape in the Russian Steppes. The Samara Valley Project,* edited by D. W. Anthony, D. R. Brown, O. D. Mochalov, A. A. Khokhlov and P. F. Kuznetsov. Los Angeles: The Cotsen Institute of Archaeology Press at UCLA.

Lazaridis, I., D. Nadel, G. Rollefson, D. C. Merrett, N. Rohland, S. Mallick, D. Fernandes, M. Novak, B. Gamarra, K. Sirak, S. Connell, K. Stewardson, E.

Harney, Q. Fu, G. Gonzalez-Fortes, E. R. Jones, S. A. Roodenberg, G. Lengyel, F. Bocquentin, B. Gasparian, J. M. Monge, M. Gregg, V. Eshed, A. S. Mizrahi, C. Meiklejohn, F. Gerritsen, L. Bejenaru, M. Bluher, A. Campbell, G. Cavalleri, D. Comas, P. Froguel, E. Gilbert, S. M. Kerr, P. Kovacs, J. Krause, D. McGettigan, M. Merrigan, D. A. Merriwether, S. O'Reilly, M. B. Richards, O. Semino, M. Shamoon-Pour, G. Stefanescu, M. Stumvoll, A. Tonjes, A. Torroni, J. F. Wilson, L. Yengo, N. A. Hovhannisyan, N. Patterson, R. Pinhasi, and D. Reich. 2016. Genomic insights into the origin of farming in the ancient Near East. *Nature* 536 (7617):419-24.

Lazaridis, I., N. Patterson, A. Mittnik, G. Renaud, S. Mallick, K. Kirsanow, P. H. Sudmant, J. G. Schraiber, S. Castellano, M. Lipson, B. Berger, C. Economou, R. Bollongino, Q. Fu, K. I. Bos, S. Nordenfelt, H. Li, C. de Filippo, K. Prufer, S. Sawyer, C. Posth, W. Haak, F. Hallgren, E. Fornander, N. Rohland, D. Delsate, M. Francken, J. M. Guinet, J. Wahl, G. Ayodo, H. A. Babiker, G. Bailliet, E. Balanovska, O. Balanovsky, R. Barrantes, G. Bedoya, H. Ben-Ami, J. Bene, F. Berrada, C. M. Bravi, F. Brisighelli, G. B. Busby, F. Cali, M. Churnosov, D. E. Cole, D. Corach, L. Damba, G. van Driem, S. Dryomov, J. M. Dugoujon, S. A. Fedorova, I. Gallego Romero, M. Gubina, M. Hammer, B. M. Henn, T. Hervig, U. Hodoglugil, A. R. Jha, S. Karachanak-Yankova, R. Khusainova, E. Khusnutdinova, R. Kittles, T. Kivisild, W. Klitz, V. Kucinskas, A. Kushniarevich, L. Laredj, S. Litvinov, T. Loukidis, R. W. Mahley, B. Melegh, E. Metspalu, J. Molina, J. Mountain, K. Nakkalajarvi, D. Nesheva, T. Nyambo, L. Osipova, J. Parik, F. Platonov, O. Posukh, V. Romano, F. Rothhammer, I. Rudan, R. Ruizbakiev, H. Sahakyan, A. Sajantila, A. Salas, E. B. Starikovskaya, A. Tarekegn, D. Toncheva, S. Turdikulova, I. Uktveryte, O. Utevska, R. Vasquez, M. Villena, M. Voevoda, C. A. Winkler, L. Yepiskoposyan, P. Zalloua, T. Zemunik, A. Cooper, C. Capelli, M. G. Thomas, A. Ruiz-Linares, S. A. Tishkoff, L. Singh, K. Thangaraj, R. Villems, D. Comas, R. Sukernik, M. Metspalu, M. Meyer, E. E. Eichler, J. Burger, M. Slatkin, S. Paabo, J. Kelso, D. Reich, and J. Krause. 2014. Ancient human genomes suggest three ancestral populations for present-day Europeans. *Nature* 513 (7518):409-13.

Lazaridis, Iosif. 2018. The evolutionary history of human populations in Europe. *Current Opinion in Genetics & Development* 53:21-27.

Lazaridis, Iosif, Anna Belfer-Cohen, Swapan Mallick, Nick Patterson, Olivia Cheronet, Nadin Rohland, Guy Bar-Oz, Ofer Bar-Yosef, Nino Jakeli, Eliso Kvavadze, David Lordkipanidze, Zinovi Matzkevich, Tengiz Meshveliani, Brendan J. Culleton, Douglas J. Kennett, Ron Pinhasi, and David Reich. 2018. Paleolithic DNA from the Caucasus reveals core of West Eurasian ancestry. *bioRxiv.*

Lazaridis, Iosif, Alissa Mittnik, Nick Patterson, Swapan Mallick, Nadin Rohland, Saskia Pfrengle, Anja Furtwängler, Alexander Peltzer, Cosimo Posth, Andonis Vasilakis, P. J. P. McGeorge, Eleni Konsolaki-Yannopoulou, George Korres, Holley Martlew, Manolis Michalodimitrakis, Mehmet Özsait, Nesrin Özsait, Anastasia Papathanasiou, Michael Richards, Songül Alpaslan Roodenberg, Yannis Tzedakis, Robert Arnott, Daniel M. Fernandes, Jeffery R. Hughey, Dimitra M. Lotakis, Patrick A. Navas, Yannis Maniatis, John A. Stamatoyannopoulos, Kristin Stewardson, Philipp Stockhammer, Ron Pinhasi,

David Reich, Johannes Krause, and George Stamatoyannopoulos. 2017. Genetic origins of the Minoans and Mycenaeans. *Nature* 548 (7666):214-218.

Lehmann, W.P. 1992. *Historical Linguistics: An Introduction*. London: Routledge.

Leonardi, Michela, Francesco Boschin, Konstantinos Giampoudakis, Robert M. Beyer, Mario Krapp, Robin Bendrey, Robert Sommer, Paolo Boscato, Andrea Manica, David Nogues-Bravo, and Ludovic Orlando. 2018. Late Quaternary horses in Eurasia in the face of climate and vegetation change. *Science Advances* 4 (7).

Linden, Marc Vander. 2015. What linked the Bell Beakers in third millennium BC Europe? *Antiquity* 81 (312):343-352.

Lipiński, Edward. 2001. *Semitic Languages Outline of a Comparative Grammar*. 2nd ed, *Orientalia Lovaniensia Analecta*. Leuven - Paris - Sterling: Peeters.

Lipson, Mark, Anna Szécsényi-Nagy, Swapan Mallick, Annamária Pósa, Balázs Stégmár, Victoria Keerl, Nadin Rohland, Kristin Stewardson, Matthew Ferry, Megan Michel, Jonas Oppenheimer, Nasreen Broomandkhoshbacht, Eadaoin Harney, Susanne Nordenfelt, Bastien Llamas, Balázs Gusztáv Mende, Kitti Köhler, Krisztián Oross, Mária Bondár, Tibor Marton, Anett Osztás, János Jakucs, Tibor Paluch, Ferenc Horváth, Piroska Csengeri, Judit Koós, Katalin Sebők, Alexandra Anders, Pál Raczky, Judit Regenye, Judit P. Barna, Szilvia Fábián, Gábor Serlegi, Zoltán Toldi, Emese Gyöngyvér Nagy, János Dani, Erika Molnár, György Pálfi, László Márk, Béla Melegh, Zsolt Bánfai, László Domboróczki, Javier Fernández-Eraso, José Antonio Mujika-Alustiza, Carmen Alonso Fernández, Javier Jiménez Echevarría, Ruth Bollongino, Jörg Orschiedt, Kerstin Schierhold, Harald Meller, Alan Cooper, Joachim Burger, Eszter Bánffy, Kurt W. Alt, Carles Lalueza-Fox, Wolfgang Haak, and David Reich. 2017. Parallel palaeogenomic transects reveal complex genetic history of early European farmers. *Nature* 551:368.

Major, Candace O., Steven L. Goldstein, William B. F. Ryan, Gilles Lericolais, Alexander M. Piotrowski, and Irka Hajdas. 2006. The co-evolution of Black Sea level and composition through the last deglaciation and its paleoclimatic significance. *Quaternary Science Reviews* 25 (17–18):2031-2047.

Mallick, S., H. Li, M. Lipson, I. Mathieson, M. Gymrek, and F. Racimo. 2016. The Simons Genome Diversity Project: 300 genomes from 142 diverse populations. *Nature* 538.

Mallory, J., and D.Q. Adams. 2007. Reconstructing the Proto-Indo-Europeans. In *The Oxford Introduction to Proto-Indo-European and the Proto-Indo-European World*. Oxford: Oxford University Press.

Mallory, J.P. 2013. The Indo-Europeanization of Atlantic Europe. In *Celtic From the West 2: Rethinking the Bronze Age and the Arrival of Indo-European in Atlantic Europe*, edited by J. T. Koch and B. Cunliffe. Oxford: Oxbow Books.

Repeated Author. 2014. Indo-European dispersals and the Eurasian Steppe. In *Reconfiguring the Silk Road: New Research on East-West Exchange in Antiquity*, edited by V. H. Mair and J. Hickman. Philadelphia: University of Pennsylvania Museum of Archaeology and Anthropology.

Mallory, J.P., and D. Q. Adams. 2007. A Place in Time. In *The Oxford Introduction to Proto-Indo-European and the Proto-Indo-European World*, edited by J. Mallory and R. B. Adams. Oxford: Oxford University Press.

Malmström, H., M.T.P. Gilbert, M.G. Thomas, M. Brandström, J. Storå, P. Molnar, P.K. Andersen, C. Bendixen, G. Holmlund, A. Götherström, and E. Willerslev. 2009. Ancient DNA reveals lack of continuity between neolithic hunter-gatherers and contemporary Scandinavians. *Curr Biol* 19:1758–1762.

Manzura, Igor. 2005. Steps to the Steppe: Or, how the North Pontic region was colonised. *Oxford Journal of Archaeology* 24 (4):313-338.

Repeated Author. 2016. North Pontic Steppes at the End of the 4th Millennium BC: the Epoch of Broken Borders. In *Man, culture, and society from the Copper Age until the Early Iron Age in Northern Eurasia (Contributions in honour of the 60th anniversary of Eugen Sava)*. Chişinău: Bons Offices.

Marcus, Joseph H., Cosimo Posth, Harald Ringbauer, Luca Lai, Robin Skeates, Carlo Sidore, Jessica Beckett, Anja Furtwängler, Anna Olivieri, Charleston Chiang, Hussein Al-Asadi, Kushal Dey, Tyler A. Joseph, Clio Der Sarkissian, Rita Radzevičiūtė, Maria Giuseppina Gradoli, Wolfgang Haak, David Reich, David Schlessinger, Francesco Cucca, Johannes Krause, and John Novembre. 2019. Population history from the Neolithic to present on the Mediterranean island of Sardinia: An ancient DNA perspective. *bioRxiv*:583104.

Martínez, Maria Pilar Prieto, and Laure Salanova. 2015. Concluding remarks. The Bell Beaker Transition: The end of the neolithisation of Europe; the starting point of a new order. In *The Bell BeakerTransition in Europe. Mobility and local evolution during the 3rd millenium BC*, edited by M. P. P. Martínez and L. Salanova. Oxford & Philadelphia: Oxbow Books.

Martínez Sánchez, Rafael M., Juan Carlos Vera Rodríguez, Leonor Peña-Chocarro, Youssef Bokbot, Guillem Pérez Jordà, and Salvador Pardo-Gordó. 2018. The Middle Neolithic of Morocco's North-Western Atlantic Strip: New Evidence from the El-Khil Caves (Tangier). *African Archaeological Review* 35 (3):417-442.

Martiniano, Rui, Lara M. Cassidy, Ros Ó'Maoldúin, Russell McLaughlin, Nuno M. Silva, Licinio Manco, Daniel Fidalgo, Tania Pereira, Maria J. Coelho, Miguel Serra, Joachim Burger, Rui Parreira, Elena Moran, Antonio C. Valera, Eduardo Porfirio, Rui Boaventura, Ana M. Silva, and Daniel G. Bradley. 2017. The population genomics of archaeological transition in west Iberia: Investigation of ancient substructure using imputation and haplotype-based methods. *PLOS Genetics* 13 (7):e1006852.

Mathieson, I., I. Lazaridis, N. Rohland, S. Mallick, N. Patterson, S. A. Roodenberg, E. Harney, K. Stewardson, D. Fernandes, M. Novak, K. Sirak, C. Gamba, E. R. Jones, B. Llamas, S. Dryomov, J. Pickrell, J. L. Arsuaga, J. M. de Castro, E. Carbonell, F. Gerritsen, A. Khokhlov, P. Kuznetsov, M. Lozano, H. Meller, O. Mochalov, V. Moiseyev, M. A. Guerra, J. Roodenberg, J. M. Verges, J. Krause, A. Cooper, K. W. Alt, D. Brown, D. Anthony, C. Lalueza-Fox, W. Haak, R. Pinhasi, and D. Reich. 2015. Genome-wide patterns of selection in 230 ancient Eurasians. *Nature* 528 (7583):499-503.

Mathieson, Iain, Songül Alpaslan-Roodenberg, Cosimo Posth, Anna Szécsényi-Nagy, Nadin Rohland, Swapan Mallick, Iñigo Olalde, Nasreen Broomandkhoshbacht, Francesca Candilio, Olivia Cheronet, Daniel Fernandes, Matthew Ferry, Beatriz Gamarra, Gloria González Fortes, Wolfgang Haak, Eadaoin Harney, Eppie Jones, Denise Keating, Ben Krause-Kyora, Isil

Kucukkalipci, Megan Michel, Alissa Mittnik, Kathrin Nägele, Mario Novak, Jonas Oppenheimer, Nick Patterson, Saskia Pfrengle, Kendra Sirak, Kristin Stewardson, Stefania Vai, Stefan Alexandrov, Kurt W. Alt, Radian Andreescu, Dragana Antonović, Abigail Ash, Nadezhda Atanassova, Krum Bacvarov, Mende Balázs Gusztáv, Hervé Bocherens, Michael Bolus, Adina Boroneanţ, Yavor Boyadzhiev, Alicja Budnik, Josip Burmaz, Stefan Chohadzhiev, Nicholas J. Conard, Richard Cottiaux, Maja Čuka, Christophe Cupillard, Dorothée G. Drucker, Nedko Elenski, Michael Francken, Borislava Galabova, Georgi Ganetsovski, Bernard Gély, Tamás Hajdu, Veneta Handzhyiska, Katerina Harvati, Thomas Higham, Stanislav Iliev, Ivor Janković, Ivor Karavanić, Douglas J. Kennett, Darko Komšo, Alexandra Kozak, Damian Labuda, Martina Lari, Catalin Lazar, Maleen Leppek, Krassimir Leshtakov, Domenico Lo Vetro, Dženi Los, Ivaylo Lozanov, Maria Malina, Fabio Martini, Kath McSweeney, Harald Meller, Marko Menđušić, Pavel Mirea, Vyacheslav Moiseyev, Vanya Petrova, T. Douglas Price, Angela Simalcsik, Luca Sineo, Mario Šlaus, Vladimir Slavchev, Petar Stanev, Andrej Starović, Tamás Szeniczey, Sahra Talamo, Maria Teschler-Nicola, Corinne Thevenet, Ivan Valchev, Frédérique Valentin, Sergey Vasilyev, Fanica Veljanovska, Svetlana Venelinova, Elizaveta Veselovskaya, Bence Viola, Cristian Virag, Joško Zaninović, Steve Zäuner, Philipp W. Stockhammer, Giulio Catalano, Raiko Krauß, David Caramelli, Gunita Zariņa, Bisserka Gaydarska, Malcolm Lillie, Alexey G. Nikitin, Inna Potekhina, Anastasia Papathanasiou, Dušan Borić, Clive Bonsall, Johannes Krause, Ron Pinhasi, and David Reich. 2018. The genomic history of southeastern Europe. *Nature* 555:197.

Mathieson, Iain, Songül Alpaslan Roodenberg, Cosimo Posth, Anna Szécsényi-Nagy, Nadin Rohland, Swapan Mallick, Iñigo Olade, Nasreen Broomandkhoshbacht, Olivia Cheronet, Daniel Fernandes, Matthew Ferry, Beatriz Gamarra, Gloria González Fortes, Wolfgang Haak, Eadaoin Harney, Ben Krause-Kyora, Isil Kucukkalipci, Megan Michel, Alissa Mittnik, Kathrin Nägele, Mario Novak, Jonas Oppenheimer, Nick Patterson, Saskia Pfrengle, Kendra Sirak, Kristin Stewardson, Stefania Vai, Stefan Alexandrov, Kurt W. Alt, Radian Andreescu, Dragana Antonović, Abigail Ash, Nadezhda Atanassova, Krum Bacvarov, Mende Balázs Gusztáv, Hervé Bocherens, Michael Bolus, Adina Boroneanţ, Yavor Boyadzhiev, Alicja Budnik, Josip Burmaz, Stefan Chohadzhiev, Nicholas J. Conard, Richard Cottiaux, Maja Čuka, Christophe Cupillard, Dorothée G. Drucker, Nedko Elenski, Michael Francken, Borislava Galabova, Georgi Ganetovski, Bernard Gely, Tamás Hajdu, Veneta Handzhyiska, Katerina Harvati, Thomas Higham, Stanislav Iliev, Ivor Janković, Ivor Karavanić, Douglas J. Kennett, Darko Komšo, Alexandra Kozak, Damian Labuda, Martina Lari, Catalin Lazar, Maleen Leppek, Krassimir Leshtakov, Domenico Lo Vetro, Dženi Los, Ivaylo Lozanov, Maria Malina, Fabio Martini, Kath McSweeney, Harald Meller, Marko Menđušić, Pavel Mirea, Vyacheslav Moiseyev, Vanya Petrova, T. Douglas Price, Angela Simalcsik, Luca Sineo, Mario Šlaus, Vladimir Slavchev, Petar Stanev, Andrej Starović, Tamás Szeniczey, Sahra Talamo, Maria Teschler-Nicola, Corinne Thevenet, Ivan Valchev, Frédérique Valentin, Sergey Vasilyev, Fanica Veljanovska, Svetlana Venelinova, Elizaveta Veselovskaya, Bence Viola, Cristian Virag, Joško Zaninović, Steve Zäuner, Philipp W. Stockhammer,

Giulio Catalano, Raiko Krauß, David Caramelli, Gunita Zariņa, Bisserka Gaydarska, Malcolm Lillie, Alexey G. Nikitin, Inna Potekhina, Anastasia Papathanasiou, Dušan Borić, Clive Bonsall, Johannes Krause, Ron Pinhasi, and David Reich. 2017. The Genomic History Of Southeastern Europe. *bioRxiv.*

Mazurkevich, A. N., B. N. Korotkevich, P. M. Dolukhanov, A. M. Shukurov, Kh A. Arslanov, L. A. Savel'eva, E. N. Dzinoridze, M. A. Kulkova, and G. I. Zaitseva. 2009. Climate, subsistence and human movements in the Western Dvina – Lovat River Basins. *Quaternary International* 203 (1-2):52-66.

Mazurkevich, Andrey, and Ekaterina Dolbunova. 2015. The oldest pottery in hunter-gatherer communities and models of Neolithisation of Eastern Europe. *Documenta Praehistorica* XLII:13-66.

Meid, W. 1975. Probleme der räumlichen und zeitlichen Gliederung der Indogermanischen. In *Flexion und Wortbildung*, edited by H. Rix. Wiesbaden: Reichert.

Meier-Brügger, Michael. 2003. *Indo-European Linguistics*. Berlin, New York: Walter de Gruyter.

Melchert, Craig. 2011. Indo-Europeans. In *The Oxford Handbook of Ancient Anatolia 10,000-323 B.C.E.*, edited by S. R. Steadman and G. McMahon. Oxford: Oxford University Press.

Melchert, H. Craig. 1998. The dialectal position of Anatolian within Indo-European. In *Annual Meeting of the Berkeley Linguistics Society*.

Meyer, Christian, Corina Knipper, Nicole Nicklisch, Angelina Münster, Olaf Kürbis, Veit Dresely, Harald Meller, and Kurt W. Alt. 2018. Early Neolithic executions indicated by clustered cranial trauma in the mass grave of Halberstadt. *Nature Communications* 9 (1):2472.

Midgley, Magdalena S. 2004. Consequences of Farming in Southern Scandinavia. In *Ancient Europe, 8000 B.C. to A.D. 1000: An Encyclopedia of the Barbarian World*, edited by P. I. Bogucki and P. J. Crabtree. New York: Charles Scribner & Sons.

Mikhailova, Tatyana A. 2015. Celtic origin: location in time and space? Reconsidering the "East-West Celtic" debate. *Journal of Language Relationship* 13 (3):257-279.

Mileto, Simona. 2018. Diet and subsistence practices in the Dnieper area of the North-Pontic region (4th - 3rd millennium BC). An integrated archaeological, molecular and isotopic approach. Dissertation, History and Cultural Studies, Freie Universität Berlin.

Mileto, Simona, Elke Kaiser, Yuri Rassamakin, and Richard P. Evershed. 2017. New insights into the subsistence economy of the Eneolithic Dereivka culture of the Ukrainian North-Pontic region through lipid residues analysis of pottery vessels. *Journal of Archaeological Science: Reports* 13:67-74.

Mittnik, Alissa, Chuan-Chao Wang, Saskia Pfrengle, Mantas Daubaras, Gunita Zariņa, Fredrik Hallgren, Raili Allmäe, Valery Khartanovich, Vyacheslav Moiseyev, Mari Tõrv, Anja Furtwängler, Aida Andrades Valtueña, Michal Feldman, Christos Economou, Markku Oinonen, Andrejs Vasks, Elena Balanovska, David Reich, Rimantas Jankauskas, Wolfgang Haak, Stephan Schiffels, and Johannes Krause. 2018. The genetic prehistory of the Baltic Sea region. *Nature Communications* 9 (1):442.

Monroy Kuhn, Jose Manuel, Mattias Jakobsson, and Torsten Günther. 2017. Estimating genetic kin relationships in prehistoric populations. *bioRxiv*.

Mooder, K. P., T. G. Schurr, F. J. Bamforth, V. I. Bazaliiski, and N. A. Savel'ev. 2005. Population affinities of Neolithic Siberians: A snapshot from prehistoric Lake Baikal. *American Journal of Physical Anthropology* 129 (3):349-361.

Moorjani, P., K. Thangaraj, N. Patterson, M. Lipson, P. R. Loh, and P. Govindaraj. 2013. Genetic evidence for recent population mixture in India. *Am J Hum Genet* 93.

Morgunova, N. L. 2002. Yamnaya (Pit-Grave) Culture in the Southern Urals Area. In *Complex Societies of Central Eurasia from the 3rd to the 1st Millennium BC: Regional Specifics in Light of Global Models*, edited by K. Jones-Bley and D. Zdanovich. Washington, D.C.: Institute for Study of Man.

Morgunova, Nina. 2014. О ХАРАКТЕРЕ КУЛЬТУРНОГО ВЗАИМОДЕЙСТВИЯ НАСЕЛЕНИЯ ЯМНОЙ КУЛЬТУРЫ СТЕПНОГО ВОЛГО-УРАЛЬЯ И АФАНАСЬЕВСКОЙ КУЛЬТУРЫ АЛТАЕ-САЯНСКОГО РЕГИОНА. *АРХЕОЛОГИЯ* 3 (26):4-13.

Morgunova, Nina L. 2015. Pottery from the Volga area in the Samara and South Urals region from Eneolithic to Early Bronze Age. *Documenta Prehistorica* XLII:311-320.

Morgunova, Nina L., and A. A. Fayzullin. 2018. The Social Structure of the Yamnaya (Pit-Grave) Culture of the Volga-Ural Interfluve. *Stratum plus* (2):35-60.

Morgunova, Nina L., and Mikhail A. Turetskij. 2016. Archaeological and natural scientific studies of Pit-Grave culture barrows in the Volga-Ural interfluve. *Estonian Journal of Archaeology* 20 (2):128-149.

Moussa, N. M., V. I. Bazaliiskii, O. I. Goriunova, F. Bamforth, and A. W. Weber. 2016. Y-chromosomal DNA analyzed for four prehistoric cemeteries from Cis-Baikal, Siberia. *Journal of Archaeological Science: Reports*.

Müller, J. 2014. 4100–2700 B.C.: Monuments and Ideologies in the Neolithic Landscape. In *Approaching Monumentality in Archaeology*, edited by J. F. Osborne. New York: Suny.

Müller, J., V. P. J. Arponen, R. Hofmann, and R. Ohlrau. 2015. The Appearance of Social Inequalities: Cases of Neolithic and Chalcolithic Societies. *Origini* 38 (2):65-86.

Müller, J., and S. VanVilligen. 2001. New radiocarbon evidence for European Bell Beakers and the consequences for the diffusion of the Bell Beaker Phenomenon. *Nicolis (ed.)*:59-80.

Müller, Johannes, and Aleksandr Diachenko. 2019. Tracing long-term demographic changes: The issue of spatial scales. *PLOS ONE* 14 (1):e0208739.

Müller, Johannes, Timo Seregély, Anne-Mette Christensen, Ullrich Schüssler, Cornelia Becker, Helmut Kroll, and Markus Fuchs. 2009. A Revision of Corded Ware Settlement Pattern–New Results from the Central European Low Mountain Range. *Proceedings of the Prehistoric Society* 75:125-142.

Murillo-Barroso, Mercedes, and Ignacio Montero-Ruiz. 2017. The Social Value of Things. In *Key Resources and Socio-Cultural Developments in the Iberian Chalcolithic*, edited by M. Bartelheim, P. B. Ramírez and M. Kunst. Tübingen: Tübingen Library Publishing.

Murphy, Eileen M., and Aleksandr A. Khokhlov. 2016. A Bioarchaeological Study of Prehistoric Populations from the Volga Region. In *A Bronze Age Landscape in the Russian Steppes. The Samara Valley Project*, edited by D. W. Anthony, D. R. Brown, O. D. Mochalov, A. A. Khokhlov and P. F. Kuznetsov. Los Angeles: Cotsen Institute of Archaeology Press.

Myres, N. M., S. Rootsi, A. A. Lin, M. Jarve, R. J. King, I. Kutuev, V. M. Cabrera, E. K. Khusnutdinova, A. Pshenichnov, B. Yunusbayev, O. Balanovsky, E. Balanovska, P. Rudan, M. Baldovic, R. J. Herrera, J. Chiaroni, J. Di Cristofaro, R. Villems, T. Kivisild, and P. A. Underhill. 2011. A major Y-chromosome haplogroup R1b Holocene era founder effect in Central and Western Europe. *Eur J Hum Genet* 19 (1):95-101.

Narasimhan, Vagheesh M, Nick J Patterson, Priya Moorjani, Iosif Lazaridis, Lipson Mark, Swapan Mallick, Nadin Rohland, Rebecca Bernardos, Alexander M. Kim, Nathan Nakatsuka, Inigo Olalde, Alfredo Coppa, James Mallory, Vyacheslav Moiseyev, Janet Monge, Luca M. Olivieri, Nicole Adamski, Nasreen Broomandkhoshbacht, Francesca Candilio, Olivia Cheronet, Brendan J. Culleton, Matthew Ferry, Daniel Fernandes, Beatriz Gamarra, Daniel Gaudio, Mateja Hajdinjak, Eadaoin Harney, Thomas K. Harper, Denise Keating, Ann-Marie Lawson, Megan Michel, Mario Novak, Jonas Oppenheimer, Niraj Rai, Kendra Sirak, Viviane Slon, Kristin Stewardson, Zhao Zhang, Gaziz Akhatov, Anatoly N. Bagashev, Baurzhan Baitanayev, Gian Luca Bonora, Tatiana Chikisheva, Anatoly Derevianko, Enshin Dmitry, Katerina Douka, Nadezhda Dubova, Andrey Epimakhov, Suzanne Freilich, Dorian Fuller, Alexander Goryachev, Andrey Gromov, Bryan Hanks, Margaret Judd, Erlan Kazizov, Aleksander Khokhlov, Egor Kitov, Elena Kupriyanova, Pavel Kuznetsov, Donata Luiselli, Farhad Maksudov, Chris Meiklejohn, Deborah C. Merrett, Roberto Micheli, Oleg Mochalov, Zahir Muhammed, Samridin Mustafakulov, Ayushi Nayak, Rykun M. Petrovna, Davide Pettner, Richard Potts, Dmitry Razhev, Stefania Sarno, Kulyan Sikhymbaevae, Sergey M. Slepchenko, Nadezhda Stepanova, Svetlana Svyatko, Sergey Vasilyev, Massimo Vidale, Dima Voyakin, Antonina Yermolayeva, Alisa Zubova, Vasant S. Shinde, Carles Lalueza-Fox, Matthias Meyer, David Anthony, Nicole Boivin, Kumarasmy Thangaraj, Douglas Kennett, Michael Frachetti, Ron Pinhasi, and David Reich. 2018. The Genomic Formation of South and Central Asia. *bioRxiv*.

Nikitin, Alexey G., Inna Potekhina, Nadin Rohland, Swapan Mallick, David Reich, and Malcolm Lillie. 2017. Mitochondrial DNA analysis of eneolithic trypillians from Ukraine reveals neolithic farming genetic roots. *PLOS ONE* 12 (2):e0172952.

Norberg, Erik. 2019. The Meaning of Words and the Power of Silence. In *The Indigenous Identity of the South Saami : Historical and Political Perspectives on a Minority within a Minority*, edited by H. Hermanstrand, A. Kolberg, T. R. Nilssen and L. Sem. Cham: Springer International Publishing.

Nordqvist, Kerkko. 2018. The Stone Age of north-eastern Europe 5500–1800 calBC : bridging the gap between the East and the West, Human Sciences, University of Oulu, Oulu.

Nordqvist, Kerkko, and Piritta Häkälä. 2014. Distribution of Corded Ware in the Areas North of the Gulf of Finland – An Update. *Estonian Journal of Archaeology* 18 (1):3-29.

Nordqvist, Kerkko, Vesa-Pekka Herva, Janne Ikäheimo, and Antti Lahelma. 2012. Early Copper Use in Neolithic North-Eastern Europe: An Overview. *Estonian Journal of Archaeology* 16 (1):3-25.

Nordqvist, Kerkko, and Teemu Mökkönen. 2016. New Radiocarbon Dates for Early Pottery in North-Eastern Europe. In *Традиции и инновации в изучении древнейшей керамики*, edited by Л. Б. Вишняцкий and Е. Л. Костылёва. Санкт-Петербург, Россия: ИИМК РАН.

Novak, Marek. 2017. Do 14C dates always turn into an absolute chronology? The case of the Middle Neolithic in western Lesser Poland. *Documenta Praehistorica* 44:240-271.

Novozhenov, Victor A. 2012. *Communications and the Earliest Wheeled Transport of Eurasia.* Edited by E. E. Kuzmina. Moscow: Taus Publishing.

Nowak, M. 2014. Późny etap rozwoju cyklu lendzielskopolgarskiego w Zachodniej Małopolsce,. In *Szkice neolityczne. Księga poświęcona pamięci Profesor Anny Kulczyckiej-Leciejewiczowej*, edited by K. Czarniak, J. Kolenda and M. Markiewicz. Wrocław.

Oettinger, Norbert. 1997. Grundsätzliche Überlegungen zum Nordwest-Indogermanischen. *Incontri Linguistici* 20 (93-111).

Repeated Author. 2003. Neuerungen in Lexikon und Wortbildung des Nordwest-Indogermanischen. In *Languages in prehistoric Europe*, edited by A. Bammesberger, M. Bieswanger, J. Grzega and T. Venneman. Heidelberg: Winter.

Oinonen, Markku, Petro Pesonen, Teija Alenius, Volker Heyd, Elisabeth Holmqvist-Saukkonen, Sanna Kivimäki, Tuire Nygrén, Tarja Sundell, and Päivi Onkamo. 2014. Event reconstruction through Bayesian chronology: Massive mid-Holocene lake-burst triggered large-scale ecological and cultural change. *The Holocene* 24 (11):1419-1427.

Ökse, A. Tuba. 2017. Salat Tepe: Overview of the Stratigraphic Sequence and the Early EBA Levels Excavated in the Last Two Seasons. In *The Archaeology of Anatolia Volume II: Recent Discoveries (2015-2016)*, edited by S. R. Steadman and G. McMahon. Newcastle: Cambridge Scholars Publishing.

Olalde, I., H. Schroeder, M. Sandoval-Velasco, L. Vinner, I. Lobon, O. Ramirez, S. Civit, P. Garcia Borja, D. C. Salazar-Garcia, S. Talamo, J. Maria Fullola, F. Xavier Oms, M. Pedro, P. Martinez, M. Sanz, J. Daura, J. Zilhao, T. Marques-Bonet, M. T. Gilbert, and C. Lalueza-Fox. 2015. A Common Genetic Origin for Early Farmers from Mediterranean Cardial and Central European LBK Cultures. *Mol Biol Evol* 32 (12):3132-42.

Olalde, Iñigo, Selina Brace, Morten E. Allentoft, Ian Armit, Kristian Kristiansen, Thomas Booth, Nadin Rohland, Swapan Mallick, Anna Szécsényi-Nagy, Alissa Mittnik, Eveline Altena, Mark Lipson, Iosif Lazaridis, Thomas K. Harper, Nick Patterson, Nasreen Broomandkhoshbacht, Yoan Diekmann, Zuzana Faltyskova, Daniel Fernandes, Matthew Ferry, Eadaoin Harney, Peter de Knijff, Megan Michel, Jonas Oppenheimer, Kristin Stewardson, Alistair Barclay, Kurt Werner Alt, Corina Liesau, Patricia Ríos, Concepción Blasco,

Jorge Vega Miguel, Roberto Menduiña García, Azucena Avilés Fernández, Eszter Bánffy, Maria Bernabò-Brea, David Billoin, Clive Bonsall, Laura Bonsall, Tim Allen, Lindsey Büster, Sophie Carver, Laura Castells Navarro, Oliver E. Craig, Gordon T. Cook, Barry Cunliffe, Anthony Denaire, Kirsten Egging Dinwiddy, Natasha Dodwell, Michal Ernée, Christopher Evans, Milan Kuchařík, Joan Francès Farré, Chris Fowler, Michiel Gazenbeek, Rafael Garrido Pena, María Haber-Uriarte, Elżbieta Haduch, Gill Hey, Nick Jowett, Timothy Knowles, Ken Massy, Saskia Pfrengle, Philippe Lefranc, Olivier Lemercier, Arnaud Lefebvre, César Heras Martínez, Virginia Galera Olmo, Ana Bastida Ramírez, Joaquín Lomba Maurandi, Tona Majó, Jacqueline I. McKinley, Kathleen McSweeney, Balázs Gusztáv Mende, Alessandra Modi, Gabriella Kulcsár, Viktória Kiss, András Czene, Róbert Patay, Anna Endrődi, Kitti Köhler, Tamás Hajdu, Tamás Szeniczey, János Dani, Zsolt Bernert, Maya Hoole, Olivia Cheronet, Denise Keating, Petr Velemínský, Miroslav Dobeš, Francesca Candilio, Fraser Brown, Raúl Flores Fernández, Ana-Mercedes Herrero-Corral, Sebastiano Tusa, Emiliano Carnieri, Luigi Lentini, Antonella Valenti, Alessandro Zanini, Clive Waddington, Germán Delibes, Elisa Guerra-Doce, Benjamin Neil, Marcus Brittain, Mike Luke, Richard Mortimer, Jocelyne Desideri, Marie Besse, Günter Brücken, Mirosław Furmanek, Agata Hałuszko, Maksym Mackiewicz, Artur Rapiński, Stephany Leach, Ignacio Soriano, Katina T. Lillios, João Luís Cardoso, Michael Parker Pearson, Piotr Włodarczak, T. Douglas Price, Pilar Prieto, Pierre-Jérôme Rey, Roberto Risch, Manuel A. Rojo Guerra, Aurore Schmitt, Joël Serralongue, Ana Maria Silva, Václav Smrčka, Luc Vergnaud, João Zilhão, David Caramelli, Thomas Higham, Mark G. Thomas, Douglas J. Kennett, Harry Fokkens, Volker Heyd, Alison Sheridan, Karl-Göran Sjögren, Philipp W. Stockhammer, Johannes Krause, Ron Pinhasi, Wolfgang Haak, Ian Barnes, Carles Lalueza-Fox, and David Reich. 2018. The Beaker phenomenon and the genomic transformation of northwest Europe. *Nature* 555:190.

Olalde, Iñigo, Swapan Mallick, Nick Patterson, Nadin Rohland, Vanessa Villalba-Mouco, Marina Silva, Katharina Dulias, Ceiridwen J. Edwards, Francesca Gandini, Maria Pala, Pedro Soares, Manuel Ferrando-Bernal, Nicole Adamski, Nasreen Broomandkhoshbacht, Olivia Cheronet, Brendan J. Culleton, Daniel Fernandes, Ann Marie Lawson, Matthew Mah, Jonas Oppenheimer, Kristin Stewardson, Zhao Zhang, Juan Manuel Jiménez Arenas, Isidro Jorge Toro Moyano, Domingo C. Salazar-García, Pere Castanyer, Marta Santos, Joaquim Tremoleda, Marina Lozano, Pablo García Borja, Javier Fernández-Eraso, José Antonio Mujika-Alustiza, Cecilio Barroso, Francisco J. Bermúdez, Enrique Viguera Mínguez, Josep Burch, Neus Coromina, David Vivó, Artur Cebrià, Josep Maria Fullola, Oreto García-Puchol, Juan Ignacio Morales, F. Xavier Oms, Tona Majó, Josep Maria Vergès, Antònia Díaz-Carvajal, Imma Ollich-Castanyer, F. Javier López-Cachero, Ana Maria Silva, Carmen Alonso-Fernández, Germán Delibes de Castro, Javier Jiménez Echevarría, Adolfo Moreno-Márquez, Guillermo Pascual Berlanga, Pablo Ramos-García, José Ramos-Muñoz, Eduardo Vijande Vila, Gustau Aguilella Arzo, Ángel Esparza Arroyo, Katina T. Lillios, Jennifer Mack, Javier Velasco-Vázquez, Anna Waterman, Luis Benítez de Lugo Enrich, María Benito Sánchez, Bibiana Agustí, Ferran Codina, Gabriel de Prado, Almudena Estalrrich, Álvaro

Fernández Flores, Clive Finlayson, Geraldine Finlayson, Stewart Finlayson, Francisco Giles-Guzmán, Antonio Rosas, Virginia Barciela González, Gabriel García Atiénzar, Mauro S. Hernández Pérez, Armando Llanos, Yolanda Carrión Marco, Isabel Collado Beneyto, David López-Serrano, Mario Sanz Tormo, António C. Valera, Concepción Blasco, Corina Liesau, Patricia Ríos, Joan Daura, María Jesús de Pedro Michó, Agustín A. Diez-Castillo, Raúl Flores Fernández, Joan Francès Farré, Rafael Garrido-Pena, Victor S. Gonçalves, Elisa Guerra-Doce, Ana Mercedes Herrero-Corral, Joaquim Juan-Cabanilles, Daniel López-Reyes, Sarah B. McClure, Marta Merino Pérez, Arturo Oliver Foix, Montserrat Sanz Borràs, Ana Catarina Sousa, Julio Manuel Vidal Encinas, Douglas J. Kennett, Martin B. Richards, Kurt Werner Alt, Wolfgang Haak, Ron Pinhasi, Carles Lalueza-Fox, and David Reich. 2019. The genomic history of the Iberian Peninsula over the past 8000 years. *Science* 363 (6432):1230-1234.

Özbal, Rana. 2011. The Chalcolithic of Southeast Anatolia. In *The Oxford Handbook of Ancient Anatolia (10,000-323 BCE)*, edited by S. R. Steadman and G. McMahon. Oxford: Oxford University Press.

Özbaşaran, Mihriban. 2011. The Neolithic in the Plateau. In *The Oxford Handbook of Ancient Anatolia (10,000-323 BCE)*, edited by S. R. Steadman and G. McMahon. Oxford: Oxford University Press.

Özdoğan, M. 2008. An alternative approach in tracing changes in demographic composition. In *The Neolithic Demographic Transition and Its Consequences*, edited by Bar-Yosef O. and B.-A. J.P.: Springer.

Özdoğan, Mehmet. 2011. Eastern Thrace: The Contact Zone between Anatolia and the Balkans. In *The Oxford Handbook of Ancient Anatolia 10,000-323 B.C.E.*, edited by S. R. Steadman and G. McMahon. Oxford: Oxford University Press.

Pagani, Luca, Toomas Kivisild, Ayele Tarekegn, Rosemary Ekong, Chris Plaster, Irene Gallego Romero, Qasim Ayub, S. Qasim Mehdi, Mark G Thomas, Donata Luiselli, Endashaw Bekele, Neil Bradman, David J Balding, and Chris Tyler-Smith. 2012. Ethiopian Genetic Diversity Reveals Linguistic Stratification and Complex Influences on the Ethiopian Gene Pool. *The American Journal of Human Genetics* 91 (1):83-96.

Pagel, Mark, Quentin D. Atkinson, Andreea S. Calude, and Andrew Meade. 2013. Ultraconserved words point to deep language ancestry across Eurasia. *Proceedings of the National Academy of Sciences* 110 (21):8471.

Palumbi, Giulio. 2011. The Chalcolithic of Eastern Anatolia. In *The Oxford Handbook of Ancient Anatolia (10,000-323 BCE)*, edited by S. R. Steadman and G. McMahon. Oxford: Oxford University Press.

Parpola, Asko. 2013. Formation of the Indo-European and Uralic (Finno-Ugric) language families in the light of archaeology: Revised and integrated 'total' correlations. In *A Linguistic Map of Prehistoric Northern Europe*. Helsinki: Société Finno-Ougrienne.

Parzinger, Hermann. 2013. Ukraine and South Russia in the Bronze Age. In *The Oxford Handbook of the European Bronze Age*, edited by H. Fokkens and A. Harding. Oxford: Oxford University Press.

Patton, Henry, Alun Hubbard, Karin Andreassen, Amandine Auriac, Pippa L. Whitehouse, Arjen P. Stroeven, Calvin Shackleton, Monica Winsborrow,

Jakob Heyman, and Adrian M. Hall. 2017. Deglaciation of the Eurasian ice sheet complex. *Quaternary Science Reviews* 169:148-172.

Pereltsvaig, Asya, and Martin W. Lewis. 2015. *The Indo-European Controversy. Facts and Fallacies in Historical Linguistics*. Cambridge: Cambridge University Press.

Petrova, VIktoria. 2016. Varna culture: an autonomous phenomenon or a local version of the Kodzhadermen-Gumelnitsa-Karanovo VI cultural complex. In *Der Schwarzmeerraum vom Neolithikum bis in die Früheisenzeit (6000-600 v. Chr.). Kulturelle Interferenzen in der zirkumpontischen Zone und Kontakte mit ihren Nachbargebieten*, edited by B. Hänsel and W. Schier. Rahden/Westfalen: Verlag Marie Leidorf.

Pickrell, Joseph K., Nick Patterson, Po-Ru Loh, Mark Lipson, Bonnie Berger, Mark Stoneking, Brigitte Pakendorf, and David Reich. 2014. Ancient west Eurasian ancestry in southern and eastern Africa. *Proceedings of the National Academy of Sciences of the United States of America* 111 (7):2632-2637.

Piezonka, Henny. 2015. Older than the farmers' pots? Hunter-gatherer ceramics east of the Baltic Sea. In *The Dąbki Site in Pomerania and the Neolithisation of the North European Lowlands (c. 5000-3000 calBC)*, edited by J. Kabaciński, S. Hatz, R. D. C. M. and T. Terberger. Rahden/Westf.: Marie Leidorf.

Repeated Author. 2016. Die frühe Keramik Eurasiens: Aktuelle Forschungsfragen und methodische Ansätze, in Multidisciplinary approach to archaeology: Recent achievements and prospects. Proceedings of the International Symposium "Multidisciplinary approach to archaeology: Recent achievements and prospects", June 22-26, 2015, Novosibirsk, Eds. V. I. Molodin, S. Hansen. In *Multidisciplinary approach to archaeology: Recent achievements and prospects. Proceedings of the International Symposium "Multidisciplinary approach to archaeology: Recent achievements and prospects", June 22-26, 2015, Novosibirsk*, edited by V. I. Molodin and S. Hansen.

Piličiauskas, Gytis, Vitali Asheichyk, Grzegorz Osipowicz, Raminta Skipitytė, Liivi Varul, Justina Kozakaitė, Mikola Kryvaltsevich, Aliaksandra Vaitovich, Vadzim Lakiza, Justina Šapolaitė, Žilvinas Ežerinskis, Mikalai Pamazanau, Alexandre Lucquin, Oliver E. Craig, and Harry K. Robson. 2018. The Corded Ware culture in the Eastern Baltic: New evidence on chronology, diet, beaker, bone and flint tool function. *Journal of Archaeological Science: Reports* 21:538-552.

Poljakov, A. V. , S. V. Svjatko, and N. F. Stepanova. 2018. Поляков А.В., Святко С.В., Степанова Н.Ф. Современное состояние радиоуглеродного датирования афанасьевской и окуневской культур. *Научное обозрение Саяно-Алтая* 21 (5):14-22.

Poska, A., and L. Saarse. 2002. Vegetation development and introduction of agriculture to Saaremaa Island, Estonia: the human response to shore displacement. *The Holocene* 12 (5):555-568.

Pospieszny, Ł. 2015. Freshwater reservoir effect and the radiocarbon chronology of the cemetery in Ząbie. *Journal of Archaeological Science* 53:264-276.

Prendergast, Mary E., Mark Lipson, Elizabeth A. Sawchuk, Iñigo Olalde, Christine A. Ogola, Nadin Rohland, Kendra A. Sirak, Nicole Adamski, Rebecca Bernardos, Nasreen Broomandkhoshbacht, Kimberly Callan, Brendan J.

Culleton, Laurie Eccles, Thomas K. Harper, Ann Marie Lawson, Matthew Mah, Jonas Oppenheimer, Kristin Stewardson, Fatma Zalzala, Stanley H. Ambrose, George Ayodo, Henry Louis Gates, Agness O. Gidna, Maggie Katongo, Amandus Kwekason, Audax Z. P. Mabulla, George S. Mudenda, Emmanuel K. Ndiema, Charles Nelson, Peter Robertshaw, Douglas J. Kennett, Fredrick K. Manthi, and David Reich. 2019. Ancient DNA reveals a multistep spread of the first herders into sub-Saharan Africa. *Science*:eaaw6275.

Prescott, C., and E. Walderhaug. 1995. The Last Frontier? Processes of Indo-Europeanization in Northern Europe. The Norwegian Case. *JIES* 23 (2):257-278.

Prescott, Christopher. 2012. No longer north of the Beakers. Modeling an interpretative platform for third millennium transformations in Norway. In *Background to Beakers: Inquiries in Regional Cultural Backgrounds of the Bell Beaker Complex* edited by H. Fokkens and F. Nicolis. Leiden: Sidestone Press.

Pustovalov, S. Z. 2000. The «Tyagunova Mogila» Burial Mound and the Problem of Wheeled Transport of the Pit Grave and Catacomb Cultures in Eastern Europe. *Stratum plus* 2:296-321.

Quade, Jay, Elad Dente, Moshe Armon, Yoav Ben Dor, Efrat Morin, Ori Adam, and Yehouda Enzel. 2018. Megalakes in the Sahara? A Review – ADDENDUM. *Quaternary Research* 90 (2):435-435.

Quiles, Carlos. 2012. Part I: Language and culture. In *A Grammar of Modern Indo-European*. Badajoz: The Indo-European Language Association.

Repeated Author. 2017. *Indo-European demic diffusion model*. 3rd ed. Badajoz: Universidad de Extremadura.

Raghavan, M., P. Skoglund, K. E. Graf, M. Metspalu, A. Albrechtsen, I. Moltke, S. Rasmussen, T. W. Stafford, Jr., L. Orlando, E. Metspalu, M. Karmin, K. Tambets, S. Rootsi, R. Magi, P. F. Campos, E. Balanovska, O. Balanovsky, E. Khusnutdinova, S. Litvinov, L. P. Osipova, S. A. Fedorova, M. I. Voevoda, M. DeGiorgio, T. Sicheritz-Ponten, S. Brunak, S. Demeshchenko, T. Kivisild, R. Villems, R. Nielsen, M. Jakobsson, and E. Willerslev. 2014. Upper Palaeolithic Siberian genome reveals dual ancestry of Native Americans. *Nature* 505 (7481):87-91.

Rahmani, Noura, and David Lubell. 2012. Early Holocene Climat Change and the Adoption of Pressure Technique in the Maghreb: The Capsian Sequence at Kef Zoura D. In *The Emergence of Pressure Blade Making: From Origin to Modern Experimentation*, edited by P. M. Desrosiers. New York: Springer.

Rascovan, Nicolás, Karl-Göran Sjögren, Kristian Kristiansen, Rasmus Nielsen, Eske Willerslev, Christelle Desnues, and Simon Rasmussen. 2018. Emergence and Spread of Basal Lineages of Yersinia pestis during the Neolithic Decline. *Cell*.

Rasmussen, Simon, Morten Erik Allentoft, Kasper Nielsen, Ludovic Orlando, Martin Sikora, Karl-Göran Sjögren, Anders Gorm Pedersen, Mikkel Schubert, Alex Van Dam, Christian Moliin Outzen Kapel, Henrik Bjørn Nielsen, Søren Brunak, Pavel Avetisyan, Andrey Epimakhov, Mikhail Viktorovich Khalyapin, Artak Gnuni, Aivar Kriiska, Irena Lasak, Mait Metspalu, Vyacheslav Moiseyev, Andrei Gromov, Dalia Pokutta, Lehti Saag, Liivi Varul, Levon Yepiskoposyan, Thomas Sicheritz-Pontén, Robert A Foley, Marta Mirazón Lahr, Rasmus

Nielsen, Kristian Kristiansen, and Eske Willerslev. 2015. Early Divergent Strains of *Yersinia pestis* in Eurasia 5,000 Years Ago. *Cell* 163 (3):571-582.

Rassamakin, Yuri. 1999. The Eneolithic of the Black Sea Steppe: Dynamics of Cultural and Economic Development 4500-2300 BC. In *Late prehistoric exploitation of the Eurasian steppe*, edited by M. Levine, Y. Rassamakin, A. Kislenko and N. Tatarintseva. Cambridge: McDonald Inst. Monogr.

Rassamakin, Yuri Ya., and Alla V. Nikolova. 2008. *Carpathian Imports in the Graves of the Yamnaya Culture on the Lower Dnieper. Some Problems of Chronology and Connections in the Black Sea Steppes During the Early Bronze Age.* Edited by F. Bertemes and A. Furtwängler, *Import and Imitation in Archaeology*. Langenweissbach: Beier & Beran.

Reich, David. 2018. *Who We Are and How We Got Here: Ancient DNA and the New Science of the Human Past* New York: Pantheon.

Reich, David, Kumarasamy Thangaraj, Nick Patterson, Alkes L. Price, and Lalji Singh. 2009. Reconstructing Indian Population History. *Nature* 461 (7263):489-494.

Reingruber, Agathe, and Yuri Rassamakin. 2016. Zwischen Donau und Kuban: Das nordpontische Steppengebiet im 5. Jt. v. Chr. In *Der Schwarzmeerraum vom Neolithikum bis in die Früheisenzeit (6000-600 v. Chr.). Kulturelle Interferenzen in der zirkumpontischen Zone und kontakte mit ihren Nachbargebieten.*, edited by V. Nikolov and W. Schier. Rahden/Westf.: Verlag Marie Leidorf.

Reinhold, Sabine, Julia Gresky, Natalia Berezina, Anatoly R. Kantorovich, Corina Knipper, Vladimir E. Maslov, Vladimira G. Petrenko, Kurt W. Alt, and Andrey B. Belinsky. 2017. Contextualising Innovation: Cattle Owners and Wagon Drivers in the North Caucasus and Beyond. In *Appropriating Innovations: Entangled Knowledge in Eurasia, 5000-150 BCE*, edited by J. Maran and P. Stockhammer. Oxford: Oxbow Books.

Renfrew, Colin. 1987. *Archaeology and Language: The Puzzle of Indo-European Origins*. London: Jonathan Cape.

Rigaud, Solange, Claire Manen, and Iñigo García-Martínez de Lagrán. 2018. Symbols in motion: Flexible cultural boundaries and the fast spread of the Neolithic in the western Mediterranean. *PLOS ONE* 13 (5):e0196488.

Ringe, D. 2006. *A Linguistic History of English: Volume I, From Proto-Indo-European to Proto-Germanic. Oxford Scholarship Online, 2006.* Edited by D. Ringe. 2 vols. Vol. 1, *A Linguistic History of English*. Oxford: Oxford University Press.

Ringe, D., T. Warnow, and A. Taylor. 2002. Indo-European and computational cladistics. *Trans. Philol. Soc.* 100 (1):59-129.

Risch, Roberto, Vicente Lull, Rafael Micó, and Cristina Rihuete. 2015. Transitions and conflict at the end of the 3rd millennium BC in south Iberia. In *2200 BC – A climatic breakdown as a cause for the collapse of the old world?*, edited by H. Meller, H. Arz, R. Jung and R. Risch. Halle: Landesmuseum für Vorgeschichte Halle.

Rivollat, Maïté, Fanny Mendisco, Marie-Hélène Pemonge, Audrey Safi, Didier Saint-Marc, Antoine Brémond, Christine Couture-Veschambre, Stéphane Rottier, and Marie-France Deguilloux. 2015. When the Waves of European

Neolithization Met: First Paleogenetic Evidence from Early Farmers in the Southern Paris Basin. *PLOS ONE* 10 (4):e0125521.

Robb, John. 2009. People of Stone: Stelae, Personhood, and Society in Prehistoric Europe. *Journal of Archaeological Method and Theory* 16 (3):162-183.

Rodríguez-Varela, Ricardo, Torsten Günther, Maja Krzewińska, Jan Storå, Thomas H. Gillingwater, Malcolm MacCallum, Juan Luis Arsuaga, Keith Dobney, Cristina Valdiosera, Mattias Jakobsson, Anders Götherström, and Linus Girdland-Flink. 2017. Genomic Analyses of Pre-European Conquest Human Remains from the Canary Islands Reveal Close Affinity to Modern North Africans. *Current Biology*.

Rosenberg, Michael, and Asli Erim-Özdoğan. 2011. The Neolithic in Southeastern Anatolia. In *The Oxford Handbook of Ancient Anatolia (10,000-323 BCE)*, edited by S. R. Steadman and G. McMahon. Oxford: Oxford University Press.

Ryan, J., J. Desideri, and M. Besse. 2018. Bell Beaker Archers: Warriors or an Ideology? . *Journal of Neolithic Archaeology* 4:95-120.

Ryan, William B. F. 2007. Status of the Black Sea flood hypothesis. In *The Black Sea Flood Question: Changes in Coastline, Climate, and Human Settlement*, edited by V. Yanko-Hombach, A. S. Gilbert, N. Panin and P. M. Dolukhanov. Dordrecht: Springer Netherlands.

Saag, Lehti, Liivi Varul, Christiana Lyn Scheib, Jesper Stenderup, Morten E Allentoft, Lauri Saag, Luca Pagani, Maere Reidla, Kristiina Tambets, Ene Metspalu, Aivar Kriiska, Eske Willerslev, Toomas Kivisild, and Mait Metspalu. 2017. Extensive farming in Estonia started through a sex-biased migration from the Steppe. *bioRxiv*.

Sagona, Antonio. 2011. Anatolia and the Transcaucasus: Themes and Variations ca. 6400-1500 B.C.E. In *The Oxford Handbook of Ancient Anatolia 10,000-323 B.C.E.*, edited by S. R. Steadman and G. McMahon. Oxford: Oxford University Press.

Repeated Author. 2017. Dolmens for the Dead: The Western Caucasus in the Bronze Age (3250–1250 BC). In *The Archaeology of the Caucasus: From Earliest Settlements to the Iron Age*. Cambridge: Cambridge University Press.

Repeated Author. 2017. Far-Flung Networks: The Chalcolithic (5000/4800–3500 BC). In *The Archaeology of the Caucasus: From Earliest Settlements to the Iron Age*. Cambridge: Cambridge University Press.

Repeated Author. 2017. Trailblazers: The Palaeolithic and Mesolithic Foundations. In *The Archaeology of the Caucasus: From Earliest Settlements to the Iron Age*. Cambridge: Cambridge University Press.

Repeated Author. 2017. Transition to Settled Life: The Neolithic (6000–5000 BC). In *The Archaeology of the Caucasus: From Earliest Settlements to the Iron Age*. Cambridge: Cambridge University Press.

Sahala, Aleksi. 2009-2013. Sumero-Indo-European language contacts. *University of Helsinki*.

Salugina, N. P. 2005. Ceramic technology of the Repin type burials within the Pit-Grave culture of the Volga-Ural. *Российская археология* 3:85-92.

Sánchez-Quinto, Federico, Helena Malmström, Magdalena Fraser, Linus Girdland-Flink, Emma M. Svensson, Luciana G. Simões, Robert George, Nina Hollfelder, Göran Burenhult, Gordon Noble, Kate Britton, Sahra Talamo, Neil

Curtis, Hana Brzobohata, Radka Sumberova, Anders Götherström, Jan Storå, and Mattias Jakobsson. 2019. Megalithic tombs in western and northern Neolithic Europe were linked to a kindred society. *Proceedings of the National Academy of Sciences* 116 (19):9469-9474.

Scerri, Eleanor M. L., Maria Guagnin, Huw S. Groucutt, Simon J. Armitage, Luke E. Parker, Nick Drake, Julien Louys, Paul S. Breeze, Muhammad Zahir, Abdullah Alsharekh, and Michael D. Petraglia. 2018. Neolithic pastoralism in marginal environments during the Holocene Humid Period, northern Saudi Arabia. *Antiquity* 92 (365):1180-1194.

Schlebusch, Carina M, Helena Malmström, Torsten Günther, Per Sjödin, Alexandra Coutinho, Hanna Edlund, Arielle R Munters, Maryna Steyn, Himla Soodyall, Marlize Lombard, and Mattias Jakobsson. 2017. Ancient genomes from southern Africa pushes modern human divergence beyond 260,000 years ago. *bioRxiv*.

Schmidt, Isabell, and Andreas Zimmermann. 2019. Population dynamics and socio-spatial organization of the Aurignacian: Scalable quantitative demographic data for western and central Europe. *PLOS ONE* 14 (2):e0211562.

Schoop, Ulf-Dietrich. 2011. The Chalcolithic on the Plateau. In *The Oxford Handbook of Ancient Anatolia (10,000-323 BCE)*, edited by S. R. Steadman and G. McMahon. Oxford: Oxford University Press.

Schrijver, Peter. 2015. Talking Neolithic: the case for Hatto-Minoan and its relationship to Sumerian. *Article accepted, still awaiting publication*.

Schroeder, Hannes, Ashot Margaryan, Marzena Szmyt, Bertrand Theulot, Piotr Włodarczak, Simon Rasmussen, Shyam Gopalakrishnan, Anita Szczepanek, Tomasz Konopka, Theis Z. T. Jensen, Barbara Witkowska, Stanisław Wilk, Marcin M. Przybyła, Łukasz Pospieszny, Karl-Göran Sjögren, Zdzisław Belka, Jesper Olsen, Kristian Kristiansen, Eske Willerslev, Karin M. Frei, Martin Sikora, Niels N. Johannsen, and Morten E. Allentoft. 2019. Unraveling ancestry, kinship, and violence in a Late Neolithic mass grave. *Proceedings of the National Academy of Sciences*:201820210.

Schuenemann, Verena J., Alexander Peltzer, Beatrix Welte, W. Paul van Pelt, Martyna Molak, Chuan-Chao Wang, Anja Furtwängler, Christian Urban, Ella Reiter, Kay Nieselt, Barbara Teßmann, Michael Francken, Katerina Harvati, Wolfgang Haak, Stephan Schiffels, and Johannes Krause. 2017. Ancient Egyptian mummy genomes suggest an increase of Sub-Saharan African ancestry in post-Roman periods. 8:15694.

Schulting, Rick J., and Michael P. Richards. 2016. Stable Isotope Analysis of Neolithic to Late Bronze Age Populations in the Samara Valley. In *A Bronze Age Landscape in the Russian Steppes. The Samara Valley Project*, edited by D. W. Anthony, D. R. Brown, O. D. Mochalov, A. A. Khokhlov and P. F. Kuznetsov. Los Angeles: The Cotsen Institute of Archaeology at UCLA.

Schulz Paulsson, B. 2019. Radiocarbon dates and Bayesian modeling support maritime diffusion model for megaliths in Europe. *Proceedings of the National Academy of Sciences* 116 (9):3460-3465.

Sherratt, A. 2004. Material resources, capital, and power: the coevolution of society and culture. Archaeological Perspectives on

Political Economies. University of Utah Press, Salt Lake City, pp. 79e103. In *Archaeological Perspectives on Political Economies*, edited by G. Feinman and L. Nicholas. Salt Lake City: University of Utah Press.

Sherratt, Andrew. 1981. Plough and pastoralism: aspects of teh secondary products revolution. In *Pattern of the Past: studies in honour of David Clarke*, edited by N. Hammond, I. Hodder and G. Isaac. Cambridge: Cambridge University Press.

Shishlina, N. I. 2008. *Reconstruction of the Bronze Age of the Caspian Steppes: Life Styles and Life Ways of Pastoral Nomads*. Oxford: British Archaeological Reports International Series 1876.

Shishlina, Natalia I., Johannes van der Plicht, and Elya P. Zazovskaya. 2011. Radiocarbon dating of the bronze age bone pins from Eurasian steppe. *Geochronometria* 38 (2):107-115.

Sikora, Martin, Vladimir Pitulko, Vitor Sousa, Morten E Allentoft, Lasse Vinner, Simon Rasmussen, Ashot Margaryan, Peter de Barros Damgaard, Constanza de la Fuente Castro, Gabriel Renaud, Melinda Yang, Qiaomei Fu, Isabelle Dupanloup, Konstantinos Giampoudakis, David Bravo Nogues, Carsten Rahbek, Guus Kroonen, Michael Peyrot, Hugh McColl, Sergey Vasilyev, Elizaveta Veselovskaya, Margarita Gerasimova, Elena Pavlova, Vyacheslav Chasnyk, Pavel Nikolskiy, Pavel Grebenyuk, Alexander Fedorchenko, Alexander Lebedintsev, Boris Malyarchuk, Morten Meldgaard, Rui Martiniano, Laura Arppe, Jukka Palo, Tarja Sundell, Kristiina Mannermaa, Mikko Putkonen, Verner Alexandersen, Charlotte Primeau, Ripan Mahli, Karl-Göran Sjögren, Kristian Kristiansen, Anna Wessman, Antti Sajantila, Marta Mirazohn Lahr, Richard Durbin, Rasmus Nielsen, David Meltzer, Laurent Excoffier, and Eske Willerslev. 2018. The population history of northeastern Siberia since the Pleistocene. *bioRxiv*.

Sikora, Martin, Andaine Seguin-Orlando, Vitor C. Sousa, Anders Albrechtsen, Thorfinn Korneliussen, Amy Ko, Simon Rasmussen, Isabelle Dupanloup, Philip R. Nigst, Marjolein D. Bosch, Gabriel Renaud, Morten E. Allentoft, Ashot Margaryan, Sergey V. Vasilyev, Elizaveta V. Veselovskaya, Svetlana B. Borutskaya, Thibaut Deviese, Dan Comeskey, Tom Higham, Andrea Manica, Robert Foley, David J. Meltzer, Rasmus Nielsen, Laurent Excoffier, Marta Mirazon Lahr, Ludovic Orlando, and Eske Willerslev. 2017. Ancient genomes show social and reproductive behavior of early Upper Paleolithic foragers. *Science*.

Silva, Fabio, and Marc Vander Linden. 2018. Author Correction: Amplitude of travelling front as inferred from 14C predicts levels of genetic admixture among European early farmers. *Scientific Reports* 8 (1):12809.

Sjogren, K. G., T. D. Price, and K. Kristiansen. 2016. Diet and Mobility in the Corded Ware of Central Europe. *PLoS One* 11 (5):e0155083.

Skoglund, P., H. Malmstrom, M. Raghavan, J. Stora, P. Hall, E. Willerslev, M. T. Gilbert, A. Gotherstrom, and M. Jakobsson. 2012. Origins and genetic legacy of Neolithic farmers and hunter-gatherers in Europe. *Science* 336 (6080):466-9.

Skoglund, Pontus, Helena Malmström, Ayça Omrak, Maanasa Raghavan, Cristina Valdiosera, Torsten Günther, Per Hall, Kristiina Tambets, Jüri Parik, Karl-Göran Sjögren, Jan Apel, Eske Willerslev, Jan Storå, Anders Götherström, and

Mattias Jakobsson. 2014. Genomic Diversity and Admixture Differs for Stone-Age Scandinavian Foragers and Farmers. *Science* 344 (6185):747.

Skoglund, Pontus, Jessica C. Thompson, Mary E. Prendergast, Alissa Mittnik, Kendra Sirak, Mateja Hajdinjak, Tasneem Salie, Nadin Rohland, Swapan Mallick, Alexander Peltzer, Anja Heinze, Iñigo Olalde, Matthew Ferry, Eadaoin Harney, Megan Michel, Kristin Stewardson, Jessica I. Cerezo-Román, Chrissy Chiumia, Alison Crowther, Elizabeth Gomani-Chindebvu, Agness O. Gidna, Katherine M. Grillo, I. Taneli Helenius, Garrett Hellenthal, Richard Helm, Mark Horton, Saioa López, Audax Z. P. Mabulla, John Parkington, Ceri Shipton, Mark G. Thomas, Ruth Tibesasa, Menno Welling, Vanessa M. Hayes, Douglas J. Kennett, Raj Ramesar, Matthias Meyer, Svante Pääbo, Nick Patterson, Alan G. Morris, Nicole Boivin, Ron Pinhasi, Johannes Krause, and David Reich. 2017. Reconstructing Prehistoric African Population Structure. *Cell* 171 (1):59-71.e21.

Skourtanioti, E., J. Choongwon, Y. S. Erdal, M. Frangipane, P. W. Stockhammer, M. Burri, J. Krause, and W. Haak. 2018. Population dynamics at Late Chalcolithic and Early Bronze Age Arslantepe, Anatolia. Paper read at 8th International Symposium on Biomolecular Archaeology ISBA 2018. 18th – 21st September, at Jena, Germany.

Slatkin, M., and F. Racimo. 2016. Ancient DNA and human history. *Proc Natl Acad Sci U S A* 113 (23):6380-7.

Smolyaninov, Roman, Andrey Skorobogatov, and Aleksey Surkov. 2017. Chronology of Neolithic sites in the forest-steppe area of the Don River. *Documenta Praehistorica* XLIV:192-202.

Smyntina, Olena. 2016. Cultural Resilience Theory as an instrument of modeling of Human response to the global climate change. A case study in the North-Western Black Sea region on the Pleistocene-Holocene boundary. *RIPARIA* 2:1-20.

Sørensen, Tim Flohr. 2017. The Two Cultures and a World Apart: Archaeology and Science at a New Crossroads. *Norwegian Archaeological Review* 50 (2):101-115.

Strahm, C. 2002. Tradition und Wandel der sozialen Strukturen vom 3. zum 2. vorchristlichen Jahrtausend. In *Vom Endneolithikum zur Fruhbronzezeit: Muster sozialen Wandels? Tagung Bamberg 14.-16. Juni 2001*, edited by J. Müller. Bonn: Habelt.

Szécsényi-Nagy, A., M. Lipson, I. Olalde, K. Oross, M. Bondár, G. Kulcsár, V. Kiss, B. Mende, K. Alt, E. Bánffy, and D. Reich. 2018. Population transformations in the 6000-2000 BC period of the Carpathian Basin. Paper read at 8th International Symposium on Biomolecular Archaeology ISBA 2018. 18th – 21st September, at Jena, Germany.

Szecsenyi-Nagy, Anna, Christina Roth, Brandt Guido, Cristina Rihuete-Herrada, Cristina Tejedor-Rodriguez, Petra Held, Inigo Garcia-Martinez-de-Lagran, Hector Arcusa Magallon, Stephanie Zesch, Corina Knipper, Eszter Banffy, Susanne Friedrich, Harald Meller, Primitiva Bueno-Ramirez, Rosa Barroso Bermejo, Rodrigo de Balbin Behrmann, Ana M. Herrero-Coral, Raul Flores Fernandez, Carmen Alonso Fernandez, Javier Jimenez Echevarria, Laura Rindlisbacher, Camila Oliart, Maria-Ines Fregeiro, Ignacio Soriano, Oriol

Vincente, Rafael Mico, Vincente Lull, Jorge Soler Diaz, Juan Antonio Lopez Padilla, Consuelo Roca de Togores Munoz, Mauro S. Hernandez Perez, Francisco Javier Jover Maestre, Joaquin Lomba Maurandi, Azucena Aviles Fernandez, Katina T. Lillios, Ana Maria Silva, Miguel Magalhaes Ramalho, Luiz Miguel Oosterbeek, Claudia Cunha, Anna J Waterman, Jordi Roig Buxo, Andres Martinez, Juana Ponce Martinez, Mark Hunt Ortiz, Juan Carlos Mejias-Gracia, Juan Carlos Pecero Espin, Rosario Cruz-Aunon Briones, Tiago Tome, Eduardo Carmona Ballestero, Joao Luis Cardoso, Ana Cristina Araujo, Corina Liesau von Lettow-Vorbeck, Conception Blasco Bosqued, Patricia Rios Mendoza, Ana Pujante, Jose I. Royo-Guillen, Marco Aurelio Esquembre Bevia, Victor Manuel Dos Santos Goncalves, Rui Parreira, Elena Moran Hernandez, Elena Mendez Izquierdo, Jorge Vega de Miguel, Roberto Menduina Garcia, Victoria Martinez Calvo, Oscar Lopez Jimenez, Johannes Krause, Sandra L. Pichler, Rafael Garrido-Pena, Michael Kunst, Roberto Risch, Manuel A. Rojo-Guerra, Wolfgang Haak, and Kurt W. Alt. 2017. The maternal genetic make-up of the Iberian Peninsula between the Neolithic and the Early Bronze Age. *bioRxiv*.

Szmyt, Marzena. 2008. Baden Patterns in the Milieu of Globular Amphorae: Transformation, Incorporation and Long Continuity. A case study from the Kujavia region, Polish Lowland. In *The Baden Complex and the Outside World*, edited by M. Furholt, M. Szmyt and A. Zastawny. Bonn: Dr. Rudolf Habelt.

Repeated Author. 2010. *Between West and East. People of the Globular Amphora Culture in Eastern Europe: 2950-2350 BC*. Edited by A. Kosko, *Baltic-Pontic Studies*. Poznań: Uniwersytet im. Adama Mickiewicza.

Repeated Author. 2013. The circulation of People and Ideas in the Baltic and Pontic Areas during 3rd millennium BC.

Tambets, Kristiina, Bayazit Yunusbayev, Georgi Hudjashov, Anne-Mai Ilumäe, Siiri Rootsi, Terhi Honkola, Outi Vesakoski, Quentin Atkinson, Pontus Skoglund, Alena Kushniarevich, Sergey Litvinov, Maere Reidla, Ene Metspalu, Lehti Saag, Timo Rantanen, Monika Karmin, Jüri Parik, Sergey I. Zhadanov, Marina Gubina, Larisa D. Damba, Marina Bermisheva, Tuuli Reisberg, Khadizhat Dibirova, Irina Evseeva, Mari Nelis, Janis Klovins, Andres Metspalu, Tõnu Esko, Oleg Balanovsky, Elena Balanovska, Elza K. Khusnutdinova, Ludmila P. Osipova, Mikhail Voevoda, Richard Villems, Toomas Kivisild, and Mait Metspalu. 2018. Genes reveal traces of common recent demographic history for most of the Uralic-speaking populations. *Genome Biology* 19 (1):139.

Tassi, Francesca, Stefania Vai, Silvia Ghirotto, Martina Lari, Alessandra Modi, Elena Pilli, Andrea Brunelli, Roberta Rosa Susca, Alicja Budnik, Damian Labuda, Federica Alberti, Carles Lalueza-Fox, David Reich, David Caramelli, and Guido Barbujani. 2017. Genome diversity in the Neolithic Globular Amphorae culture and the spread of Indo-European languages. *Proceedings of the Royal Society B: Biological Sciences* 284 (1867).

Telegin, D. Ya., M. Lillie, Inna Potekhina, and M. M. Kovaliukh. 2015. Settlement and economy in Neolithic Ukraine: A new chronology. *Antiquity* 77 (297):456-470.

Terberger, Thomas, Joachim Burger, Friedrich Lüth, Johannes Müller, and Henny Piezonka. 2018. Step by step – The neolithisation of Northern Central Europe in the light of stable isotope analyses. *Journal of Archaeological Science* 99:66-86.

Thangaraj, Kumarasamy, Lalji Singh, Alla G. Reddy, V. Raghavendra Rao, Subhash C. Sehgal, Peter A. Underhill, Melanie Pierson, Ian G. Frame, and Erika Hagelberg. 2003. Genetic Affinities of the Andaman Islanders, a Vanishing Human Population. *Current Biology* 13 (2):86-93.

Tolochko, P. P. 1997. *Davnya istoriya Ukraynu 1*. Kiyv: Naukova dumka.

Tóth, Csaba, Katalin Joó, and Attila Barczi. 2015. Lyukas Mound: One of the Many Prehistoric Tumuli in the Great Plain. In *Landscapes and Landforms of Hungary*, edited by D. n. Lóczy. Pécs: Springer.

Trager, George L., and Henry Lee Smith. 1950. A chronology of Indo-Hittite. *Studies in linguistics* 8 (3):61-70.

Trifonov, V. A., N. I. Shishlina, A. Yu. Loboda, N. N. Kolobylina, E. Yu. Tereschenko, and E. B. Yatsishina. 2018. The Production of Thin-Walled Jointless Gold Beads from the Maykop Culture Megalithic Tomb of the Early Bronze Age at Tsarskaya in the North Caucasus: Results of Analytical and Experimental Research. *Archaeometry* 0 (0).

Underhill, P. A., G. D. Poznik, S. Rootsi, M. Jarve, A. A. Lin, J. Wang, B. Passarelli, J. Kanbar, N. M. Myres, R. J. King, J. Di Cristofaro, H. Sahakyan, D. M. Behar, A. Kushniarevich, J. Sarac, T. Saric, P. Rudan, A. K. Pathak, G. Chaubey, V. Grugni, O. Semino, L. Yepiskoposyan, A. Bahmanimehr, S. Farjadian, O. Balanovsky, E. K. Khusnutdinova, R. J. Herrera, J. Chiaroni, C. D. Bustamante, S. R. Quake, T. Kivisild, and R. Villems. 2015. The phylogenetic and geographic structure of Y-chromosome haplogroup R1a. *Eur J Hum Genet* 23 (1):124-31.

Vai, Stefania, Stefania Sarno, Martina Lari, Donata Luiselli, Giorgio Manzi, Marina Gallinaro, Safaa Mataich, Alexander Hübner, Alessandra Modi, Elena Pilli, Mary Anne Tafuri, David Caramelli, and Savino di Lernia. 2019. Ancestral mitochondrial N lineage from the Neolithic 'green' Sahara. *Scientific Reports* 9 (1):3530.

Valdiosera, Cristina, Torsten Günther, Juan Carlos Vera-Rodríguez, Irene Ureña, Eneko Iriarte, Ricardo Rodríguez-Varela, Luciana G. Simões, Rafael M. Martínez-Sánchez, Emma M. Svensson, Helena Malmström, Laura Rodríguez, José-María Bermúdez de Castro, Eudald Carbonell, Alfonso Alday, José Antonio Hernández Vera, Anders Götherström, José-Miguel Carretero, Juan Luis Arsuaga, Colin I. Smith, and Mattias Jakobsson. 2018. Four millennia of Iberian biomolecular prehistory illustrate the impact of prehistoric migrations at the far end of Eurasia. *Proceedings of the National Academy of Sciences* 115 (13):3428-3433.

van den Brink, Edwin C. M., Ron Beeri, Dan Kirzner, Enno Bron, Anat Cohen-Weinberger, Elisheva Kamaisky, Tamar Gonen, Lilly Gershuny, Yossi Nagar, Daphna Ben-Tor, Naama Sukenik, Orit Shamir, Edward F. Maher, and David Reich. 2017. A Late Bronze Age II clay coffin from Tel Shaddud in the Central Jezreel Valley, Israel: context and historical implications AU - van den Brink, Edwin C. M. *Levant* 49 (2):105-135.

van Dorp, Lucy, David Balding, Simon Myers, Luca Pagani, Chris Tyler-Smith, Endashaw Bekele, Ayele Tarekegn, Mark G. Thomas, Neil Bradman, and Garrett Hellenthal. 2015. Evidence for a Common Origin of Blacksmiths and Cultivators in the Ethiopian Ari within the Last 4500 Years: Lessons for Clustering-Based Inference. *PLOS Genetics* 11 (8):e1005397.

Vander Linden, Marc. 2015. An impossible dialogue? On the interface between archaeology, historical linguistics and comparative philology. In *The Linguistic Roots of Europe. Origin and development of European languages*, edited by R. Mailhammer, T. Vennemann and B. A. Olsen. Copenhagen: Museum Tusculanum Press.

Vanhanen, Santeri, Stefan Gustafsson, Håkan Ranheden, Niclas Björck, Marianna Kemell, and Volker Heyd. 2019. Maritime Hunter-Gatherers Adopt Cultivation at the Farming Extreme of Northern Europe 5000 Years Ago. *Scientific Reports* 9 (1):4756.

Varul, Liivi, Ravil M. Galeev, Anna A. Malytina, Mari Tõrv, Sergey V. Vasilyev, Lembi Lõugas, and Aivar Kriiska. 2019. Complex mortuary treatment of a Corded Ware Culture individual from the Eastern Baltic: A case study of a secondary deposit in Sope, Estonia. *Journal of Archaeological Science: Reports* 24:463-472.

Vasilyeva, I. N. 2002. About technology of ceramics of the I Hvalynsk Eneolithic burial ground. *Вопросы археологии Поволжья. Самара* 2:15-49.

Veeramah, Krishna R. 2018. The importance of fine-scale studies for integrating paleogenomics and archaeology. *Current Opinion in Genetics & Development* 53:83-89.

Villalba-Mouco, Vanessa, Marieke S. van de Loosdrecht, Cosimo Posth, Rafael Mora, Jorge Martínez-Moreno, Manuel Rojo-Guerra, Domingo C. Salazar-García, José I. Royo-Guillén, Michael Kunst, Hélène Rougier, Isabelle Crevecoeur, Héctor Arcusa-Magallón, Cristina Tejedor-Rodríguez, Iñigo García-Martínez de Lagrán, Rafael Garrido-Pena, Kurt W. Alt, Choongwon Jeong, Stephan Schiffels, Pilar Utrilla, Johannes Krause, and Wolfgang Haak. 2019. Survival of Late Pleistocene Hunter-Gatherer Ancestry in the Iberian Peninsula. *Current Biology* 29 (7):1169-1177.e7.

Villar Liébana, Francisco. 2014. *Indoeuropeos, iberos, vascos y sus parientes. Estratigrafía y cronología de las poblaciones prehistóricas*. Salamanca: Ediciones Universidad de Salamanca.

Vybornov, Aleksandr. 2016. Initial stages of two Neolithisation models in the Lower Volga basin. *Documenta Prehistorica* 43:161-166.

Vybornov, Aleksandr A., Markku Oinonen, Natalia S. Doga, Marianna A. Kulkova, and Aleksandr S. Popov. 2016. On the chronological aspect of productive economy origin in the Lower Volga region. *Science Journal of Volgograd State University* 21 (3):6-13.

Vybornov, Alexander, Pavel Kosintsev, and Marianna Kulkova. 2015. The origin of farming in the Lower Volga Region. *Documenta Praehistorica* XLII:67-76.

Vybornov, A.A., A.I. Yudin, I.N. Vasilyeva, P.A. Kosintsev, N.S. Doga, A.S. Popov, V.I. Platonov, and N.V. Roslyakova. 2018. New results of studies on the Oroshaemoe site in the Lower Volga regionn. *Археология и этнография* 20 (3):215-222.

Wakabayashi, Ken, Ryan Schmidt, Takashi Gakuhari, Kae Koganebuchi, Motoyuki Ogawa, Jordan Karsten, Mykhailo Sokhatsky, and Hiroki Oota. 2017. Analysis of ancient human mitochondrial DNA from Verteba Cave, Ukraine: insights into the origins and expansions of the Late Neolithic-Chalcolithic Cututeni-Tripolye Culture. *bioRxiv*.

Wang, Chuan-Chao, Sabine Reinhold, Alexey Kalmykov, Antje Wissgott, Guido Brandt, Choongwon Jeong, Olivia Cheronet, Matthew Ferry, Eadaoin Harney, Denise Keating, Swapan Mallick, Nadin Rohland, Kristin Stewardson, Anatoly R. Kantorovich, Vladimir E. Maslov, Vladimira G. Petrenko, Vladimir R. Erlikh, Biaslan Ch Atabiev, Rabadan G. Magomedov, Philipp L. Kohl, Kurt W. Alt, Sandra L. Pichler, Claudia Gerling, Harald Meller, Benik Vardanyan, Larisa Yeganyan, Alexey D. Rezepkin, Dirk Mariaschk, Natalia Berezina, Julia Gresky, Katharina Fuchs, Corina Knipper, Stephan Schiffels, Elena Balanovska, Oleg Balanovsky, Iain Mathieson, Thomas Higham, Yakov B. Berezin, Alexandra Buzhilova, Viktor Trifonov, Ron Pinhasi, Andrej B. Belinskij, David Reich, Svend Hansen, Johannes Krause, and Wolfgang Haak. 2019. Ancient human genome-wide data from a 3000-year interval in the Caucasus corresponds with eco-geographic regions. *Nature Communications* 10 (1):590.

Wang, Chuan-Chao, Sabine Reinhold Reinhold, Alexey Kalmykov, Antje Wissgott, Guido Brandt, Choongwon Jeong, Olivia Cheronet, Matthew Ferry, Eadaoin Harney, Denise Keating, Swapan Mallick, Nadin Rohland, Kristin Stewardson, Anatoly R. Kantorovich, Vladimir E. Maslov, Vladimira G. Petrenko, Vladimir R. Erlikh, Biaslan C. Atabiev, Rabadan G. Magomedov, Philipp L. Kohl, Kurt W. Alt, Sandra L. Pichler, Claudia Gerling, Harald Meller, Benik Vardanyan, Larisa Yeganyan, Alexey D. Rezepkin, Dirk Mariaschk, Natalia Y. Berezina, Julia Gresky, Katharina Fuchs, Corina Knipper, Stephan Schiffels, Elena Balanovska, Oleg Balanovsky, Iain Mathieson, Thomas Higham, Yakov B. Berezin, Alexandra P. Buzhilova, Viktor Trifonov, Ron Pinhasi, Andrej B. Belinskiy, David Reich, Svend Hansen, Johannes Krause, and Wolfgang Haak. 2018. The genetic prehistory of the Greater Caucasus. *bioRxiv*.

Welton, Megan Lynn. 2010. Mobility and Social Organization on the Ancient Anatolian Black Sea Coast: An Archaeological, Spatial and Isotopic Investigation of the Cemetery at İkiztepe, Turkey, Near & Middle Eastern Civilizations, University of Toronto, Toronto.

West, M.L. 2007. *Indo-European Poetry and Myth*. Oxford: Oxford University Press.

Whittaker, Gordon. 2008. The Case for Euphratic. *Bull. Georg. Natl. Acad. Sci.* 2 (3):156-168.

Repeated Author. 2012. Euphratic: A phonological sketch. In *The Sound of Indo-European: Phonetics, Phonemics, and Morphophonemics*, edited by B. N. Whitehead, T. Olander, B. A. Olsen and J. E. Rasmussen. Copenhagen: Museum Tusculanum Press.

Whittle, Alasdair. 1996. *Europe in the Neolithic. The Creation of New Worlds*. Edited by N. Yoffee, *Cambridge World Archaeology*. Cambridge: Cambridge University Press.

Wilk, Stanisław. 2016. New data about chronology of the impact of the Hunyadihalom-Lažňany horizon on Younger Danubian cultures north of the Carpathian Mountains. *Recherches Archéoloqiques* 8:7–27.

Repeated Author. 2018. Can we talk about the Copper Age in Lesser Poland? Contribution to the discussion. In *Multas per gentes et multa per saecula. Amici magistro et collegae suo Ioanni Christopho Kozłowski dedicant.*, edited by P. Valde-Nowak, K. Sobczyk, M. Nowak and J. Źrałka. Kraków: Institute of Archaeology, Jagiellonian University in Kraków.

Włodarczak, Piotr. 2001. The absolute chronology of the Corded Ware Culture in the south-eastern Poland. In *Die absolute Chronologie in Mitteleuropa 3000-2000 v. Chr.*, edited by J. Czebreszuk and J. Müller. Poznań/Bamberg/Rahden.

Repeated Author. 2008. Corded Ware and Baden Cultures. Outline of Chronological and Genetic Relations based on the Finds from Western Little Poland. In *The Baden Complex and the Outside World. Proceedings of the 12th Annual Meeting of the EAA in Cracow 19-24th September 2006*, edited by M. Furhold, M. Szmyt and A. Zastawny. Bonn: Dr. Rudolf Habelt.

Repeated Author. 2017. Battle-axes and beakers. The Final Eneolithic societies. In *The Past Societies: Polish lands from the first evidence of human presence to the Early Middle Ages 2 (5500-2000 BC)*, edited by P. Urbańczyk. Warszawa: Institute of Archaeology and Ethnology, Polish Academy of Science.

Repeated Author. 2017. Kurgan rites in the Eneolithic and Early Bronze age Podolia in light of materials from the funerary ceremonial centre at Yampil. In *Podolia "Barrow Culture" Communities: 4th/3rd-2nd Mill. BC. The Yampil Barrow Complex: Interdisciplinary Studies*. Poznan.

Wutke, Saskia, Edson Sandoval-Castellanos, Norbert Benecke, Hans-Jürgen Döhle, Susanne Friederich, Javier Gonzalez, Michael Hofreiter, Lembi Lõugas, Ola Magnell, Anna-Sapfo Malaspinas, Arturo Morales-Muñiz, Ludovic Orlando, Monika Reissmann, Alexandra Trinks, and Arne Ludwig. 2018. Decline of genetic diversity in ancient domestic stallions in Europe. *Science Advances* 4 (4):eaap9691.

Yang, Melinda A., Xing Gao, Christoph Theunert, Haowen Tong, Ayinuer Aximu-Petri, Birgit Nickel, Montgomery Slatkin, Matthias Meyer, Svante Pääbo, Janet Kelso, and Qiaomei Fu. 2017. 40,000-Year-Old Individual from Asia Provides Insight into Early Population Structure in Eurasia. *Current Biology* 27 (20):3202–3208.E9.

Yanko-Hombach, Valentina, Allan S. Gilbert, and Pavel Dolukhanov. 2007. Controversy over the great flood hypotheses in the Black Sea in light of geological, paleontological, and archaeological evidence. *Quaternary International* 167–168:91-113.

Yoon, Sook Hee, Wonseok Lee, Hyeonju Ahn, Kelsey Caetano-Anolles, Kyoung-Do Park, and Heebal Kim. 2018. Origin and spread of Thoroughbred racehorses inferred from complete mitochondrial genome sequences: Phylogenomic and Bayesian coalescent perspectives. *PLOS ONE* 13 (9):e0203917.

Zaitseva, G., V. Skripkin, N. Kovaliukh, G. Possnert, P. Dolukhanov, and A. Vybornov. 2009. Radiocarbon dating of Neolithic pottery. *Radiocarbon* 51 (2):795-801.

Zakościelna, A. 2010. *Studium obrządku pogrzebowego kultury lubelsko-wołyńskiej*. Lublin.

Zalizniak, Leonid L. 2016. Mesolithic origins of the first Indo-European cultures in Europe according to the archaeological data. *Ukrainian Archaeology*:26-42.

Zaliznyak, L.L. 1999. Tanged point cultures in the Western part of Eastern Europe. In *Tanged Points Cultures in Europe*, edited by S. K. Koslowski, J. Gurba and L. L. Zaliznyak. Lublin: Maria Curie-Sklodowska University Press.

Zastawny, Albert. 2015. The Baden complex in Lesser Poland - Horizons of cultural influences. In *The Baden culture around the Western Carpathians*, edited by M. Nowak and A. Zastawny. Kraków: Krakowski Zespół do Badań Autostrad.

Zeng, Tian Chen, Alan J. Aw, and Marcus W. Feldman. 2018. Cultural hitchhiking and competition between patrilineal kin groups explain the post-Neolithic Y-chromosome bottleneck. *Nature Communications* 9 (1):2077.

Zhang, Cheng, Pan Ni, Hafiz Ishfaq Ahmad, M Gemingguli, A Baizilaitibei, D Gulibaheti, Yaping Fang, Haiyang Wang, Akhtar Rasool Asif, Changyi Xiao, Jianhai Chen, Yunlong Ma, Xiangdong Liu, Xiaoyong Du, and Shuhong Zhao. 2018. Detecting the Population Structure and Scanning for Signatures of Selection in Horses (Equus caballus) From Whole-Genome Sequencing Data. *Evolutionary Bioinformatics* 14:1176934318775106.

Zhilin, Mikhail. 2017. Mesolithic bone arrowheads from Ivanovskoye 7 (central Russia): Technology of the manufacture and use-wear traces. *Quaternary International* 427:230-244.

Zimmermann, Thomas. 2007. Anatolia and the Balkans, once again? Ring-shaped idols from Western Asia and a critical reassessment of some Early Bronze Age items from İkiztepe, Turkey. *Oxford Journal of Archaeology* 26 (1):25-33.

Zwyns, N., and L. V. Lbova. 2018. The Initial Upper Paleolithic of Kamenka site, Zabaikal region (Siberia): A closer look at the blade technology. *Archaeological Research in Asia*.